LIVING DOWNSTREAM

LIVING DOWNSTREAM

An Ecologist's Personal Investigation of Cancer and the Environment

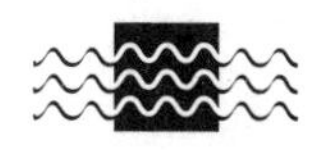

SANDRA STEINGRABER

A Merloyd Lawrence Book
Da Capo Press
A Member of the Perseus Books Group

 Printed in the United States of America. For information, address Da Capo Press, 11 Cambridge Center, Cambridge, MA 02142.

Designed by Trish Wilkinson
Set in 11 point Minion by the Perseus Books Group

Library of Congress Cataloging-in-Publication Data

Steingraber, Sandra.
Living downstream : an ecologist's personal investigation of cancer and the environment / Sandra Steingraber. — 2nd ed.
p. cm.
"A Merloyd Lawrence Book."
Includes bibliographical references and index.
ISBN 978-0-306-81869-1 (alk. paper)
1. Cancer—Environmental aspects. 2. Environmental toxicology. I. Title.
RC268.25.S74 2010
616.99'4071—dc22 2009039502

First Da Capo Press edition 2010

A Merloyd Lawrence Book
Published by Da Capo Press
A Member of the Perseus Books Group
www.dacapopress.com

Da Capo Press books are available at special discounts for bulk purchases in the U.S. by corporations, institutions, and other organizations. For more information, please contact the Special Markets Department at the Perseus Books Group, 2300 Chestnut Street, Suite 200, Philadelphia, PA, 19103, or call (800) 810-4145, ext. 5000, or e-mail special.markets @perseusbooks.com.

10 9 8 7 6 5 4 3 2 1

For Jeannie Marshall
And for Rita Arditi
And for my mother,
Whose original plan was to build
A laboratory in the north bedroom

contents

There was once a village along a river. The people who lived there were very kind. These residents, according to parable, began noticing increasing numbers of drowning people caught in the river's swift current. And so they went to work devising ever more elaborate technologies to resuscitate them. So preoccupied were these heroic villagers with rescue and treatment that they never thought to look upstream to see who was pushing the victims in.

This book is a walk up that river.

foreword to the second edition

Thirty years ago, in between my sophomore and junior years of college, I was diagnosed with bladder cancer. Those are amazing words to write: "Thirty years ago I had cancer." I had just turned twenty. I was hoping that I would live long enough to have sex with someone; I hadn't done that yet. I could not have imagined, while lying in my hospital bed, exhaling anesthesia, that someday I could write, "Thirty years ago I had cancer."

Last September, on a sunny afternoon, the phone rang while I was trying to meet a writing deadline. It was the nurse in my urologist's office. She was calling to say that the pathologist had found, in the urine collected from my last cystoscopic check-up, abnormal cell clusters. And traces of blood.

After I hung up, I looked out the window of my small house where the sun still shone on the last of the marigolds. I looked down at my computer screen where the cursor still blinked on the same paragraph. I noticed the crayons on the floor, cast aside in the morning rush for the school bus, and could hear in the kitchen the tomatoes still bobbing in the stockpot that was steaming away on the stove. The world was still the same, but it felt to me a suddenly altered place.

I provided a second urine sample for further testing, and based on the results of that, a third sample that was sent out for genetic analysis. I began living within that period of time known as watchful waiting. This is a familiar place to me. *Watch* means screening tests, imaging, blood work, self-advocacy, second opinions, and hours logged in hospital parking garages. *Wait* means you go back to your half-finished essay, to the tomatoes on the stove. You lay plans and carry on within the confines of ambiguity. You meet deadlines and make grocery lists. And sometimes you jump when the phone rings on a sunny afternoon. Bladder cancer recurs in 50–70 percent of patients. There are evidence-based reasons for feeling jumpy.

Ten days later, I got a call from the urology nurse. The results were normal. A few months later I had a second cystoscope and a renal ultrasound. Normal. No explanation for the blood. It might mean nothing. Let's look again in six months.

Thirty years ago I had cancer. When I left the hospital, I went back to my college dormitory, resumed my life as a biology major with a side interest in poetry, and began mucking around in the medical literature. I was curious about a series of questions my young, new-to-the-area urologist had asked me a few days after my surgery. Had I ever worked in a tire factory? Any exposure to textile dyes? What about employment in the aluminum industry? As I lay there, still tethered to catheter tubes, these queries had seemed surreal to me. I was the clean-living winner of the local Elks Club scholarship, a high-achieving college student with plans for graduate research. Of course I wasn't out vulcanizing tires or smelting aluminum. But why had he asked?

It didn't require many hours in the university library to learn that bladder cancer is considered a quintessential environmental cancer, meaning that more evidence exists for a link between toxic chemical exposure and bladder cancer risk than for almost any other kind of cancer, with data going back a hundred years. I also learned that the identification of bladder carcinogens does not preclude their ongoing use in commerce. Just because researchers discover, through careful scientific study, that a chemical contributes to cancer doesn't mean it's automatically banished from our economy.

On all these fronts, not much has changed in the three decades since my diagnosis. Of the 80,000 synthetic chemicals now in use, only about 2 percent have been tested for carcinogenicity and, since 1976, exactly five have been outlawed under the Toxics Substances Control Act. Our environmental regulatory system requires no rigorous toxicological testing of chemicals as a precondition for marketing them. It promulgates legal limits on chemical releases, largely overlooking that we are all exposed to trace amounts of many contaminants, and not just one at a time. It is still no one's job to make sure that the total burden of toxic exposures is not too much for any one of us.

A 2007 investigation published by the American Cancer Society identified 216 chemicals known to cause breast cancer in animals. Of these, 73 are found in food or consumer products; 35 are air pollutants; and 29 of them are produced in the United States in large amounts every year.

In 1981, I went off to graduate school, pursuing first a degree in creative writing and then another in field biology. Both offered opportunities to travel far from my hometown in central Illinois. Wherever I was, I dutifully submitted to cancer check-ups. I also started a collection of pamphlets on bladder cancer, gathered from the various waiting rooms and hospitals where I spent time. I noticed that they seldom contained the words "carcinogen" or "environment." (More on these in Chapter Twelve.) Nor were these words used anymore in conversations I had with my various health care providers. There seemed to be a disconnect between the evidence that medical researchers had compiled about the environmental origins of bladder cancer and what patients heard about that evidence. To judge by the medical intake forms, the more relevant variable was genetics: I was asked again and again about my family medical history. I was happy enough to provide it. There is a lot of cancer in my family. My mother was diagnosed with breast cancer at age 44. I have uncles with colon cancer, prostate cancer, stromal cancer. My aunt died of the same kind of bladder cancer—transitional cell carcinoma—that I had.

But here's the punch line to my family story: I am adopted.

And when I looked at the literature on cancer among adoptees, I learned that, in fact, the chance of an adopted person dying of cancer is more closely related to whether or not her adoptive parents had died of cancer and far less related to whether or not her biological parents had met such a fate. What runs in families does not necessarily run in genes. And while knowledge of one's genetic history is important for understanding health risks, so too is knowledge about one's environment. (More on this in Chapter Eleven.)

The environmental questions posed by my first urologist became the seeds of this book. The research it required began in the stacks of the Harvard Medical School Library, where I spent a postdoctoral year, and continued in my Midwestern hometown. As a biologist, my goal with this book was to bring together two categories of information—data on environmental contamination and data on cancer—to see what patterns might exist, to identify questions for further inquiry, and to urge precautionary action, even in the face of incomplete answers. To explore the extent to which toxic chemicals, including cancer-causing agents, have trespassed into our air, food, water, and soil, I drew heavily upon databases available under federal right-to-know laws. Cancer registry data provided a view of cancer's trajectory through time and its distribution across space. Various published studies, gathered from far-flung corners of the biological and medical literature, offered other glimpses of the connection between cancer and the environment. Informing my discussion throughout this book, these range from reports on pesticides, river sediments, and trash incinerators to surveys of farmers, sport anglers, and nursing mothers. They include investigations of laboratory animals, wildlife, and pets, as well as examinations of human tissues and cellular machinery. They range from atmospheric science to neuroendocrinology.

But this book is also a deeply personal story. Woven throughout the various scientific descriptions is a memoir set on the east bluff of the Illinois River where I grew up. As a biologist, I will tell you that my Illinois home is utterly unexceptional: as in many other communities, the dramatic transformation of its industrial and agricultural practices that followed World War II had unintended environmental consequences. This place nevertheless receives my devotional attention because central Illinois is the source of my ecological roots and my search for these roots is the other half of this book.

Bladder cancer rates among women are rising. I am one data point in that statistical story. Bladder carcinogens have turned up in my hometown aquifer and in the sediments of the river that runs by it. (How did they get there? What shall we do about them?) I am one voice in that human story. Both of these stories are told here.

In January 2004, the phone rang while I was trying to meet a writing deadline. It was a film director (not a nurse!), and the conversation we had with each other that day led directly to this new edition. Chanda Chevannes, the director, wished to option *Living Downstream* for a documentary. In so doing, she intended to represent cinematically both of its stories—the scientific one and the personal one. This happy plan created for me three tasks. One of them was to accompany her and her Canadian film crew to central Illinois.

And so I did. I introduced them to the river barges (Chapter Nine), the ethanol plant (Chapter Five), and the wind turbines across the road from my cousin John's farm (Chapter Seven). I introduced them to John's cornfields—and to the combines advancing across them, the afternoon thunderheads building over them, and the way you can navigate from them using a grain elevator as a compass point (Chapter Eight). I introduced them to the toxic waste landfill (Chapter Five again) and to my mom's backyard swing and my Aunt Ann's pear tree (Chapter Ten). When they commissioned a helicopter to explore the Illinois River valley from above, however, they were on their own.

The second task was to introduce them to my private life as a medical patient (Chapters Six and Eleven). This was trickier. It meant bringing a film crew to a cystoscopic examination during which a fiber-optic tube would be inserted into my bladder. It meant that men carrying a movie camera and a boom mike would follow me into a room with a stack of backless, blue cotton gowns (one of which I would change into) and up onto an examination table equipped with stirrups (into which I would place my feet). Here, all eyes but mine would stare into the large-screen monitor (onto which the interior walls of my bladder would then be displayed). Meanwhile, I would be lying quietly, pondering the ceiling tiles. Taped to the walls around us all would be posters of enlarged prostates and penile dysfunctions (the anatomical details of which I ritualistically study in the moments before the doctor steps through the door). And the camera would be rolling through it all.

I was determined to carry this off. Cystoscopies save lives. However ghastly a cystoscopic exam might sound to the uninitiated, it is brief, minimally painful, and can locate cancers at an early stage. As a tool for screening and early detection, it is unsurpassed. (From a medical point of view, I give cystoscopies the *sine qua non* award along with colonoscopies and Pap smears.) No one who finds blood in their urine should delay seeking help out of fear of a tube with a flashlight on the end of it. So, if I could bear witness to the value of cystoscopes, I would do it. If I had the chance to pull back the curtain of silence surrounding urological exams, I would take it. As someone who has undergone upwards of seventy cystoscopic exams, who better than me?

And so I did, and in so doing, discovered something unexpected: cystoscopies are actually better when you bring a camera crew along. Whether or not I successfully demystified the cystoscopy for the movie's audience, I certainly demystified it for myself that afternoon. *The procedure room*—that chamber I had always entered with solemn ceremony—now seemed dinky and ordinary. The penis posters were suddenly hilarious. And the quiet, reassuring voice of my urologist during the exam itself—which I have always appreciated—seemed to me, more than ever, a sign of steadfast human compassion. The relationship between doctor and long-time patient can be an intimate one.

My third task was to explore with the film director the science of the book (Chapters One through Twelve plus nearly 100 pages of source notes). Chanda had to figure out how to make cancer data visible to her film audience—a challenge that took her to laboratories and field stations across the continent to shoot, for example, whale autopsies in Quebec, frog studies in California's Salinas River, tissue micrographs in federal offices in North Carolina, and DNA extraction in cancer laboratories in Vancouver. Meanwhile, I went to work updating the scientific research in the book itself.

The result of those revisions is this second edition. The time interval between this edition and the first represents a period of rapid growth in our understanding of the environmental links to human cancer. For the most part, new published findings support the evidence I had compiled in 1997. I was thus able to add a few more pieces to the big jigsaw puzzle of cancer causation and answer some questions that earlier studies had raised. Where I needed to make corrections or shift the emphasis, I did. Happily, my residency at Cornell University's Program on Breast Cancer and Environmental Risk Factors and subsequent advisory role in the California Breast Cancer Research Program have offered me, over the past ten years, a front-row seat from which to observe scientific research in action.

Providing up-to-the-minute insights into the ongoing encroachment of industrial chemicals into our communities proved more problematic. Databases that disclose the routine release of 650 toxic substances from industrial facilities were first made available under federal right-to-know laws passed by Congress in 1986. They allowed the public to identify polluters within their communities and researchers to track pollution and cross-reference with cancer patterns. In the mid-1990s, when I was drafting the first edition of this book, the Toxics Release Inventory went up on the Internet. Between 2001 and 2008, however, the inventory was scaled back, and thousands of facilities were no longer required to report. In 2009, some of the original requirements for reporting were reinstated. However, because of the changing criteria for reporting, right-to-know data available now are less comprehensive than in years past. Therefore, I let stand much of my previous reportage on toxic chemical contamination, which draws on data I gathered in the mid-1990s when the databases were more robust.

The personal story of *Living Downstream* is also unchanged. I wrote this book as a single woman in my thirties who lived with my dog in a

Boston apartment. In those days, I ignored national holidays and read cancer registry data in the bathtub. That solitary woman is still the narrator of this book. And this means that its autobiographical scenes are set in the recent past while the scientific descriptions are *au courant.* Thus, in Chapter Ten, the drama on the farm takes place in fall 1994, but the passages that describe the behavior of dioxin include evidence published in subsequent years.

By contrast, my own life has been altered in many ways since I wrote the first edition. I am now an almost-50 mother of two who is married to the father of my children—he is also their art teacher—and we all live in a small village in upstate New York. I am seldom allowed reading time in the bathtub, and I not only observe Valentine's Day, I have baked heart-shaped pizzas for the entire population of a nursery school. For descriptions of my embedded life as a mother, I gladly refer the reader to *Having Faith: An Ecologist's Journey to Motherhood* and my forthcoming book on the environmental life of children.

Over the last decade, six clear trends have emerged in our understanding of the environment's contribution to cancer. The first is a growing acknowledgment that **cancer causation is complex.** The old way of thinking was to imagine cancer risk factors as independent agents that could be boxed up into three neat categories: genes, lifestyle, and environment. Of the three, genes and lifestyle were thought to be the dominant players with only a small fraction of cancers attributable to the environment. That kind of simplistic accounting is increasingly seen as naive. Cancer is now believed to result from a web of interwoven variables, any one of which can modify another. For example, breastfeeding is protective against breast cancer. It is considered a classic lifestyle factor: you can choose to nurse your baby or not, and if you do, you may lower your later risk of breast cancer. But evidence also suggests that exposure to certain organochlorine chemicals may impair a woman's ability to lactate and breastfeed successfully. Thus, environmental contaminants can affect a lifestyle choice that, in turn, affects breast cancer risk. In short, cancer risk factors can interact with each other to exert direct and indirect effects.

The second trend is an emerging awareness of the **importance of epigenetics.** The old way of thinking saw DNA—the bricks and mortar of our genes—as a master molecule. Cancer was thought to arise through the inheritance of bad genes or by damage to good genes (mutations). The new thinking acknowledges that cancer can arise through a third route: by changing the behavior of genes. The study of how substances alter gene expression is part of the field of epigenetics. Some chemical exposures appear to turn on and turn off genes in ways that disregulate cell growth and predispose for cancer. From this perspective, our genes are less the command-and-control masters of our cells and more like the keys of a piano, with the environment as the hands of the pianist.

The third trend is a mounting appreciation for the **role of endocrine disruption** in the story of cancer. If there were ever a contest for Most Easily Duped biological system, I would nominate our endocrine system—the hormonal messaging service that guides our development, runs our metabolism, and allows us to reproduce. Many chemicals, at vanishingly small concentrations, have the ability to interfere with hormonal signals, sometimes by crude imitation. The endocrine system is impressively incapable of distinguishing between real hormones and environmental chemicals that act like hormones. It is a patsy for sabotage. When I wrote the first edition of this book, my focus was on chemicals that had the ability to imitate estrogen. But the simple mimicry of sex hormones is now only part of the story of endocrine disruption. Hormonally active chemicals can infiltrate the signaling circuitry throughout our bodies. There is even a newly identified category of endocrine disruptors known as *obesogens*: chemicals that perturb the suite of hormonal messages that oversee fat deposition.

An ancient principle of toxicology posits that the dose makes the poison: "Solely the dose determines that a thing is not a poison." That axiom dates back to the sixteenth century, and it still appears on the opening page of my copy of *Casarett and Doull's Toxicology*, 6th edition, expressing the prevailing belief that our risk for harm from exposure to an inherently toxic substance is proportional to how much we were exposed to. There is still a lot of truth to this old chestnut. But what's becoming increasingly evident is that the risk posed by a toxic substance depends as well on *when* we were exposed. Timing matters. Especially if the exposure in question involves an endocrine disruptor. The fourth trend, then, is an expanding recognition that **the timing makes the poison.** The search for environmental links to breast cancer, in particular, is increasingly focused on exposures early in life that influence the course of breast development. Altered breast development may increase susceptibility to breast cancer in later life. As is discussed further in Chapter Six, the majority of breast cancers arise from structures within the ducts of glands called terminal end buds. Any chemical that increases the number of cells in the end buds or delays their maturation, according to the new thinking, may raise the risk for breast cancer.

The fifth trend is a recognition that combinations of chemicals may have consequences not predicted by one-chemical-at-a-time analyses. **Chemical mixtures need attention.** Real-life exposures seldom involve single agents. And yet when testing chemicals for their potential to cause cancer or when deciding what the acceptable limit of exposure to suspected carcinogens should be, our regulatory system considers them in isolation from each other. Some chemicals operate down similar cellular pathways; their effects may be additive. Others may interact in more complex ways, as when exposure to one pesticide alters the activity of enzymes in ways that cause a second pesticide to be metabolized into a more powerful toxicant. Mixtures of chemical exposures *with* other stressors—like obesity or poverty—may also create cancer risks not predictable by examination of each variable by itself.

The sixth trend is a shift toward embracing **the precautionary principle as a normative guide** to environmental decision making. This idea was first articulated in Germany in the 1970s when scientists realized they needed to halt the ongoing death of their nation's forests in advance of working out all the details of how exactly air pollution was contributing to the problem. Now enshrined into the Treaty of the European Union, the precautionary principle urges us to take action to prevent harm in situations where substantive proof is unavailable and where delays caused by waiting for proof may create irreversible, catastrophic damage. In so doing, the precautionary principle grants the benefit of the doubt to public health rather than to the things that threaten it. More on this in Chapter Twelve.

The most frequent question I am asked by my readers is, how do you have hope? I have two responses, one personal and the other evidence-based. The personal one: I'm a cancer survivor. I learned, early on in my life, how to have hope in times of desperation. I am also now a mother. I would like my children to live in a world without carcinogens in the groundwater. I would like them not to fear that a phone ringing on a sunny afternoon is bringing them bad news from a pathology lab. In other words, despair feels like a luxury I cannot afford right now.

My other answer goes like this. The mounting evidence that our environment is playing a bigger role in the story of cancer than previously supposed is good news because we can do something about it. We can choose, for example, to change our antiquated chemicals policy. We can resolve, collectively, to divorce our economy from its current dependencies on toxic chemicals known to trespass inside our bodies. We can decide that the presence of cancer-causing substances in our air, water, and food is too expensive. A 2009 study, for example, has found that coal mining in Appalachia costs the region five times more in premature deaths, including from cancer, than it provides to the region in jobs, taxes, and economic benefits. In California, the production and use of hazardous chemicals cost the state $2.6 billion in 2004 alone in lost wages and health-care expenses to treat workers and children with pollution-linked diseases. (As a percentage of U.S. health-care spending, which has tripled since 1970, cancer is the third most costly condition. For an individual person, cancer is the most costly.)

We can change our thinking. Rather than viewing the chemical adulteration of our environment and our bodies as the inevitable price of convenience and progress, we can decide that cancer is inconvenient and toxic pollution archaic and primitive. We can start seeing the creation of carcinogens as the result of outmoded technologies. We can demand green engineering and green chemistry. We can let our systems of industry and agriculture know that they are suffering from a design flaw. (See Chapter Five.)

By contrast, none of us (adopted or not) can change our ancestors. If the science had instead pointed to genes as the kingpins of cancer, if nothing could be done but wait for the ticking time bombs inside our cells to detonate at random, then I would feel depressed. Happily, that is not our situation.

The even better news is that the synthetic chemicals linked to cancer largely derive from the same two sources as those responsible for climate change: petroleum and coal. Finding substitutes for these two substances is already on the collective to-do list. The U.S. petroleum industry alone accounts for one-quarter of toxic pollutants released each year in North America. This does not include the air pollutants generated from cars and trucks burning the products that the petroleum industry makes. (As is described in Chapter Eight, vehicle emissions are linked to lung, breast, and bladder cancers.) Coal-burning electric utilities are also among the nation's top generators of toxic chemical releases, as are mining operations. Investments in green energy are therefore also investments in cancer prevention. In this, it feels to me that we are standing at a historic confluence, a place where two rivers meet: a stream of emerging knowledge about what the combustion of fossil fuels is doing to our planet is joining a stream of emerging knowledge about what synthetic chemicals derived from fossil fuels are doing to our bodies.

The War on Cancer, declared by President Nixon in 1971, has savored few victories. The idea of a cure, presumed just around the corner for decades, seems almost fanciful. With a few notable exceptions, improvements in existing treatment have not translated into significant numbers of lives saved. Indeed, the death rate from cancer is only 6 percent lower than it was in 1950. In 1999, cancer surpassed heart disease as the leading killer of Americans under 85. At present, 45 percent of men and 40 percent of women will be diagnosed with cancer at some point during their lives, a far higher proportion than 50 years ago. And as our population ages, the number of people suffering from cancer is expected to jump by 45 percent in the next two decades.

But data from cancer registries—which receive my close attention in Chapter Three—also contain another message: eliminating exposures to carcinogens saves lives. The death rate from cancer is now falling. That decline is largely attributable to the success of smoking cessation programs and changing attitudes about the glamour quotient of cigarettes. Overall cancer incidence has also dropped, slowly but steadily, over the last decade, likewise driven by declines in lung cancer diagnoses and, to a lesser extent, colon cancer. (Colonoscopies: *sine qua non.*)

With bans on smoking in public places now enacted in many states and tobacco under the regulatory control of the Food and Drug Administration, U.S. smoking rates will almost surely decline further in the years to come. The lives saved will include people who might otherwise have started to smoke or continued smoking, as well as those of us nonsmokers who would otherwise have breathed in the carcinogens that our smoking compatriots breathed out. We will not know which among us owe our continued existence to the collective decision to denormalize tobacco, but the lives spared will be visible in the descending slope of the line that expresses trends in death from tobacco-related cancers over time. None of us aspires to become a data point on that graph.

Here in upstate New York, smoking was banned from public places in 2003. The ironic result for my young children was that they saw cigarettes for the first time, as the tobacco-addicted took to the sidewalks and alleyways. In the winter months, smokers were easily identifiable by their hunched-over posture. As we looked out of the window of our village coffee shop one blizzardy afternoon, my son, then three years old, whispered to me with alarm, "Mama, there's a man in the snow trying to light his face on fire!"

To my children, smoking doesn't look glamorous. It looks grotesque. And their perception is a direct result of changes in public policy that were put in first motion during my own early childhood when, in 1964, the U.S. Surgeon General warned, on the basis of good but partial evidence, that smoking causes lung cancer. That was a courageous decision and an example of the precautionary principle in action. Proof for a link between smoking and lung cancer was not demonstrated until 1996, three decades later.

In *Living Downstream*, I advocate that we bring the same precautionary approach to other carcinogens, known and suspected. In so doing, I fully agree with the conclusion of a consensus statement, signed by many members of the cancer research and advocacy community and submitted to the President's Cancer Panel in October 2008:

> The most direct way to prevent cancer is to stop putting cancer-causing agents into our indoor and outdoor environments in the first place.

This task is made urgent by ascending rates of cancers unrelated to tobacco. Among U.S. men, age-adjusted incidence rates of multiple myeloma and cancers of the kidney, liver, and esophagus are rising. Among women, the cancers of increasing frequency include melanoma, non-Hodgkin lymphoma, leukemia, and tumors of the bladder, thyroid, and kidney. As is explained in Chapter Three, improvements in diagnostic techniques cannot explain away these trends. Many of the cancers that are now increasing in incidence are those with links to environmental exposures.

Most troubling: childhood cancer has increased steadily since 1975. Cancers among teenagers and young adults are also more prevalent. Indeed, support groups now abound for young adults with cancer, who have their own nonprofit organization (The I'm Too Young for This! Foundation), their own radio show ("The Stupid Cancer Show—The Voice of Young Adults with Cancer"), and a signature alcoholic beverage (the cancertini). Rising rates of cancer among college students have spawned the birth of a new social movement that includes lapel pins,

T-shirts, Visa cards, networking sites, retreat centers, and the slogan, "Stupid Cancer. Survivors Rule."

I am inspired by activism that destigmatizes cancer and breaks silence about its presence among us, old and young alike. Ultimately, though, I would prefer that cancer among twenty-year-olds return to levels of startling uncommonness. And I believe this goal is attainable. *Living Downstream* is my best attempt as a biologist and a cancer survivor to lay out the case for cancer prevention through environmental change. There are individuals who claim, as a form of dismissal, that links between cancer and environmental contamination are unproven and unprovable. There are others who believe that we are obligated to act, as did the surgeon general in 1964, on the basis of the evidence we have before us now. "To ignore the scientific evidence is to knowingly permit thousands of unnecessary illnesses and deaths each year." This was the conclusion of a recent state-of-the-science review of the links between cancer and the environment.

I have copied that sentence onto the outside of a file folder on my desk. In it are published papers documenting links between bladder cancer and a group of synthetic chemicals called aromatic amines. The earliest report comes from a German surgeon in 1895 who noticed bladder cancer among textile dye workers exposed to the color magenta during a period of time when coal tar–derived pigments—aromatic amines—were replacing plant-based pigments in the European textile industry. Another paper recounts that all fifteen workers in a British mill had succumbed to bladder cancer. A series of papers in the 1950s painstakingly documented increased rates of bladder cancer among chemical industry workers exposed to aromatic amines. Nearly identical findings continued to be published in the 1960s and 1980s. In 1991, the National Institute for Occupational Safety and Health uncovered bladder cancer rates among aromatic amine–exposed workers that are twenty-seven-fold higher than normal. The most recent paper I have was published in 2009. It reports elevated bladder cancer rates among farmers who use imazethapyr, a pesticide containing aromatic amines. Imazethapyr was registered for use in 1989—more than 100 years after the German surgeon's early warning.

This is a file folder of madness. *To ignore the scientific evidence is to knowingly permit thousands of unnecessary illnesses and deaths each year.* Or as my son would say, we don't have to keep lighting our faces on fire, right?

Sandra Steingraber
July 2009

one

trace amounts

On a clear night after the harvest, central Illinois becomes a vast and splendid planetarium. This transformation amazed me as a child. In one of my earliest memories, I wake up in the back seat of the car on just such a night. When I look out the window, the black sky is so inseparable from the plowed, black earth—which dots are stars and which are farmhouse lights?—that it seems I am floating in a great, dark, glittering bowl.

Rural central Illinois still amazes me. Buried under the initial appearance of ordinariness are great mysteries. At least, I attempt to convince newcomers of that.

Were you to visit this countryside for the first time, its apparent flatness is probably what would impress you first—and indeed, for almost half the year, the landscape seems to consist of a simple plain of bare earth overlain by sky. But Illinois is not flat at all, I would insist, as I unfold geological survey maps that make visible the surprisingly contoured lay of the land. Parallel arcs of scalloped moraines slant across the state, each ridge representing the retreating edge of a glacier as it melted back into Lake Michigan and surrendered the tons of granulated rock and sand it had churned into itself.

Better than maps is a ground fog on a summer night when I drive you across these moraines and basins. Now you see how the shrouded bottomlands are distinguished from the uplands, the floodplains from the ridges, how the daytime perception of flatness belies a great depth. Out of

the car and walking, I encourage you to feel, as we traverse land that appears to be utterly level, the slight tautness in the thighs that comes with ascending a long grade versus the looseness in our feet that indicates descent.

Then there is the issue of water. Consider your own body, how the blood does not pulse through your tissues in great tidal surges—as was presumed before the English physician William Harvey discovered circulation in 1628—but instead flows within a diffuse net of permeable vessels. So too in Illinois, a capillary bed of creeks, streams, forks, and tributaries lies over the land. Your newly found skill of walking downhill will help you locate it.

And this is only the water that is visible. Under your feet lie pools of groundwater held in shallow aquifers—interbedded lenses of sand and gravel—and in the bedrock valleys of ancient rivers that lie below. One of these is the Mahomet, part of a river system that once ran west across Ohio, Indiana, and Illinois. Thousands of tons of debris, let loose by melting glaciers, completely buried the Mahomet River at the end of the last ice age. It now flows underground. In Mason County you can stand over a place where the Mahomet once joined the Illinois River. Here, in an area called the Havana Lowlands, the groundwater lies just below the earth's surface. In times of heavy rain, lakes brim up from under the earth and reclaim whole fields and neighborhoods.

In the eastern half of my county, Tazewell, the ancestral Mississippi River cut a valley 3 miles wide and 450 feet deep before glaciers exiled it to the western border of Illinois, its current channel. Buried by soil, clay, silt, and stones, the old Mississippi River valley is still down there, connected to the same ancient tributaries, its fractures and pores full of water. Islands still rise from the bedrock channel. If you could see through dirt, imagine the dramatic view you would have.

Of course, what you do see are corn and soybean fields. About 87 percent of Illinois is cropland, meaning that if you fell to earth in Illinois, chances are good you would land in a farm field. Illinois grows more soybeans and corn than any other state but Iowa. Read any supermarket la-

hol. Corn syrup, corn gluten, cornstarch, dextrose, soy oil, and soy proteins are found in almost every processed food from soft drinks to sliced bread to salad dressing. These are also the ingredients of the food we feed to the animals we eat. Thus, you could say that we are standing at the beginning of a human food chain. The molecules of water, earth, and air that rearrange themselves to form these beans and kernels are the molecules that eventually become the tissues of our own bodies. You have eaten food that was grown here. You *are* the food that is grown here. You are walking on familiar ground.

Illinois is called the Prairie State, but, to find prairie, you must really know where to look. Most of it vanished after John Deere invented the self-scouring steel plow in 1836. To be exact, 99.99 percent went under the plow. The .01 percent that escaped occupies odd and neglected places: along railroad tracks, encircling gravestones in old pioneer cemeteries, on hillsides too awkward to plow. Of the original 281,900 acres of tallgrass prairie in my home county, an official 4.7 fragmented acres remain (.0017 percent). I have never found them. Illinois conceals not only its topography but its ecological past as well, and even though I went on to become a plant ecologist, I have no real relationship to the native plants of my native state.

Truthfully, the closest I have felt to the prairie is when looking at plain, unadorned dirt. There are plenty of opportunities to do this in central Illinois—although the fields look less naked between October and April than they did when I was a child, thanks to low-till and no-till farming. These practices have largely replaced the habit of turning the field completely over after the harvest. The newer techniques leave on the surface a certain fraction of stalks, leaves, and stems to serve as a thin blanket against the wind. It's a tricky business: Too much residue leaves the soil compressed, without air, and unable to warm up in time for spring planting; water puddles on the surface. Too little residue, and the soil refuses to clump up at all, is prone to blow away or run with meltwater into the nearest creek bed.

Then, each September at the Farm Progress Show, farm equipment representatives demonstrate all the latest technology for striking the perfect balance between these two states. Popular among farmers is the disc and chisel plow combination: parallel rows of sliding silver plates like large pizza cutters, alternating with rows of beveled metal claws. These grids of disks and chisels are pulled, one by one, through an exhibition field as an announcer extols the virtues of each particular model. Onlookers, including me and my uncle, stand on either side of the tractor as it cuts a wide swath through corn stubble. We then step into the black wake and bend down to take a look. To assess a depth of penetration, we are encouraged to poke yardsticks into the chiseled furrows. We heft clumps of dirt in our hands to check diameter and ease of crumbliness. We then walk 10 yards over and form two lines on either side of the next tractor in the queue of tractors to cut a path through this field of stubble. We step in, bend down, heft clumps, stand up, walk over. And so on. It's a peculiar kind of country line dance. Each plowed strip is subtly different from the others.

There is no reason I should participate in this ritual except that my mother's family still farms the Illinois prairie and watching the earth being tilled offers me a connection to the past. Even though I live out of state now, it's important to me to maintain a relationship with both Illinoises—the present and familiar one as well as the Illinois that has vanished and is barely discernible. What remains of the 22 million acres of tallgrass prairie that once covered this state is the deep black soil that those grasses produced from layers of sterile rock, clay, and silt dumped here by wind and glaciers. The molecules of earth contained in each plowed clod are the same molecules that once formed roots and runners of countless species unfamiliar to me now. They died and became soil. This most obvious of realizations occurs to me every September as though for the first time. When I am touching Illinois soil I am touching prairie grass.

~~~~~~

Illinois soil holds darker secrets as well. To the 87 percent of Illinois that is farmland, an estimated 54 million pounds of synthetic pesticides are
~~~~~~

applied each year. Introduced into Illinois at the end of World War II, these chemical poisons quickly familiarized themselves with the landscape. In 1950, less than 10 percent of cornfields were chemically treated. Fifty-five years later, 99 percent were sprayed with pesticides. The most abundantly used is the weedkiller atrazine, which in 2007 was applied to 81 percent of Illinois cornfields—nearly 10 million acres of soil. With so much acreage now planted in field corn, fungus, which breeds on corn stubble, has emerged as a significant pest. The use of fungicides is now sharply up, refamiliarizing rural folk with an icon of the past: the low-flying crop duster droning above the fields in midsummer.

Pesticides do not always stay on the fields where they are sprayed. They evaporate and drift in air. They dissolve in water and flow downhill into streams and creeks. They bind to soil particles and rise into the air as dust. They migrate into glacial aquifers and thereby enter groundwater. They fall in the rain. They are found in snowflakes. And fog. And wind. And clouds. And backyard swimming pools. Little is known about how much goes where. By 1993, 91 percent of Illinois's rivers and streams showed pesticide contamination. Ten years later, the streams and rivers within my childhood watershed contained 31 different pesticides, and atrazine was in all samples. These chemicals travel in pulses: pesticide levels in surface water during the months of spring planting—April through June—are sevenfold those during winter and often contain levels of atrazine that exceed legal limits for drinking water. Even less is known about pesticides in groundwater. About 18 percent of all samples of groundwater surveyed in Illinois in 2006 contained atrazine byproducts, while a 1992 study found that one-quarter of private wells tested in central Illinois contained agricultural chemicals of some type. Drinking water wells in the Havana Lowlands region of Mason County showed some of the most severe contamination. A 2009 report identified two public drinking water systems in Illinois with running annual averages for atrazine in tap water that exceeded legal limits. In the same year, the wind blowing across my home county was so full of weedkiller that the air itself withered grape vines in a local vineyard.

Some of the pesticides inscribed into the Illinois landscape have been linked to cancer. One of these is DDT. Banned for use decades ago, DDT is so chemically stable that it remains the most common pesticide in fish in North American rivers and streams. A 2009 national survey of pesticide residues in homes across the United States found traces of DDT on 42 percent of kitchen floors. Like islands in preglacial river valleys, its presence endures. DDT has been variously linked, in human studies, to low sperm count, premature birth, diabetes, brain damage, pancreatic cancer, impaired breastfeeding, and breast cancer. Some of the pesticides inscribed into the Illinois landscape are hormonally active—even at vanishingly small concentrations. One of these is atrazine, which has been variously linked, in animal studies, to increased estrogen production, birth defects, sexual ambiguity, disrupted ovulation, and altered breast development.

A lot goes on in the 13 percent of Illinois that is not farmland. In 2007, 1,102 different industries released more than 114 million pounds of toxic chemicals into air, water, and soil, making Illinois the nation's thirteenth biggest polluter. In the same year, 763 chemical spills occurred—more than two a day—making Illinois ninth among states in number of reported toxic accidents.

Like pesticides, industrial chemicals have filtered into the groundwater and surface waters of streams and rivers. Metal degreasers and dry-cleaning fluids are among the most common contaminants of glacial aquifers. Both have been linked to cancer in humans. At last count, 415 dry cleaners throughout Illinois have poisoned soil, and at least 30 represent a threat to groundwater. An assessment of the Illinois environment concluded that chemical contamination "has become increasingly dispersed and dilute," leaving residues that are "increasingly chemically exotic and whose health effects are not yet clearly understood."

~~~~~~

I was born in 1959 and so share a birthdate with atrazine, which was first registered for market that year. In the same year DDT—dichloro diphenyl trichloroethane—reached its peak usage in the United States. The 1950s
~~~~~~

were also banner years for the manufacture of PCBs—polychlorinated biphenyls—the oily fluids used in electrical transformers, [illegible] carbonless copy paper, and small electronic parts. DDT was outlawed the year I turned thirteen and PCBs a few years later. Both have been linked to cancer.

I am compelled to learn what I can about the chemicals that presided over the industrial and agricultural transformations into which I was born. Certainly, all of these substances have an ongoing biological presence in my life. Atrazine remains the most frequently detected pesticide in water throughout the United States, found in three of every four American streams and rivers and 40 percent of all groundwater samples. PCBs still lace the sediments of the river I grew up next to as well as the flesh of the fish that inhabit it. PCBs are why I'm unfamiliar with the taste of smallmouth bass and channel catfish. In fact, I have never eaten fish from my own river. State fish advisories warn women and children against doing so. DDT also continues to separate people from fish. The coastal waters of Palos Verdes, California became unfishable after 100 tons of DDT were drizzled into the sea between 1947 and 1971. The nine-mile stretch of ocean floor where the poison lies is considered one of the most hazardous places in the nation. The current plan for remediation is to cover it over with 18 inches of silt. Work is expected to begin in 2011, nearly four decades after DDT was banned.

I honestly have no memories of DDT. Instead, my images come from archival photographs and old film clips. In one shot, children splash in a swimming pool while DDT is sprayed above the water. In another, a picnicking family eats sandwiches, their heads engulfed in clouds of DDT fog. Old magazine ads are even more surreal: an aproned housewife in stiletto heels and a pith helmet aims a spray gun at two giant cockroaches standing on her kitchen counter. They raise their front legs in surrender. The caption reads, "Super Ammunition for the Continued Battle on the Home Front." DDT is a ruthless assassin. In another ad, the aproned woman appears in a chorus line of dancing farm animals who sing, "DDT is good for me!" DDT is a harmless pal.

During the 1940s and 1950s, this chemical of multiple personalities found its way into all kinds of civic campaigns and household products. One Illinois town not far from where I grew up conducted aerial fumigations of DDT in an attempt to control polio, mistakenly thought to be spread by flies. Meanwhile, a paint company advertised a formulation that could be brushed onto porches, window screens, and baseboards. When dry, DDT crystals would rise to the surface, forming "a lethal film." Perfect for summer cottages and trailers. Perhaps I spent childhood vacations in some of them. And perhaps, while there, I slept soundly between pesticide-impregnated blankets. In 1952, researchers proudly announced that woolens could now be mothproofed by adding DDT to the dry-cleaning process.

Fellow baby boomers just a few years older do not rely on old magazine ads to recall DDT. From memory, they can describe the fogging trucks that rolled through their suburban neighborhoods as part of mosquito, Dutch elm disease, or gypsy moth control programs. Some can even describe childhood games that involved chasing these trucks. "Whoever could stay in the fog the longest was the winner," remembers one friend. "You had to drop back when you got too dizzy. I was good at it. I was almost always the winner." Says another, "When the pesticide trucks used to come through our neighborhood, the guys would haul their hoses into our backyard and spray our apple trees. Mostly we kids would throw the apples at each other. Sometimes we would eat them."

Hazards that are universally common or repetitive assume "the harmless aspect of the familiar," observed the wildlife biologist Rachel Carson in her book *Silent Spring*, published when I was three years old. "It is not my contention that chemical insecticides never be used," Carson emphasized. "I do contend we have put poisonous and biologically potent chemicals indiscriminately into the hands of persons wholly ignorant of their potentials for harm. We have subjected enormous numbers of people to contact with these poisons, without their consent and often without their knowledge." She went on to predict that future generations would not condone this lack of prudent concern.

Reading *Silent Spring* as a member of this generation, across a distance of more than three decades, I gain another view of DDT. What impresses me most is just how much was known about the harmful aspects of this familiar and seemingly harmless substance. As Carson made clear, the scientific case against DDT—even by the late 1950s—was damning. It was not objective science, nor was it blissful ignorance, that created the impression that DDT was somehow both our most lethal weapon against undesirable life forms ("killer of killers," "the atomic bomb of the insect world") and a completely benign helpmate. In fact, scientific study after scientific study showed that DDT was failing at both roles. It triggered population explosions in insect pests who evolved resistance and whose natural enemies were killed by the spray. It poisoned birds and fish. It disrupted sex hormones in laboratory and domestic animals. It showed signs of contributing to cancer. By 1951, it had become a contaminant of human breast milk and was known to pass from mother to child.

Nevertheless, people continued using DDT until Carson's preliminary damning evidence was supplemented with more and more corroborating damning evidence, producing a great accumulation of damning evidence, and its registration was finally revoked in 1972. I find this phenomenon boundlessly fascinating. Across my desk are spread forty years of toxicological profiles, congressional testimonies, laboratory studies, field reports, and public health investigations of toxic chemicals both officially outlawed and officially permitted. Like crossing and recrossing the same field, I move back and forth between *Silent Spring* and the scientific literature that preceded it, between *Silent Spring* and the scientific literature published in the decades since. At what point does preliminary evidence of harm become definitive evidence of harm? When someone says, "We were not aware of the dangers of these chemicals back then," whom do they mean by *we*?

DDT, lindane, aldrin, dieldrin, chlordane, heptachlor. These names, unfamiliar to us now, are a roll call of the pesticides Rachel Carson featured in *Silent Spring*. All have links to cancer in at least some studies. All are now prohibited or heavily restricted for domestic use. Lindane was banned for most uses in 1983 and banned entirely in 2006, although a controversial exemption allows its ongoing use as a treatment for lice and scabies. And [illegible] several pounds of lindane into the air in [illegible] and dumped several more pounds into the sewer system. I know this because federal right-to-know laws make such events public information. Thus, lindane appears in the 1992 federal government's Toxics Release Inventory for Tazewell County. I was stunned to discover it there as I scanned the electronic list that documents emissions, dumpings, and transfers of toxic chemicals. Lindane has been associated in several studies with cancers of the lymph system.

Aldrin and dieldrin were banned in 1975, although aldrin was allowed as a termite poison until 1987. Aldrin converts to dieldrin in soil and inside our tissues. Dieldrin suppresses the immune system and produces abnormal brain waves in mammals. As late as 1986, dieldrin was still turning up in milk supplies because the soils of hayfields sprayed more than a decade earlier remained contaminated. Most agricultural uses of chlordane in the United States were ended in 1980 and heptachlor in 1983. Both have been linked to leukemia and certain childhood cancers.

For those of us born in the 1940s, 1950s, and 1960s, the time between the widespread dissemination of these pesticides and their subsequent prohibition represent our prenatal periods, infancies, childhoods, and teenage years. We were certainly the first generation to eat synthetic pesticides in our pureed vegetables. By 1950, residue-free produce was so scarce that the Beech-Nut Packing Company began allowing detectable levels of residue in baby food.

~~~

At what point does preliminary evidence of harm become definitive evidence of harm? When someone says, "We were not aware of the dangers of these chemicals back then," who is the *we*?
~~~

With a focus on breast cancer let's look at the evidence of harm for three chemicals: DDT, PCBs, and atrazine.

In 1976, four years after DDT was banned, researchers reported that women with breast cancer had significantly higher levels of DDE (dichloro diphenyl dichloroethylene) and PCBs in their tumors than in the surrounding healthy tissues of their breasts. (DDT is metabolized in the human body into DDE, a chemical that acts like estrogen.) The study was small but the finding provocative because DDT and PCBs were already linked to breast cancer in rodents.

Other studies followed. Some showed an association between breast cancer and residues of pesticides or PCBs. Some did not. In 1993—seventeen years after the first study—the biochemist Mary Wolff and her colleagues conducted the first carefully designed, major study on this issue. They analyzed DDE and PCB levels in the stored blood specimens of 14,290 New York City women who had attended a mammography screening clinic. On average, they reported, the blood of breast cancer patients contained 35 percent more DDE than that of healthy women, but PCB levels were only slightly higher. The most stunning discovery was that the women with the highest DDE levels in their blood were four times more likely to have breast cancer than the women with the lowest levels. The authors concluded that residues of DDE "are strongly associated with breast cancer risk."

By now, breast cancer activists were paying attention. Throughout the 1990s, as breast cancer rates continued rising, they urged scientists to direct more research dollars down lines of inquiry that would reveal, once and for all, whether exposure to pesticides and industrial chemicals was contributing to breast cancer. They pointed out that pesticide use in the United States had doubled since Rachel Carson wrote *Silent Spring* and that women born in the United States between 1947 and 1958 had almost three times the rate of breast cancer that their great-grandmothers had when they were the same age. Women cancer activists marched in the streets carrying signs proclaiming, "Rachel Carson was right!" Taking a page from the playbook of AIDS activists, these women demanded a seat at the table where research proposals were reviewed and funding decisions made.

Studies were funded and papers published. Yet the results were maddeningly inconsistent. For every finding of a positive association, another showed no association or yielded a complicated picture. One study found that African American women with breast cancer showed more past exposure to PCBs than their counterparts without breast cancer. Mysteriously, however, the trend for white women went in the opposite direction: the highest levels of blood PCBs tended to occur in women *without* the disease. The largest and best-designed investigation within this suite of studies, published in the *New England Journal of Medicine* in 1997, found no association at all between risk of breast cancer and blood levels of PCBs and DDT. Interpretation of these contradictory results sparked considerable debate, but the majority opinion within the scientific community was that women with breast cancer, as a group, do not appear to have higher body burdens of DDE and PCBs than women without breast cancer.

Some researchers found these results reassuring. Others worried that these studies had not considered the underlying genetic differences among women nor taken into account the timing of exposure. What if, they asked, some genetic subgroups were more susceptible to environmentally induced breast cancers than others? Furthermore, most studies had measured levels of DDE or PCBs in adult women—after much of their residues had been eliminated from the body and the chemicals themselves long banned. What if contemporary measures do not accurately reflect historical exposures? What if the important variable is DDT exposure during childhood or adolescence—when the developing breast is most vulnerable? Animal studies clearly demonstrate the importance of toxic exposures that occur in early life when the breast is most sensitive to damage.

The ideal study would be designed like this: go back in time to a year of peak DDT usage—say, 1963—gather blood from U.S. girls, and then follow them through their adult lives to see if those exposed to the highest levels of DDT as children went on to suffer higher rates of breast cancer as adults.

And then someone did just that. More or less. As described in a paper published in 2007, Barbara Cohn and her colleagues at the University of California unearthed medical records and banked blood samples of women who had visited a clinic between 1959 and 1967 to seek routine prenatal care. Knowing that DDT came on the market in [illegible], Cohn was able to calculate how old each woman was when she was first exposed to DDT. And she was also able to trace these women and learn their current breast cancer status. The results were clarifying: Women exposed to DDT after age 14—those born in 1931 or before—showed no association between exposure to DDT and breast cancer. But among women exposed to DDT when they were younger than 14, a significant relationship existed: women with high DDT levels were five times more likely to be diagnosed with breast cancer by age 50 than those with the lowest levels. In other words, this study showed a fivefold increase in breast cancer risk among women who had experienced high exposures to DDT before puberty but not in women so exposed after their breasts had already developed. Thanks to hundreds of test tubes that stood silently in the back of a freezer in Oakland, California for a half century, and thanks to breast cancer activists who insisted that environment studies go forward, we now know that DDT exposure in childhood can significantly increase breast cancer risk in adulthood. And we gained this knowledge nearly forty years after DDT was banned.

Meanwhile, other researchers went to work categorizing women genetically. They looked closely at women who had inherited a variation of CYP1A1, a gene that is involved in metabolizing hormones and that is known to be influenced by PCB exposure. About 10–15 percent of white women in the United States are thought have the variant gene. The proportion of black women who have it is not yet known. When data on women with the variant gene were examined in isolation, a picture began to emerge: women who possessed both the genetic variation as well as a high PCB body burden had an elevated rate of breast cancer. Indeed, their rate of breast cancer was two to three times higher than that of women with lower levels and without this genetic trait. The evidence to date now supports an association between breast cancer and PCB exposure for subpopulations of women who have inherited this particular genetic variation. And we gained this knowledge nearly thirty years after PCBs were banned.

The story of atrazine today is much like the story of DDT and PCBs as it was told decades ago. Worrisome findings followed by equivocal ones. Inconsistencies. Contradictions. Balls of confusion. The difference is that atrazine is not banned. It is the second most abundantly used pesticide in the United States, and its manufacturer plays an aggressive role in defending its product. A proven endocrine disruptor, atrazine causes breast cancer in one strain of rat. Some argue that it does so by a mechanism not relevant to humans. The human studies themselves are inconclusive and, while a few show possible associations, most do not report a link between adult exposure to atrazine and breast cancer. However, no human studies have looked at early-life exposures to atrazine, which is when atrazine exerts its strongest effects in lab animals. A 2009 study called for an investigation into how atrazine might be affecting the pace and tempo of sexual maturation in girls. (Early puberty is, by itself, a known risk factor for breast cancer.) Other human studies have found suggestive evidence for an association between atrazine exposures and several other cancers, including lymphoma and cancer of the prostate, ovary, testes, and brain. There is also suggestive evidence for unique toxicities arising from mixtures of atrazine with other farm chemicals. Laboratory studies report possible synergistic effects: among invertebrate animals, atrazine induces an enzyme that makes a second pesticide, chlorpyrifos, more toxic. In this way, exposure to one contaminant can turn another into a more powerful poison. Are these results applicable to humans? It's not yet clear.

By 1994, the evidence against atrazine was troubling enough that the U.S. Environmental Protection Agency (EPA) initiated a special review of its registration. Nine years passed. Meanwhile, across the Atlantic, regulators in Europe announced the results of their own review: atrazine was banned throughout the European Union. Finally, in 2003, the EPA announced its decision: continued use of atrazine was approved. This was an intensely controversial decision. One researcher pointed out in disgust that DDT was abolished on the basis of less evidence than we now had for atrazine.

In October 2009, the EPA announced a plan—and a timetable—for a new evaluation of atrazine.

~~~

Ten thousand years of tallgrass prairie have left a fainter trace on the place I call home than twenty-seven years of DDT, forty-six years of PCBs, and fifty years of atrazine. Because it is my home, I am driven to pursue the question of the past and ongoing contamination of Illinois and its possible link to the increasing frequency of cancer there. I believe that all of us, wherever our roots, need to examine this relationship. And I think it reasonable to ask—nearly a half century after *Silent Spring* alerted us to a possible problem—why so much silence still surrounds questions about cancer's connection to the environment and why so much scientific inquiry into this issue is still considered "preliminary."

From dry-cleaning fluids to pesticides, harmful substances have trespassed into the landscape and have also woven themselves, in trace amounts, into the fibers of our bodies. This much we know with certainty. It is not only reasonable but essential that we should understand the lifetime effects of these incremental accumulations.
~~~

two

silence

The very modern Beinecke Library at Yale University is the resting place for Rachel Carson's papers. The cool, gray archival boxes that contain her correspondence, lecture notes, and personal writings must be requested one at a time from the librarian's assistant. The special room for viewing them is hushed and spacious. A wall of windows looks out over a green collegiate lawn. One enters after a ritual of giving over all personal possessions to the librarian. No ink is allowed in the viewing room—only pencils or laptop computers.

Alone in this room with the first box I sift slowly through the pages it holds as though I were sorting botanical specimens. It is an automatic reflex, although I have not worked in a botanical herbarium for years. Herbarium sheets, onto which the delicate skeletons of dried plants are pressed, must never be flipped over like pages in a book but rather are to be laid gently in reverse order to the left of the stack one is looking through. When finished, the examiner places the sheaves, one at a time, on top of the stack to the right, and they thus assume their original position. At least, this is the method I was taught. Something about the ceremony of my current task has triggered this old behavior. I can only hope it approximates correct archival technique.

The sight of Rachel Carson's handwriting is exhilarating. I uncover a note to Carson from Jacqueline Kennedy. Deep in another file is a letter of complaint Carson sent to a music company after receiving an erroneous

In a nation where guarantees of free speech are carved into the heart of our legal system, we are very often baffled by those who claim they have been silenced. I myself have never feared my mail would arrive with passages blacked out by a censor's invisible hand. I have never wondered if the police would stop me on the way to class to announce that the content of my lecture was unacceptable. And yet perhaps we have all witnessed certain subtle codes of silence in operation—an unspoken agreement in the workplace or a family secret that everyone knows but does not discuss.

Rachel Carson was interested in three forms of silence. As a government scientist—she rose through the ranks of the U.S. Fish and Wildlife Service—Carson became concerned that the noise of important ecological debates carried on within federal agencies seldom reached the public. The long-running quarrel over the claim that pesticides were harmless was one she followed most closely. By virtue of her position, she had access to field reports clearly indicating that attempts to eradicate insect pests through massive chemical spraying programs had many unintended consequences for people and wildlife alike. This view, although denied vociferously by some in the government, was shared by many of Carson's colleagues. Yet the citizenry heard little of this debate. The problem was not so much that those questioning the wisdom of eradication programs were spirited away in the middle of the night but that much of their data remained soundproofed in internal documents and technical journals, that follow-up research was sorely underfunded, and that government officials turned a deaf ear to bearers of bad news.

By 1952, Carson had become a best-selling author of nature books and was able to retire from government service. However, she continued to follow the pesticide debate as it clamored through the halls of the U.S.

Department of Agriculture and the National Academy of Sciences. In 1958, a gardener named Olga Owens Huckins sent Carson a letter full of painful details about a mosquito control campaign that had resulted in a mass death of songbirds near her home. Those that lay scattered around her DDT-contaminated birdbath had perished in a posture of grotesque convulsion: legs drawn up to their breasts, beaks gaping open.

This letter prompted Carson to begin a comprehensive investigation of pesticides. In letters to friends about this project, she referred often to her need to speak out in defense of the natural world: "Knowing what I do, there would be no future peace for me if I kept silent." Having documented a cavalcade of problems attributable to pesticides—from blindness in fish to blood disorders in humans—she could find no magazine or periodical willing to publish her work. Carson decided to write a book.

Its title, *Silent Spring*, refers to the eerier kind of silence: the absence of birdsong in a world poisoned by chemicals. Indeed, Carson argued, pesticidal warfare, waged with reckless disregard, threatens to extinguish a chorus of living voices—those of birds, bees, frogs, crickets, coyotes, and ultimately us. On this level, *Silent Spring* can be read as an exploration of how one kind of silence breeds another, how the secrecies of government beget a weirdly quiet and lifeless world.

Through this process of silencing, the interconnectedness of all life forms is revealed. Carson studied the failed attempt to prevent the Japanese beetle from invading Iroquois County, Illinois, a rural farming community located due east of my home county. After intense and repeated pesticide bombardments by air during the mid-1950s, many insect species, sickened by the spraying, became easy prey for insect-eating birds and mammals. These creatures became poisoned in turn and, in ever-widening circles of death, went on to sicken and kill those who fed on their flesh, leaving a landscape devoid of animal life—from pheasants to barnyard cats.

Meanwhile, the targeted beetle species continued its westward advance. The protracted war against this enemy had accomplished nothing, but the residues of dieldrin remaining in the water and soil—like landmines left behind by a retreating army—guaranteed further casualties for decades to come. All for the dream of a bugless world. The ecological tragedy of [illegible], said Carson, is narrated by the mute testimony of its dead ground squirrels, found with their mouths full of dirt; they had gnashed at the ground as they died.

The third kind of silence that fascinated Carson was the hushed complicity of many individual scientists who were aware of—if not directly involved in documenting—the hazards created by chemical assaults on the natural world. While dutifully publishing their research, most were reluctant about speaking out publicly, and some refused Carson's requests for more information. Writing in *Silent Spring*, Carson acknowledged the constant threat of defunding that hushed many government scientists. But she made clear in her private correspondence that she had little respect for those who knew but did not speak, a combination she saw as cowardice:

> The other day I saw a wonderful quote from [Abraham] Lincoln. . . . I told you once that if I kept silent I could never again listen to a veery's song without overwhelming self-reproach. . . . The quote is "To sin by silence when they should protest makes cowards out of men."

After *Silent Spring* was published, Carson turned her attention to the political and economic reasons behind the fearful silence of her colleagues in science. In a speech to the Women's National Press Club, she questioned the cozy relations between scientific societies and for-profit enterprises, such as chemical companies. When a scientific society acknowledges a trade organization as a "sustaining associate," Carson asked, whose voice do we hear when that society speaks—that of science or of industry?

Carson was just beginning to develop her ideas on the interlocking economic structures that bound the direction of medicine and science to the interests of industry when she herself was silenced. Leaving behind an adopted son, plans for summer fieldwork, and sketches for two more books, Rachel Carson died of breast cancer on April 14, 1964.

Sheltered from wind and waves, the Rachel Carson National Wildlife Refuge in southern Maine is essentially a salt marsh. It bears little resemblance to the rest of the Maine coastline, where the intense drama of ocean meeting rock prohibits marsh grasses from taking root. It is, therefore, a very different place from the craggy tidal pools and moonlit coves of Rachel Carson's beloved summer home farther north.

Walking along the paths of the refuge that bears her name, I realize I feel less close to Rachel Carson here than in the climate-controlled sanctum of the Beinecke Library. At the dedication site, a large plaque dutifully lists the titles of her books and then credits her for inspiring millions to greater environmental consciousness. Its brief, abstract sentences remind me how remote a figure Carson became after her death. Like Rosa Parks, Carson is a symbol, a muse, a spark that ignited a social movement, a name to be invoked before a speech. In this, she seems unknowable and unhuman.

Still, my Illinois nerve endings are stirred by the softness of the landscape here. The lay of the land feels familiar, although most of the plant species are not. Salt meadow grass knits together the higher grounds, while the lower sweeps are bound by the taller and stiffer saltwater cordgrass. The sinuous borders between them represent the reach of the tide. The trail guide boasts that these two grasses together can produce as much plant matter per acre per year as a prime midwestern cornfield. I smile. No way.

It is November 1993. I have driven here from Boston with my friend Jeannie Marshall, who patiently endures my lecture on corn productivity and then turns my attention to the weather. "Doesn't it feel like a different season?" Jeannie asks. On the dry uplands, a rich summery light pours through the oak trees that hang willfully onto their curled leaves.

[illegible] light me. I depend on surface water to reveal slope and direction, but poised here at the margin of the sea, these two concepts are subordinated to a larger force. At low tide, the creeks flow into the ocean. At high tide, the ocean flows into the creeks. The streambeds here pulse back and forth, flooding and draining, in a continual exchange of water and salt. There is no clear direction.

Which is exactly how I feel standing next to my friend: poised without direction in an uncertain but beautiful season. Hopeful yet unnerved.

Just diagnosed for a second time with a rare cancer of the spinal cord, Jeannie is in between surgery and radiation treatments. She is recovering quickly—getting well in preparation for becoming sick in an attempt to get well. She moves so nimbly along the paths looping through the refuge that I scarcely need to modify my own movements. If not for her cane, we could be mistaken for any two young day-trippers escaping from the city. But we are on an escape of another kind, and I feel protective and scan the path ahead for rocks, roots, and sinkholes.

Although our friendship is a recent one, the many parallels in our lives promote intense conversations whenever we are together. Both of us are writers in our thirties. Both of us became cancer patients in our twenties. Both of us grew up in communities with documented environmental contamination, high cancer rates, and suspicions that these two factors are related to each other. Both of us grew up in families constructed through adoption (Jeannie's mother was adopted, as I was), and we each have a keen curiosity about the interplay between heredity and environment in our lives.

And we have spoken at length about all of these topics. We have talked about what it means to have cancer as young women and about the relative significance of genealogy and ecology in that context. We have discussed our relationship with our doctors, our families, our hometowns, our writing, our bodies.

The depth and easiness of our talking carry us along today—through the luminous oak groves, out along the boardwalks that float over salt meadow grass, up onto the observation deck that overlooks the confluence of the Mariland River and Branch Brook, whose waters throb back and forth. It seems to me in these moments that Jeannie and I have words for everything. We have rejected the cultural taboos of the past that wrapped the topic of cancer in shrouds of silence, but we have also turned away from the happy cancer chatter that regularly arrives in our mailboxes in the form of brochures and magazines dedicated to the concepts of coping, accommodating, and adjusting to this disease. In its place, we have created a language between us that is compassionate, smart, fearless, open.

What my friend and I do not choose to talk about this afternoon are the dark days that lie ahead for her. Days of lying under the crosshairs of a proton-beam cyclotron. Fatigue, vomiting, blood tests. Continuously handing one's body over to technicians and doctors in a process that we call becoming medicalized. But between us, we have years of experience with cancer. I have no doubt that when those days arrive we will find a vocabulary for every experience.

We pause to examine some small ponded areas near the brook. These are salt pannes—low spots that hold water when the tide ebbs. Evaporation concentrates the salt to such extraordinary levels that only a few inconspicuous plants can survive. Glassworts. Sea-blite. Life thriving among bitterness.

"I like this place," I finally admit.

"I do, too. It's nice to be here."

~~~

On average, breast cancer robs the woman it kills of twenty years of life. This means that in the United States, nearly one million years of women's lives are lost each year. In 1964, Rachel Carson died at age fifty-six—twenty
~~~

years short of the average life expectancy for U.S. women at that time. Despite all the ways she [illegible], as a victim of breast cancer Carson was utterly typical.

Carson was diagnosed in 1960, in the thick of researching and writing *Silent Spring*. Her tumor spread to her lymph nodes and to her bones, eventually including her spine, pelvis, and shoulder. She continued writing, even though surgery left her exhausted and radiation treatments, nauseated. Other ailments—joint and heart problems that were exacerbated, if not caused, by the radiation—brought crippling and immobility. The tumors in her cervical vertebrae caused her writing hand to go numb.

Carson lived for eighteen months after finishing *Silent Spring*—long enough to smoke out a hornet's nest of ridicule and invective from the chemical industry, as well as to receive every imaginable award from the world of arts, letters, and science. Privately, Carson expressed relief and satisfaction at having lived to see *Silent Spring* complete—a reaction many of Carson's commentators and colleagues have repeatedly underscored.

But there is another story embedded in the remaining fragments of Carson's private writings. Far from viewing *Silent Spring* as her crowning achievement, Carson ached to go on to new projects as well as to seize the opportunities that her success now afforded. She did not go gently or gratefully into any good night. As her letters reveal, she died hoping for another remission, another field season, more time. And in this desire, Carson appears before us again as a typical woman with breast cancer.

From a letter to her dearest friend, Dorothy Freeman, in November 1963:

> There is still so much I want to *do*, and it is hard to accept that in all probability, I must leave most of it undone. And just when I have attained the power to achieve so much I feel is important! Strange, isn't it?

And a few months later:

But in spite of the blow yesterday, darling, [presumably, news of more cancer] I am able to feel that another reprieve will perhaps be won. . . . Now it really seems possible there might be another summer.

There was not.

~~~

The winter of 1994 let go of Boston during the second week of March. Over 100 inches of snow had fallen since December, and most of it lay in towering mounds over every inch of grass and concrete that was not a passage for car traffic or an entrance to a building. Now the ice piles were finally melting, and everything that had been lost or abandoned began to surface: mittens, shovels, coat hangers, trash cans, lumber, laundry baskets, entire automobiles. Stratified layers of sand, cat litter, and gravel, which had been trapped at various depths, redeposited themselves in swirling alluvial fans along the sidewalks as rivulets of meltwater streamed toward the storm sewers.

Jeannie and I move through this landscape on our way from the Massachusetts General Hospital to her apartment in the North End. Neither of us speaks. The sound of our boots on the gravelly outwash seems deafening. Jeannie is not using a cane today, and we are walking even faster than we did four months ago at the salt marsh. In my mind's eye, I am tossing all obstacles out of our way—chunks of ice, orange traffic cones, parked cars, cement barricades. I am aiming a wrecking ball at every building.

Neither of us can believe what we have just heard. After eight miserable weeks of radiation treatments to the tumor in her lower back, the original tumor in her neck—successfully removed and treated six years ago—has returned. "Massive recurrence," to quote the neurologist who had just received the scans from the radiologist.

In fact, he said these words to us as soon as we walked into his office and closed the door. We were still standing in our winter coats and had
~~~

not yet found our chairs. "Massive recurrence." I struggled with my buttons, my scarf, the [illegible]. My hands refused to work correctly. It had become my job in [illegible] to serve as the scribe and, as such, to provide complete documentation of conversations between patient and doctor.

This ritual could not withstand the current assault. I am a crack note-taker, but my hands did not want to write the words being spoken. All my attention was trained on overriding my desire to lay down the pen. The doctor spoke quickly and relentlessly as he described the tissues that were being "destroyed" or "strangled" by the chordoma's advance. He was clearly upset but seemed unable to blend his despair with a demonstration of compassion or hope.

Jeannie remained calm. She asked him to conduct a neurological exam; her symptoms, after all, were improving. Her body seemed to be telling a different story. He refused. What would be the point? The scans told the whole story. He asked her to look at them. She refused. They each accused the other of not listening. I focused on writing faster. It was a battle of narrative. Which told the true story? The radiologist's report? Or Jeannie's body? Finally, the meeting ended.

"Don't shoot the messenger," he said flatly as we were once again standing and struggling with our coats.

Now we are back in Jeannie's apartment. A garbage truck backing down the street sets off a car alarm. I imagine setting fire to them both. Jeannie lies on the bed, saying nothing. I make tea.

Say something, I order myself. The words I have just transcribed in the doctor's office are the same ones I have dreaded since my own diagnosis. Now I have heard them spoken—by a doctor who was looking into the eyes of the person sitting next to me. Not mine. Not me.

Say something.

On the day of my diagnosis, I was hospitalized and friends from college came to visit. They politely stepped into the hallway when the doctor came in. He gently told me the results of the pathology reports and the treatment plan he had in mind. We sat together for a while. After he left, my friends gingerly reentered the room. They were trying to be appropriate.

"I have cancer."

There was silence—and then some kind of awkward talking, but no one really acknowledged what I had said, including myself. Later, I was furious with all of us.

Say something.

But what? I sit down at Jeannie's kitchen table and begin to review the notes I have taken to make sure they are legible and complete. Were these the words that were really said? Can their meanings be trusted? Perhaps we had simply entered an unfamiliar culture where the phrase "massive recurrence" actually means "Hello, have a seat," and "don't shoot the messenger" is a way of saying "So long, take care."

You are not saying anything.

I think back to the sunlit oak grove and the salt pannes where language was so easy. How sure I was then that I could be depended on to push any situation, no matter how dire, into the bright daylight of human speech. I think back to Rachel Carson. Tumors in her cervical vertebrae caused loss of functioning in her right hand, the writing hand. Jeannie is also right-handed. It is her left hand that is becoming weak.

~~~~~~

In the four years Rachel Carson struggled with breast cancer, she worked to break silence in the public arena. Yet in her private life, she created at least two kinds of silence. One was permeable; one absolute.

The former kind was a sort of drapery Rachel periodically pulled between herself and her confidante, Dorothy Freeman. In some of her letters to Dorothy, Rachel described the progress of her disease in detailed medical terms. But in others she spoke only in code, referring elliptically to "menacing shadows." Rachel often refrained from divulging bad news, downplayed the miseries of treatment, and stated her belief that expression of fearful thoughts would only make them loom larger.
~~~~~~

[illegible] tones. She refers not to [illegible] radical mastectomy but to her "hurt side."

And yet at other times, Dorothy seemed to feel shut out by Rachel's silences. Both correspondents entreated the other not to censor her thoughts or feelings. Both correspondents also admitted they were not fully disclosing their own secret fears, out of a need to protect the other. Rachel sometimes pulled back the curtain and confided a darker story—one that admitted to pain and despair. Sometimes she followed these communications with retractions and apologies. And sometimes the letters containing the dark confessions were, upon request, destroyed.

Confessing and recanting. Withholding and divulging. This mesh of conflicting impulses is part of a familiar script that is enacted again and again between cancer patients and those who love them. And in this familiarity, Carson emerges once more, poignantly, as an ordinary woman.

The second kind of silence was a fortress of secrecy Rachel constructed around her own diagnosis, a secrecy she expected Dorothy to collude with her in maintaining. Rachel strictly forbade any discussion, public or private, about her illness. This decision was intended to retain the appearance of scientific objectivity as she was documenting the human cost of environmental contamination. She wished to yield her enemies in industry no further ground from which to launch their attacks.

Accordingly, Rachel instructed Dorothy to say nothing of her condition to their mutual acquaintances, lest rumors take root. If need be, Dorothy was to lie. "Say you heard from me recently and that I said I was fine," she told Dorothy to tell her neighbors in Maine. "Say . . . *that you never saw me look better*. Please say that."

What personal price each of these women paid for upholding this code of silence is impossible to know. Being sworn to secrecy can be a terrible burden. Anticipating the unintentional slip of the tongue that could ruin one's career must have been equally crushing. Against this backdrop of agreed-upon silence is the fact that Carson's state of health should have been obvious to anyone who cared to look at her. But not seeing is another form of silence.

As soon as *Silent Spring* was published, Carson was thrust into the national spotlight. She spoke in front of Congress, at the National Press Club, and on national television. In the photographs and old film clips documenting these occasions, she looks for all intents and purposes like a woman in treatment for cancer. She wears an unfortunate black wig. Her face and neck exhibit the distorting puffiness characteristic of radiation. She holds herself in the ginger, upright manner of one who has undergone surgery. The alteration in her appearance that followed her cancer diagnosis is dramatic.

The newspaper clippings in the Beinecke Library that trace her various public appearances in the waning days of her life are full of elaborate descriptions of what type of elegant suit Miss Carson chose to wear and how delightfully she comported herself. The accompanying pictures tell a different story. But it is a story read in silence by a woman from a future generation who knows how it will end.

~~~

Thanksgiving morning is sunny and mild. Jeannie and I decide to walk to Waterfront Park overlooking Boston Harbor. It is now more than a year since our buoyant walk through the wildlife refuge. Jeannie has just finished another round of radiation treatment, and because her balance has been affected, our pace is much slower. Orange tail swishing, my dog circles patiently, herding us toward the water. Somehow, Jeannie has managed to finish writing two articles, one about the search for cancer genes and another on breast cancer prevention for a British medical text. Feeling triumphant, she is in the mood to talk about cancer—but not her own.

"You remind me of Rachel Carson," I laugh.

~~~

Silent Spring is remembered for the birds. When I ask people to name words, phrases, or images that Rachel Carson's book evokes for them, "thin eggshells" is among the most frequent responses. Yet this consequence of pesticide exposure—bird eggs so fragile they crush under the airy weight of their own brooding parents—is scarcely mentioned in *Silent Spring*. Perhaps we like to equate Carson with eggshell thinning because it is a problem that largely fixed itself after DDT and a handful of other pesticides were finally restricted for domestic use. In this way, Carson's predictions of disaster can be simultaneously viewed as both prophetic and successfully averted. A comfortable reckoning.

Of course, the fate of birds and other innocents caught in the chemical crossfire certainly was a central concern of *Silent Spring*. As proof of harm, their deaths were starkly visible. Who can deny the ground squirrels' cold little mouths packed with dirt? Or shrug off the pitiful sight of songbirds writhing in the grass? But *Silent Spring* makes clear that this kind of evidence, however immediate and tangible, is only one part of a much larger assemblage that also includes human cancer. Even while hiding the image of herself as a cancer patient, Carson provided many others: from farmers with bone marrow degeneration to spray gun–toting housewives stricken with leukemia.

Making visible the links between cancer and environmental contamination was challenging for Carson, and the task continues to be daunting. However agonizing their deaths, cancer patients do not collapse around the birdbath. Decades can transpire between the time of exposure to cancer-causing agents and the first outward symptoms of disease. When birds drop out of the sky in great numbers, we ask why. When someone we love is diagnosed with cancer, questions of cause are often of less immediate relevance than questions about treatment. Questions about the past are subordinated to questions about the suddenly uncertain future.

Based on all the data available to her in 1962, Carson laid out five lines of evidence linking cancer to environmental causes. While any one alone would be insufficient proof, when viewed all together, Carson asserted, a startling picture emerges that we ignore at our peril. First, although some cancer-producing substances—called carcinogens—are naturally occurring and have existed since life began, twentieth century industrial activities have created countless such substances against which we have no naturally occurring means of protection.

Second, since the arrival of the nuclear and chemical age that followed World War II, everyone—not just industrial workers—has been exposed to these carcinogens from the moment of conception until death. Industry manufactures carcinogens in such large quantities and in such diverse array that they are no longer confined to the workplace. They have seeped into the general environment, where we all come into intimate and daily contact with them.

Third, cancer is striking the general population with increasing frequency. At the time of Carson's writing, the postwar chemical era was less than two decades old—less than the time required for many cancers to manifest themselves. Carson predicted that the full maturation of "whatever seeds of malignancy have been sown" by the new lethal agents of the chemical age would occur in the years to come. She also believed that the first signs of catastrophe were already visible. At the end of the 1950s, death certificates showed that a far greater proportion of people were dying of cancer than had been true at the turn of the century. Most ominously, children's cancers, once a medical rarity, were becoming commonplace—as revealed both by vital statistics and by doctors' observations.

Carson's fourth line of evidence came from animals. Experimental tests were beginning to reveal that low doses of many pesticidal chemicals in common use caused cancer in laboratory mice, rats, and dogs. Moreover, many animals inhabiting contaminated environments develop malignant tumors; *Silent Spring* not only documents acute poisonings of songbirds but also reports on cases of sheep with nasal tumors. These incidents supported the circumstantial evidence from human populations.

Finally, Carson argued, the unseen inner workings of the cell itself corroborate the story. At the time of *Silent Spring*'s publication, the mechanisms responsible for basic cellular processes such as energy production and regulation of cell division were just beginning to be elucidated. The role and structure of the twisting DNA molecule had been discovered only recently. From the glimmers she was able to gather from widely scattered studies, Carson spotlighted three properties that she believed would ultimately explain why these contaminants were associated with cancer. They were able to damage chromosomes and thereby cause genetic mutations (a property shared with radiation, which had already been shown to cause cancer); they were able to mimic and disrupt sex hormones (high estrogen levels were already correlated with high cancer rates); and they were able to alter enzyme-directed processes of metabolism (by which we break apart molecules, including foreign chemicals that are sometimes metabolically converted into carcinogens). Carson predicted that future studies on the mysterious transformation of healthy cells into malignant ones would reveal that the roads leading to the formation of cancer are the same pathways that pesticides and other related chemical contaminants operate along once they enter the interior spaces of the human body.

Like the assembling of a prehistoric animal's skeleton, this careful piecing together of evidence can never furnish final or absolute answers. There will always be a few missing parts, first because experimenting on human beings is not, thankfully, considered ethically acceptable. Human carcinogens must, therefore, be identified through inference. One set of clues is provided by observations of people who have been inadvertently exposed to substances suspected of having cancer-causing tendencies. But often these people have been exposed to unknown quantities over unknown periods of time. Observations of laboratory animals exposed to known quantities of possible carcinogens supply a second set of clues. But different animal species can vary in their vulnerability to certain kinds of cancers and in their sensitivity to certain kinds of chemicals. Which species can serve as our surrogates in these studies? Rats? Mice? Fish? Dogs? Is there a species whose lymph nodes, bone marrow, brain tissue, prostate glands, bladders, breasts, livers, and spinal cords behave most like those in humans when exposed to particular substances?

Another reason for scientific uncertainty is that the widespread introduction of suspected chemical carcinogens into the human environment is itself a kind of uncontrolled experiment. There remains no unexposed control population to whom the cancer rates of exposed people can be compared. Moreover, the exposures themselves are uncontrolled and multiple. Each of us is exposed repeatedly to minute amounts of many different carcinogens and to any one carcinogen through many different routes. From a scientific point of view, such combinations are especially dangerous because they have the capacity to do great harm while yielding meaningless data. Science loves order, simplicity, the manipulation of a single variable against a background of consistency. The tools of science do not work well when everything is changing all at once.

~~~

It is March 1995. Winter and spring have hung together in the air for weeks, neither yielding to the other. On the phone, Jeannie is trying to describe to me a new sensation she feels across the skin of her chest. It is vague and formless. There are no real words for it. I am attempting to understand how this symptom fits together with a few other recent problems she has reported. Morning vertigo. A funny feeling when she swallows. What picture is emerging here? What does her doctor say? She turns back my questions.

"Let's talk about the chapter you're writing now. What is it called?"

"Silence."

"Let's talk about that."

~~~

Recently, I have become fascinated by the evident reciprocity between environmental activism and *Silent Spring*. I have come to believe that Carson was as influenced by activism and advocacy as the contemporary environmental movement was influenced—some would say inaugurated—by the publication of *Silent Spring*.

In her acknowledgments at the beginning of *Silent Spring*, Carson credits citizen activists as much as scientists for convincing her to speak out. "In a letter written in January 1958, Olga Owens Huckins told me of her own bitter experience of a small world made lifeless and so brought my attention sharply back to a problem with which I had long been concerned," writes Carson. "I then realized I must write this book." But Huckins, the letter writer Carson credits for [illegible] her to write *Silent Spring*, was more than a gardener. As Carson's biographer, Linda Lear, documents, Huckins was also a member of the Committee Against Mass Poisoning, which sought, through lawsuits and protests, to halt the aerial spraying of pesticides. Throughout 1957, members of this committee, including Huckins, wrote letters to the editors of many New England and Long Island newspapers. These letters did not scientifically document the harm created by the broadcasting of pesticides. Instead, they bore witness to many small tragedies, such as dead birds piling up around backyard birdbaths in the aftermath of a particular spraying episode. These letters were also quite polemical. From one: "Stop the spraying of poisons everywhere until all the evidence, biological and scientific, immediate and long run, of the effects upon wildlife and human beings is known."

The Committee Against Mass Poisoning took a human rights approach to environmental harm—as contemporary environmental justice advocates continue to do. In the parlance of today's environmental activists, the introduction of harmful chemicals into air, food, and water (and thereby into our bodies) violates the right to privacy as well as security of person and is referred to as an act of "toxic trespass." Likewise, Huckins condemned aerial spraying of pesticides as "inhumane, undemocratic, and probably unconstitutional."

The citizen lawsuit filed by the committee was useful to Carson because, as it wended its way to the Supreme Court, it became a magnet for media attention. Carson was thus able to elicit the interest of *The New Yorker*. When its editor offered Carson fifty thousand words in the magazine to write about pesticides, she knew she was on her way to a book.

In short, environmental activism in the 1950s raised awareness among editors, and that development, as much as the slow accumulation of scientific knowledge, allowed Rachel Carson to speak out against silence in all its forms.

three

Time

Like a jury's verdict or an adoption decree, a cancer diagnosis is an authoritative pronouncement, one with the power to change your identity. It sends you into an unfamiliar country where all the rules of human conduct are alien. In this new territory, you disrobe in front of strangers who are allowed to touch you. You submit to bodily invasions. You agree to the removal of body parts. You agree to be poisoned. You have become a cancer patient.

Most of the traits and skills you bring with you from your native life are irrelevant, while strange new attributes suddenly matter. Beautiful hair is irrelevant. Prominent veins along the soft skin at the fold of your arm are highly prized. The ability to cook a meal in thirty minutes is irrelevant. The ability to lie motionless for thirty minutes while your bones are scanned for signs of tumor is, conversely, quite useful.

Whether it happens at a hospital bedside, in a doctor's office, or on the phone, most of us remember the event of our diagnosis with a mixture of photographic recall and amnesia. We may be able to describe every word spoken, the arrangement of photographs on the doctor's desk, the exact color of the office draperies—but have no memory of how we got home that day. Or we may remember nothing that was said but everything about the bus ride. The scene I happen to remember most vividly—and this must have occurred weeks after my discharge from the hospital—is unlocking my door and discovering that my roommate had moved out. She did not want to live with a cancer patient. This was my [illegible] humiliation. Fifteen years later, the sight of a bare mattress can still cause me to burst into tears.

In [illegible] an estimated 1.48 million people in the United States—four thousand people a day—were told they had cancer. Each of the diagnoses is a border crossing, the beginning of an unplanned and unchosen journey. There is a story behind each one.

These diagnoses also form a collective, statistical story. When all the diagnoses of years past and present are tallied, an ongoing narrative emerges that tells us how the incidence of cancer has been and is changing. Changes in cancer incidence, in turn, provide key clues about the possible causes of cancer. For example, if heredity is suspected as the main cause of a certain kind of cancer, we would not expect to see its incidence rise rapidly over the course of a few human generations because genes cannot increase their frequency in the population that quickly. Or if a particular environmental carcinogen is suspected, we can see whether a rise in incidence corresponds to the introduction of such substances into the workplace or the general environment (taking into account the lag time between exposure and onset of disease). Such an association does not constitute absolute proof, but it gives us ground to launch additional inquiries.

The work of compiling statistics on cancer incidence is carried out at a network of cancer registries, which exist in the United States at both the state and the federal levels. Theoretically, for each new cancer diagnosis, a report is sent to a registry. How a diagnosed person has experienced, reacted to, coped with, remembered, or repressed this stunning event are aspects not included in this accounting, of course. What each report does contain is a coded description of the type of cancer; the stage to which it has advanced; and the geographic region, age, sex, and ethnicity of the newly diagnosed person.

This incoming information is then processed, analyzed, audited, graphed, and disseminated by teams of statisticians. In and of itself, a head count is not very useful. There are more people with cancer now in part, because there are more people. There are also proportionally more older people alive now than ever before, and the aged tend to get more cancers than the young. Between 1970 and 1990, for example, the U.S. population increased by 22 percent, and the number of people over sixty-five increased by 55 percent. To eliminate the effects of the changing size and age structure of the population, cancer registries standardize the data. One way of doing this is to calculate a cancer incidence rate, which is traditionally expressed as the number of new cases of cancer for every 100,000 people per year. For example, in 1973, 99 out of every 100,000 women living in the United States were diagnosed with breast cancer. By 1998, the incidence had risen to 141 out of 100,000. The rate of breast cancer has declined since then and in 2005 stood at 118. Thus, we are still standing on higher ground than we were in 1973, when the federal registry was founded, but we seem to be in a better place than we were a decade ago.

These numbers are also age-adjusted. That is, the data from all the differently aged people from any given year are weighted to match the age distribution of a particular census year. Thus standardized, the statistics from various years can be compared to each other. In this way, we know that the 43 percent rise in breast cancer between 1973 and 1998 did not happen because the population was aging. Alternatively, cancer registry data can be made age-specific: the percentage of forty-five- to forty-nine-year-olds contracting breast cancer can, for example, be compared with the percentage from a decade ago.

I have often wondered about the daily lives of tumor registrars, those souls responsible for keeping count of cancer's casualties. How strange it must be to monitor the thousands of cancer reports that flow into the registries every day. Surely I would want to pluck each one from the current and imagine the life behind the name. A seventy-five-year-old black woman from an urban area with advanced-stage breast cancer . . . or a forty-five-year-old white man from a farming community with chronic lymphocytic leukemia . . . or a seven-year-old girl with a brain tumor.

Reading the circles would make me long to ask, "What has happened to you since your diagnosis? Are you getting good care? Are you surrounded by people who love you?"

Cancer registries publish their findings in thick annual volumes and online databases replete with tables and graphs. My own reaction to these reports follows a particular evolution. At first examination, my eyes disassemble the data. In a graph displaying the age-adjusted rates for ovarian cancer, for example, I initially focus on the points rather than the lines that connect them. I wonder at the individual women whose lives are contained in the little black circles and gray squares that float in a white field of mathematical space. Gradually, as when I am looking at a picture that contains a hidden pattern, another way of seeing emerges from the page. Years of biological training kick in, and my eyes automatically begin to trace the slope of the lines, check the coordinates, imagine how the data might appear if displayed logarithmically.

In many ways, tracking the changing patterns of cancer incidence is not unlike tracking the patterns of ecological change. The statistical methods are certainly very similar—as are the vexing problems.

I once compiled old and current species inventories in order to monitor gradual changes in the composition of a Minnesota forest over several decades. During this time, some species became more common and others more rare. Sometimes I literally could not see the forest for the trees. The graphs constructed from my data showed clear trends often not apparent to me as I walked the deer paths that meandered among the pillars of the ancient canopy pines and through the green tangle of shrubs and saplings below. Without an exact count, I tended to overestimate the presence of rare plants because my delight at discovering them was more memorable than my efforts to note the prevalence of their more common neighbors. Perception can be misleading.

But I also had reasons to distrust parts of my data. To study time trends over half a century, one must rely on census counts conducted by many previous researchers, including some no longer living. If their system of coding and classifying differed significantly from mine, or if any one of us consistently misidentified certain species, then the changes indicated by my graphs were artifacts of our different techniques rather than reflections of a real biological shift. The seeming disappearance of a species that then suddenly reappeared in abundance five years later was a likely indication of a methodological snafu.

Cancer registry data are cursed with similar problems. We need the data because perception can mislead. It may seem to us that more and more people are getting brain tumors or that breast cancer is striking women at increasingly younger ages, but what do the numbers actually show? Perhaps people with cancer are now simply more outspoken than their predecessors. Yet the numbers can also deceive. Earlier detection, changes in the rate of misdiagnosis, and coding and classifying tumor types mean that apparent rises or falls in incidence rates can be artificial. How to quantify and correct such problems is a recurring question at tumor registrars' conferences.

Let's look closely at the statistics on the rise and fall of breast cancer incidence. Its swift ascendency between 1973 and 1991 corresponds to the introduction of mammography as a screening tool. This new technology changed the way many U.S. women were diagnosed with the disease, presumably because malignancies could be identified before being felt as a lump. How much of this rise can be explained by the increased use of mammograms? To answer this question, statisticians first look to see whether breast cancer incidence began to rise at the same time mammography became widely available. An internal audit of the data can also show whether groups of women with the highest rates of cancer are those receiving the most mammograms. And, since mammograms purportedly detect cancer earlier, statisticians can check whether the diagnosis of small breast tumors has been increasing faster than the diagnosis of large, advanced ones.

While still a matter of some debate, the most widely accepted estimate is that between 25 and 40 percent of the 43 percent upsurge in breast cancer incidence in the 1970s, 1980s, and 1990s is attributable to earlier detection. Underlying this acceleration there still exists a gradual, steady, and long-term increase in breast cancer incidence that began after World

War II and continued into the 1990s. This slow rise—between 1 and 2 percent each year since 1943—predated the introduction of mammograms as a common diagnostic tool. Moreover, the groups of women in whom breast cancer incidence was ascending most swiftly—blacks and the elderly—were among those least served by mammography. Therefore, the majority of the increase in breast cancer during this time period cannot be explained by mammograms.

From its peak in the 1990s, the breast cancer rate began to ebb. It stabilized in 2001–2003 and then dropped noticeably in 2003 and stabilized again in 2004. While this disease remains, by far, the most prevalent cancer among U.S. women, and while U.S. women still have the highest rates in the world, we can now say with some conviction that the constellation of factors contributing to breast cancer—whatever they are—seem to be receding. The serial killer known as breast cancer is claiming proportionally fewer victims than a decade ago.

But why?

There are four possible explanations, and they all have some merit, and they all have some defects. About them all, I remain agnostic. At this writing, the most popular hypothesis is that the ongoing decline in breast cancer is attributable to the widespread collective decision by postmenopausal women to stop taking hormone-replacement drugs in 2002. In that year, the *New England Journal of Medicine* reported trial results from the Women's Health Initiative that indicated likely excesses of breast cancer and undeniable excesses of heart problems among women taking estrogen and progestin hormones to ease the side effects of menopause. These findings made headlines around the nation and triggered a repudiation of pharmaceutical approaches to menopause. Among California women alone, the use of estrogen-progestin replacement drugs plunged by 68 percent between 2001 and 2003.

Scoring more points for this hypothesis: the fall in breast cancer rate that followed was mostly driven by a waning in the number of estrogen-dependent tumors among postmenopausal women, exactly the subpopulation that makes up the customer base for estrogen-progestin drugs.

Breast cancer among younger women and women with estrogen-negative tumors have not shown the same dramatic downward trends.

But they do show some signs of change. Breast cancer rates among young African American women—which are higher than among their white counterparts—are also now, thankfully, starting to fall. These women have, disproportionately, more estrogen-negative tumors. Changing attitudes about hormones and changing practices about coping with menopause are not an issue for this group. So what explains their receding rates?

The second hypothesis—which is not mutually exclusive with the first—is that declining rates of mammography screening explain the declining numbers of breast cancer diagnoses. (There is evidence to show that women are not as faithful as they once were about getting mammograms.) If true, this would be the saddest explanation of the four because it means that the incidence of breast cancer may not be declining after all, just temporarily hiding from the statisticians. Sooner or later, women now walking around with undiscovered breast tumors will be diagnosed, and breast cancer rates will ascend again.

The third possible explanation for falling rates of breast cancer is disproportional underreporting. It's possible that the reports of recent cases, still dribbling in, are more likely than reports of older cases to be missing from the analysis. Old data sets are always more completely analyzed than new data sets. On more than one occasion in the past, headlines trumpeting a decline in breast cancer incidence had to be retracted upon subsequent analysis. (And yet the current downward trend has persisted for several consecutive years.)

The fourth possibility is that declining breast cancer rates are caused by declining exposures to causative agents other than (or in addition to) pharmaceutical estrogen. Indeed, biomonitoring data show blood levels of some hormonally active suspected breast carcinogens, such as DDT and PCBs, are now finally falling, many years after their ban. (More on this in Chapters Five and Twelve.) This hypothesis would help explain why the beginning of the recent recession of breast cancer predates the swift and dramatic change in hormone replacement drug use. Supporting this idea is a recent Spanish study that demonstrated an association between breast cancer risk and body burden of all estrogenic chemical contaminants, including natural hormones.

I am writing in a running stream of data here. Science writers always do. If you are reading these words years in the future and know how this story turns out, you may be smiling at the silly notions that early twenty-first-century scholars put forth about the origins of breast cancer, a disease that currently kills forty-one thousand women a year in the United States. For my readers in the present epoch, I urge the following approach: the evidence today says that lowering population-wide exposures to estrogen prevents breast cancer. The abandonment of hormone replacement therapy as a treatment for menopause and the attendant drop in breast cancer incidence is a natural human experiment that serves as Exhibit A. On the grounds that we are morally obligated to act on the basis of the best information available to us now, let's initiate another natural human experiment: screen chemicals in commerce for their ability to act like estrogen and systematically phase out the estrogen mimics. Then see what happens to the breast cancer rate. Go on to publish papers with titles such as, "The Impact on Breast Cancer Incidence of Eliminating Hormonally Active Agents from Agriculture and Consumer Products."

Cancer registries provide clues about cancer's causes when many years of data are available. Unfortunately, many state cancer registries are new; they cannot look back across fifty years as I could with my tree inventories. The Illinois State Cancer Registry was created in 1985. My own diagnosis, which took place in 1979, is therefore not part of the collective story of cancer in Illinois.

Regional comparisons are often difficult because cancer registries in neighboring states can vary wildly in their length of operations. For example, Connecticut has the oldest functioning registry, one started in 1935. The Connecticut Tumor Registry provides one of the only truly long-term views of U.S. cancer incidence. Massachusetts established its cancer registry in 1982. Nearby Vermont had no cancer registry until 1992.

This patchwork of state-based registries is afflicted with another problem that we who count plants never have to worry about. People, unlike trees, move. Lifelong residents of one state, for example, may migrate to another upon retirement and become statistics in their new community. Without a comprehensive national cancer registry—which the United States does not have—state registries must rely on an elaborate system of data exchange. This is especially crucial for my elongated home state of Illinois, which shares a border with five other states. When faced with a serious health problem, many rural folk in the central and southern counties wind up being diagnosed across the Mississippi and Wabash Rivers because they would rather travel to cities in Iowa, Missouri, Indiana, or Kentucky than make the long trek north to Chicago. When Illinois began trading registry data with its neighbors, cancer incidence suddenly rose in its many east and west border counties.

A handful of state and metropolitan registries also contribute data to a partial federal cancer registry. The so-called SEER (Surveillance, Epidemiology, and End Results) Program, overseen by the National Cancer Institute, does not attempt to record all cases of cancer in the country, but instead samples the populace from seventeen geographic areas. The 26 percent of the U.S. population so represented stands in for us all. Collecting cancer diagnoses since 1973, SEER is a child of the War on Cancer as declared by President Richard Nixon and codified within the National Cancer Act of 1971. Without a nationwide registry, no one can know exactly how many new cases of cancer are diagnosed in the United States every year. Instead such numbers are estimated by applying rates from the SEER registry to the population projection for any particular year. Admittedly, it's a makeshift system.

As recently as 1992, ten states were not counting cancer at all. In that year, Congress passed the National Program of Cancer Registries to establish registries in all states and improve their quality. The Centers for Disease Control oversees these state registries and, since 2001, shares cancer

From 1950 to 2001, the overall incidence rate for cancer increased by 85 percent, peaking in the early 1990s. Since 1992, cancer incidence has fallen by 6.5 percent—about a half percent a year—and most of that decline was driven by declines in lung cancer among men and, to a lesser extent, by declines in colon cancer among both men and women, and breast cancer among women. The overall cancer incidence rate is more than twice the mortality rate and currently stands at 463/100,000.

~~~

Incidence data were not available to Rachel Carson when she first documented what she believed were the beginnings of a cancer epidemic. Instead, Carson focused on rising death rates from cancer. She was most disturbed by evidence that, within a few decades, childhood cancer had jumped from the realm of medical rarity to the most common disease killer of American schoolchildren.

Some researchers believe that mortality rates—which are also adjusted for age and population size—are still a more reliable indicator than incidence because they are less affected by changes in diagnostic technique. Death is certain and absolute. Moreover, causes of death, duly noted in all states of the union, have been tallied for far longer than tumors have been registered. We have a much deeper and wider view when we examine cancer trends over time using information gleaned from death certificates. Cancer mortality is calculated as the number of cancer deaths each year per 100,000 persons. It stands at about 200/100,000, a figure that has not budged much for sixty years. Mortality data reveal that we have failed to substantially reduce the deaths from cancer in spite of massive funding for cancer treatment research. Cancer mortality in
~~~

2005 was 10 percent lower than it was in the 1970s and a mere 8 percent lower than in 1950. Reduction in smoking rates accounts for most of the lives saved. Mortality rates reveal no substantial declines in the burden of cancer for middle-aged adults. For persons 45–64, cancer remains not only our leading cause of death, it kills more of us each year than heart disease, accidents, and stroke combined.

Nevertheless, rising incidence rates coupled with stable or falling death rates do mean that more people with cancer are surviving longer. Over 11 million people in the United States—more than ever before—are now walking around with cancer, or walking around in remission from cancer, or even, in some cases, walking around cured from cancer. This is good news: people get to live longer after being diagnosed than they did in previous decades. But the bad news that the good news carries with it is that this increase in the prevalence of cancer—the so-called cancer burden—ushers in economic, social, and psychological costs that we have not had to bear before.

Nowhere is this more true than childhood cancers, which jumped in incidence by 22 percent between 1973 and 2000 even as the death rate fell by 45 percent. Using mortality to measure the occurrence of cancer in children today would create a falsely rosy picture. Improved treatment may be saving more children from death, but every year more children are diagnosed with cancer than the year before. Increases are most apparent for leukemia (up 35 percent), non-Hodgkin lymphoma (up 33 percent), soft tissue cancers (up 50 percent), kidney cancer (up 45 percent), and brain and nervous system tumors (up 44 percent). Cancer among children provides a particularly intimate glimpse into the possible routes of exposure to contaminants in the general environment and their possible significance for rising cancer rates among adults. It's hard to blame children's cancers on dangerous lifestyle choices. The lifestyle of toddlers has not changed much over the past half century. Young children do not smoke, drink alcohol, or hold stressful jobs. Children do, however, receive a greater dose of whatever chemicals are present in air, food, and water because, pound for pound, they breathe, eat, and drink more than adults do. In proportion to their body weight, children drink 2.5 times more water, eat 3 to 4 times more food, and breathe 2 times more air. They are also affected by parental exposures before conception, as well as by exposures in the womb and in breast milk.

The night before Jeannie's death, I dreamed I traveled on a large boat with many other people. No shorelines were visible. Someone suggested I walk out onto the deck and get some sun. "It's too hot," I said. But I walked out anyway and discovered the weather very pleasant. Someone suggested I go for a swim. "Too dangerous," I said. But I dove in, and the water was cool and crystalline. Dolphins circled me protectively. Back in the boat, I asked, "Where are we?" And someone smiled and handed me a map.

Driving across the Charles River to the hospital the next morning, I took the dream as a sign that I had accepted what I understood now to be imminent. But by the time I crossed the river again that night, I knew I had not and never would.

I wanted time to stop. I wanted all the clocks unplugged and the calendars nailed flat to the walls. It was April. I wanted no leaves to emerge from the buds that blurred the outlines of the trees.

Time had become such a strange commodity in the preceding month. On the surface, it had seemed to speed up as the vague progression of Jeannie's various symptoms had suddenly accelerated. One day she found she could no longer type. A week later she could not turn doorknobs. The next week, buttons were impossible. Each loss was profound and irrevocable—the ability to write, to walk through a doorway, to undress.

But under the quick surface, in the deep water at the center of every hour and every moment, time was slowing down. Each meal, each conversation, each walk from one room to another unfolded with such deliberateness that an afternoon spent in Jeannie's apartment was the equivalent of a week.

"You understand this is a terminal event." A doctor's voice on the magnetic tape of my answering machine. The [illegible] drive to the intensive care unit. Each heartbeat visible as data on a video screen. Slow drippings in tubes. An endless night. A blue-black dawn. A nurse's voice as though from a distant room: "Okay. These are her last breaths now."

The whole concept of time was unbearable. I wanted to be back in Illinois in the middle of winter. I wanted to walk across frozen fields. No ocean. No leaves. No boats. She was gone.

~~~

During the middle of the twentieth century, a cancer diagnosis was the expected fate of one in every four Americans—a ratio Carson found so shocking that it inspired the title of one of her chapters. Today, more than 40 percent of us (38.3 percent of women and 48.2 percent of men) will contract the disease sometime within our lifespans. Cancer is now the second leading cause of death overall, and, among adult Americans younger than 85, it is the number-one killer—beating out stroke and heart disease.

More than one-fourth of all cancer deaths are from lung cancer. Because the fatality rate is so high, lung cancer incidence and lung cancer mortality are very nearly the same statistic, and, in the United States, both closely mirror historical patterns of cigarette consumption. Among U.S. men, lung cancer incidence peaked in 1984. Mortality has been declining since the early 1990s. U.S. women began smoking in large numbers later in the twentieth century than did men and took up the habit in earnest when heavily bombarded by gender-specific advertising between 1965 and 1975 (the Virginia Slims effect). Women's lung cancer incidence finally began to wane in 2001. Declines in mortality will surely follow. Overall, 85–90 percent of the deaths from lung cancer can be attributed to cigarette smoking or exposure to passive smoke from other smokers.

This statistic also means, of course, that 10–15 percent of all lung cancer deaths have nothing to do with smoking. This is not a trivial number:
~~~

between sixteen thousand and twenty-four thousand people died in 2008 from lung cancer for which smoking or passive smoking was not the cause. If lung cancers among nonsmokers were broken out and presented as a separate statistic, lung cancers not attributable to smoking would still appear on the list of the ten most common cancers in terms of death. Thus, although smoking dominates the lung cancer picture, additional mysteries need sleuthing here. (More on this in Chapter Eight.)

And, while smoking remains the largest single known preventable cause of cancer, the majority of cancers cannot be traced back to cigarettes. Testicular cancer, now the most common cancer to strike men in their twenties and thirties, has been increasing in incidence for fifty years and rose 23 percent in the last decade alone. The same time period saw marked increases in non-Hodgkin lymphoma and cancers of the esophagus, liver, pancreas, kidney, thyroid, bladder, bone marrow, and skin (melanoma). Although tobacco is indeed a risk factor for some of these cancers (kidney, bladder, pancreas, esophagus), it is not strongly implicated in the others. Nor can trends in smoking, which are going down—and dragging down lung cancer rates along the way—explain why any of these cancers continue to increase in incidence. Nor can their increases be explained away entirely by improved diagnostic practices. Thyroid cancer, for example, doubled in incidence between 1973 and 2002. A 2009 analysis of the incidence data concluded that more sensitive diagnostic procedures could not sufficiently account for the ongoing surge.

~~~

In 1964, when I was five years old, two senior scientists at the National Cancer Institute, Wilhelm Hueper and W. C. Conway, looked at the data on cancer trends available to them at the time and concluded: "Cancers of all types and all causes display even under already existing conditions, all the characteristics of an epidemic in slow motion." This unfolding crisis, they asserted, was being fueled by "increasing contamination of the human environment with chemical and physical carcinogens and
~~~

with chemicals supporting and potentiating their action." And yet the possible relationship between cancer and "the growing chemicalization of the human economy" has still not been pursued in any systematic, exhaustive way.

The environment keeps falling off the cancer screen. The circumstances surrounding the birth of the Illinois State Cancer Registry is a case in point. The registry came into being when the Illinois Health and Hazardous Substances Registry Act was signed by the governor in September 1984. As implied by its name, this state law was intended to "monitor the health effects among the citizens of Illinois related to exposures to hazardous substances in the work place and in the environment." Accordingly, the registry system was to collect information not only on the incidence of cancer among the Illinois populace but also on their "exposure to hazardous substances, including hazardous nuclear material," thus prompting public health studies that would relate "measurable health outcomes to environmental data to help identify contributing factors in the occurrence of disease."

The cancer registry was funded. The hazardous substances registry was not.

Like a thriving child with a stillborn twin, the Illinois State Cancer Registry dutifully acquires information on health outcomes, but this activity now goes on independently of any attempt to correlate health with exposure to hazardous substances.

Hueper and Conway's call to focus cancer research on environmental carcinogens may finally be heard. The National Children's Study, after vanquishing efforts to defund it, was officially launched in January 2009. An investigation of epic proportion, the NCS intends to enroll 100,000 infants to serve as its study subjects. Researchers will monitor the health and development of these children from prenatal life through age twenty-one and, as well, will gather information on the chemicals in their environment—from their mother's bodies during pregnancy to the dust in their teenage bedrooms. This prospective study has potential to reveal, in far more robust ways than previously possible, associations between early-life exposures and a host of health conditions, including pediatric cancers.

But many of the NCS's tiny subjects of study are not yet born, and the final results are more than two decades away, which is a long time for parents of children with cancer to wait.

Meanwhile, cancer epidemiologist Dr. Devra Davis [illegible] has recommended that university cancer centers throughout the nation open centers for environmental oncology. The principal goal of these centers would be a simple but noble one: to improve our ability to prevent cancer. Or more comprehensively, they will "develop specific interventions and policy recommendations regarding ways to lower the burden of cancer, based on existing information about cancer hazards in the personal, occupational, and general environment." In this, Davis already established a beachhead at the University of Pittsburgh Cancer Institute, where she directs its Center for Environmental Oncology.

~~~~~~

Two months pass before I visit the cemetery. It is June. Four days of stormy weather have pelted the last of New England's rhododendron blossoms into the grass. The just-awakening roses, however, are luminescent in the streaming rain. In fact their buds seem to be opening before my eyes.

Time still seems speeded up, as in an old movie when the wind tears the page from the calendar and the characters leap forward into another season. Cars drive too fast. People walk too fast. Food even seems to cook too fast. I've learned to avoid quickness—like dashing out to the post office before it closes—because sudden movements seem to rush time forward even faster. I'm hoping an afternoon in the cemetery will slow the world down again.

I realize immediately that I have no idea where her gravesite is. When last here, I'd noticed nothing but the flower-swathed casket and the mound of dirt draped in green plastic. There was some kind of old, severely pruned tree nearby, but I can't recall the species. In my mind, I can see the round bull's eyes of its sawed-off limbs and the humped roots that had pushed away the hurricane fence behind it. I scan the fence line.
~~~~~~

A row of old basswoods runs along the far side. Finally, I see it: nearly at the end and standing exactly as in my memory. So, it is a basswood. The tree leads me to the rectangle of earth I'm looking for. At the top is a plastic plaque coated with wet petals, seed coats, leaf bits, and stems. JeanMarie Marshall. 1958–1995. Finally, everything seems still enough.

Devra Davis and her colleagues have analyzed U.S. cancer patterns in a novel and revealing way. Rather than simply look at changes in cancer rates over calendar time, Davis grouped people according to year of birth, as well as year of diagnosis, and explored how cancer has affected successive generations. Because data on nonwhites in the early years of the SEER Program are unreliable, she restricted her view to U.S. whites and separated cancers generally believed to be associated with smoking from those not known to be so associated.

Davis found that cancer not tied to smoking has increased steadily down the generations. U.S. white women born in the 1940s have had 30 percent more non-smoking-related cancers than did women of their grandmothers' generation (women born between 1888 and 1897). Among men, the differences were even starker. White men born in the 1940s have had more than twice as much non-tobacco-related cancer than their grandfathers did at the same age. "What this is telling us," Davis says, "is that there is something going on here in addition to smoking, and we need to figure out what that is."

The grandparents of those born in the 1940s are mostly all dead now. Of all the worries they carried for their baby-boom grandchildren—those riotous offspring who opened the original generation gap—cancer, as I recall, was not high on their list. I say this as a lifelong observer of this birth cohort. At one point in my childhood, it seemed that the entire generation born in the decade before mine might die young. By eleven, we all wore metal bracelets engraved with the names of those officially missing in action in Vietnam. About the rest, we heard various dire predictions from adults: perhaps they would all be felled by police truncheons or end

Nothing slows time down as much as waiting for lab reports. This time I am the patient. In the interior waiting room, dressed in a wraparound smock identical to the ones worn by every other human being who has entered this room, I try to conjure Jeannie out of thin air. Of the ample supply of magazines provided us here, she would choose *Vanity Fair*. Of this, I feel certain. During these moments of waiting, which celebrity interview would she, in her unflagging attempt to bring me up to snuff on popular culture, read aloud to me? And when I drifted into anxious thinking, what clever thing would she say to keep me from floating off too far?

Last summer she waited with me for hours at the ultrasound clinic.

"They had a hard time seeing what they wanted to see," I reported back to her as we finally walked out the door. "And then one of the technicians looked at the image in the monitor and whistled."

She laughed. "You know that ranks right up there with 'Hey, nice tits!'"

My name is called and I follow the doctor down the corridor to her office. Like a defendant studying the faces of the jurors as they file back into the courtroom, I try to read her expression.

It seems my situation today is mostly good, but a little bit ambiguous. The specialists have conferred and would like to recommend I undergo a new type of test, which the doctor explains in clear detail.

"I know this isn't what you wanted to hear," she says, with genuine compassion. "But you don't need me to be your best friend right now."

Time lurches forward again. Where is she?

Rising cancer incidence among children is one line of evidence that implicates environmental factors. The increase in cancer incidence among successive generations of adults is another. A third line of evidence comes from a close consideration of the cancers that exhibit particularly rapid rates of increase. If we restrict our view to just one of these cancers, what patterns emerge? Let's zero in on the fifth most common cancer in women and the sixth most common cancer in men: lymphoma.

Non-Hodgkin lymphoma strikes at a tissue designed to protect us from harmful invasions: the knobby lymph nodes clustered in our throats, armpits, groins, and elsewhere. Our tonsils, the most accessible example, represent a constellation of lymph nodes wrapped in a mucous membrane.

The watery fluid that fills the microscopic spaces between all of our cells is, for all intents, lymph. It does not receive that name, however, until it flows from those spaces, like rainwater from a field, into the creekbeds called lymphatic vessels. The origin of all this fluid is the bloodstream, and when held within that system, it is known as plasma. Each day, about three quarts of blood plasma leak out of the capillaries, swirl around freely, and then drain into the lymph vessels. Eventually, lymph becomes plasma again when it is poured back into blood just at the point where the jugular vein joins the subclavian in their return to the heart. Several tasks are accomplished during the ceaseless transformation of plasma to lymph and lymph to plasma. The identification and destruction of foreign substances is one of them. Lymph nodes, scattered along the lymph vessels at various intervals, are honeycombed with a diverse array of cell types specialized for immune response. As the fluid is channeled through the nodes' intricate meshwork, alien life forms are trapped and killed. Lymph nodes can also send immune-responsive cells forth to circulate in other territories of the body.

Because the lymph system also serves as a highway for runaway cancer cells of all kinds, lymph nodes are a significant feature in the cancer landscape. Breast cancers very often spread to nearby lymph nodes, for example. Breast cancer patients are quickly categorized as node positive or node negative, a distinction that depends on whether breast cancer cells, shed from the original tumor, have lodged themselves in the lymph nodes [illegible] between the arm and the trunk of the body. Their presence there indicates the disease has likely dispersed to other, more distant locations.

As a way of measuring the extent of this cancer diaspora, node-positive women are further classified by the number of nodes containing breast cancer cells: 1 to 4 is one kind of identity; 11 to 17 is quite another. "How many nodes?" is very often the first question women in breast cancer support groups ask each other.

But a lymphoma is a different condition. In this case, the tumors derive from lymph tissue itself, not from the immigrant cells that have floated in from someplace else. Lymphoma can arise inside a node, or, because lymph tissue is diffused throughout the body, it can originate almost anywhere elsewhere—in the spleen, for example, or even in the skin. Non-Hodgkin lymphoma is therefore a collection of diseases, in contrast to the very specific and more curable lymphoma called Hodgkin lymphoma, which, in the vast majority of cases, arises from a type of white blood cell called a Reed-Sternberg cell.

While the incidence of Hodgkin lymphoma has declined, non-Hodgkin lymphoma has shot up—doubling in incidence since 1973. This increase is evident in both sexes and most age groups. Jackie Kennedy Onassis was killed by one of its most malignant incarnations. Between 1995 and 2005, its incidence increased faster among women than among men, although it still afflicts more men than women.

At its peak, the AIDS epidemic contributed to some of the increase in non-Hodgkin lymphoma. A small but significant percentage of AIDS patients were diagnosed with lymphoma, which for many caused their death. However, the steadily upward momentum of non-Hodgkin lymphoma incidence in the United States was already under way decades before the AIDS epidemic sank its teeth in, and it still continues now that AIDS has retreated.

What do we know about the people who get this disease? We know that people who work in certain occupations are overrepresented among NHL patients. Farm workers and workers in dry-cleaning shops are cer-

only two. Firefighters, airplane mechanics, and farmers may be others. We also know that lymphoma is consistently associated with exposure to synthetic chemicals, including solvents, PCBs, and pesticides of many kinds, especially a class of weedkillers known as phenoxy herbicides.

In a 2007 publication, biostatistician John Spinelli and his colleagues in British Columbia reported on a study that measured PCBs in the blood of lymphoma patients and compared them to those of matched controls. He and his team found a relationship: high blood levels of certain types of PCBs raised non-Hodgkin lymphoma risk by moderate but significant amounts. This study, which corroborates others also reporting elevated lymphoma risk with PCB exposure, makes biological sense: PCBs are known immune suppressors; immune alterations are known to increase the risk for non-Hodgkin lymphoma. In 2009, the hypothesis that PCBs contribute to non-Hodgkin lymphoma via mechanisms that undermine immunity was strengthened by two other findings. The first was the discovery that infection with the Epstein-Barr virus makes PCB exposure an even more potent risk factor for lymphoma. The second was the discovery of an interaction between PCB exposure and certain genes. Specifically, researchers found that associations between PCB exposure and lymphoma risk were limited to individuals who happened to carry particular genetic variations. And the genes that conferred this vulnerability were involved in immunity or PCB metabolism.

Like PCBs, phenoxy herbicides are chlorinated compounds. Unlike PCBs, they were invented with warfare in mind. First synthesized in 1942, phenoxys were part of a never-implemented plan by the U.S. military to destroy rice fields in Japan. The most famous phenoxy is a mixture of two chemicals, 2,4,5-trichlorophenoxyacetic acid (2,4,5-T) and 2,4-dichlorophenoxyacetic acid (2,4-D). This combination is called Agent Orange, and it was finally deployed between 1962 and 1970 by U.S. troops to clear brush, destroy crops, and defoliate rainforests in Vietnam. The military career of phenoxy herbicides was thus revived.

Linked to miscarriages and contaminated with dioxin, 2,4,5-T was eventually outlawed. By contrast, 2,4-D went on to become one of the most popular weed killers in lawns, gardens, and golf courses, as well as in farm fields and timber stands. It has been marketed under a schizophrenic collection of trade names: Ded-Weed, Lawn-Keep, [illegible], Plantgard, Miracle, Ferxone.

Evidence for a connection between phenoxy herbicides and non-Hodgkin lymphoma comes from several corners. Vietnam veterans suffer excess rates. So do golf course superintendents. So do farmers who use 2,4-D. Risk to farmers rises with the number of days per years of use, the number of acres sprayed, and the length of time they wear their application garments. In Sweden, exposure to phenoxy herbicides was found to raise the risk of lymphoma by sixfold.

Dogs also acquire lymphoma. One recent study showed that pet dogs living in households whose lawns were treated with 2,4-D were significantly more likely to be diagnosed with canine lymphoma than dogs whose owners did not use weed killers. Risk rose with number of applications: the incidence of lymphoma doubled among pet dogs whose owners applied lawn chemicals at least four times per year. A study of people, however, found no excess lymphoma among those exposed to residential pesticides.

The evidence linking phenoxy compounds to non-Hodgkin lymphoma remains preliminary. "The long, slow epidemic of the last half of the 20th century remains poorly understood." So concluded a 2006 review of lymphoma trends and possible causative agents. No one knows exactly how traces of weed killer find their way into our extracellular fluid as it is funneled back and forth between blood and lymph. Absorption through the skin is considered the most likely route of exposure. No one has explicated the exact mechanism by which these chemicals might alter the cells inside the far-flung network of nodes, canals, and lymph tissues-at-large and thereby set the stage for a lymphoma. No one knows whether phenoxys require interactions with other agents to work their damage nor what proportion of the current rise in lymphoma might be attributed to phenoxy exposure.

Most of us are probably far less exposed to phenoxy herbicides than soldiers, farmers, or even our own beloved dogs, who use our lawns for their outdoor activities. Nevertheless, the presence of disease in these specific groups is a clue to which we, as readers of a complicated mystery, need to pay attention when trying to determine why non-Hodgkin lymphoma casts an ever-longer shadow among us all.

A month before her death, Jeannie initiated a massive housecleaning project. She reorganized all her files, returned books, gave away clothing. Waiting for me on her kitchen table one morning was a stack of medical papers, department of public health reports, press releases, and newspaper clippings. They were her collection of articles about the cluster of cancer cases in southeastern Massachusetts, where she grew up.

"I thought you might want them for your research."

"You don't want to keep these?"

"You take them."

Eighteen months after Jeannie's death, I finally read them—prompted by the release of a new study confirming the patterns documented by previous ones. Jeannie's cancer is not included in any of these studies, which concern sharply rising leukemia rates in five neighboring towns during the 1980s and their possible relationship to documented radioactive releases at the Pilgrim nuclear power plant—the result of a fuel rod problem—ten years earlier. While no firm cause-and-effect relationship has been established, meteorological data indicate that coastal winds may have trapped the airborne radioactive isotopes and recycled them within a five-town area. "Individuals with the highest potential for exposure to Pilgrim emissions . . . had almost four times the risk of leukemia as compared with those having the lowest potential for exposure."

Although one of the towns is her own, Jeannie's cancer was far too rare for the case-control comparisons made in the studies she collected. Her cancer has no known cause, and cancer registries do not track its incidence. I will not find her here.

four

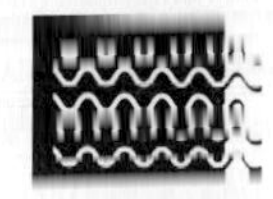

space

Pekin is the judicial seat of Tazewell County, Illinois. It is situated across the Illinois River and a few miles downstream from Peoria. Just outside of Pekin's city limits, about two miles west of the house I grew up in, is the unincorporated subdivision of Normandale. The community was created in 1926 to provide housing for factory workers, and its streets are named for the original prewar products that the residents who slept here at night toiled by day to create: Karo Street (after the syrup), Quaker Street (after the paper mill's round oatmeal boxes), Fleischmann Street (after the yeast).

Normandale is home to 480 people, a popular supper club, a beautiful brick church, and a root-beer stand where I hung out with my best friend in the summers after we learned to drive. Eating onion rings in her father's car, Gail Williamson and I debated the merits of German versus Latin, big universities versus small colleges, sex versus celibacy. In this parking lot, we decided to settle for nothing short of everything. Gail would go to medical school *and* play the violin. I would go to graduate school *and* write poetry. Our present and future boyfriends would just have to understand. Of course. And they also would have to play the guitar.

Normandale is situated on a wedge of land near Dead Lake, a dumping pond for industrial wastes near the river's east bank. It is flanked on two sides by industry: a foundry, a grain-processing plant, a couple of chemical companies, a coal-burning power plant, and an ethanol distillery. Its third side is bounded by a landfill that operated without state permits until the Illinois Pollution Control Board shut it down in 1998. Twenty rusted barrels leaking an unknown tarry substance were discovered along the blacktop just south of town. This is also Normandale.

The distribution of cancer across space, like its trajectory through time, reveals key clues about its possible causes. For example, if ethnicity played a major role in determining cancer risk, then immigrants should retain the cancer incidence of their homelands. Conversely, if the cancer rates of immigrants come to approximate those of their host country (and this is, in fact, the case), then we have good reason to suspect that environmental agents are at work. If cancer rates are elevated in certain geographic areas—within cities, for example, or in areas of intensive agriculture—we have further leads to pursue. If high rates of cancer follow the course of a river or the path of the prevailing wind or are clustered around a drinking-water well or a certain industrial site, then we have very strong clues indeed.

Paradoxically, the closer we stare at the map of cancer, the more unclear the picture becomes. On the largest scale, when cancer registry data from many nations are pooled and we are looking across whole continents, distinct areas of high and low cancer rates are clearly visible. As we narrow our view to one regional area—a single county or town or, like Normandale, a particular subdivision within a town—our power to discriminate differences decreases. Recall that cancer rates are based on the number of people annually diagnosed for each 100,000 people. Determining whether a cancer cluster exists in a small community of only a few thousand or a few hundred inhabitants is statistically difficult work, and it is at this level where the fiercest arguments fly.

At a global level, fewer arguments arise. The time trends and spatial features of cancer's occurrence around the globe clearly belie the notion that cancer is a random misfortune. Cancer associates with westernization. Whereas forty years ago, cancer was mostly a disease of wealthy nations, half of all cancers now occur in developing nations, particularly those rapidly industrializing. Some of the ballooning global burden of cancer is attributable to increasing longevity and an aging population, but age-adjusted incidence rates are also rising in many regions. In India, cancer incidence rose by 7 percent between 1983 and 1997. It rose by 12 percent in Latin America. Reporting on these trends is a little-known arm of the World Health Organization. Located in Lyon, France, the International Agency for Research on Cancer is charged with the daunting job of monitoring cancer incidence around the world. It does so by collecting registry data from as many countries as possible. Its *World Cancer Report 2008* points out that cancer is ascendant in the parts of the world where smoking rates are increasing, diets are westernizing, and rates of obesity are rising. The report expresses frustration about multiple potential exposures to chemical carcinogens, which are also associated with industrialization but about which little is known.

In another global report, the Blacksmith Institute, a New York–based environmental foundation, provides an account of health problems in the world's worst-polluted places. Cancer is among them. At the top of the unhappy list of the ten most polluted cities in the world stands Sumqayit in Azerbaijan, which had served as a center of petrochemical manufacturing in the former Soviet Union. The cancer rate in Sumqayit is as much as 51 percent higher than the national average. Two of China's cities—Tianying and Lifen—also appear on this list.

Between 2000 and 2005, coal burning increased in China by 75 percent, making this nation the world's largest coal producer and consumer. The economic burden of premature death and illness associated with the resulting air pollution—likely the worst among the world's nations—has been estimated at 3.8 percent of its GDP. What part of this is the cost of cancer is not fully known nor yet fully manifest. In China as a whole, cancer incidence increased by 33 percent between 1973 and 1997. What we know about the geography of cancer's surge in China comes mainly from the reports of a few heroic journalists. Steven Ribert provides first-hand accounts of apparent cancer clusters in and around the oilfields and petrochemical refineries in northern China. Along the riverbanks of the heavily industrialized Huai river basin, a team of Chinese journalists has identified dozens of 'cancer villages' where the incidence of cancer [illegible] the national average.

Coal and petroleum. They stand accused of marching humanity toward the cliff edge of catastrophic climate change. Along the way, as coal is burned for energy and to make petroleum-derived products and chemicals, they are killing some unknown number of us from cancer.

Migrant studies also provide clues to the origins of cancer. When immigrants arrive in their adopted country, they leave behind the cancer rates of their homelands and quickly equilibrate with the rates of their new surroundings. "The most important single conclusion to derive from migrant studies," states the International Agency for Research on Cancer, "is that, for a group as a whole, it is the new 'environment' that determines cancer risk and not the genetic component associated with the ethnic stock of the migrants." The quotation marks around that stretchy word 'environment' call attention to its many elements: dietary habits, cultural attitudes about breastfeeding, social stress, and opportunities for physical activity are all part of our environment. So, too, are chemical pollutants in air, food, and water.

Cancer patterns among migrants to Australia, Canada, Israel, and the United States all illustrate the primacy of environment—as defined expansively—in determining cancer risk. Consider Jewish women who migrate from North Africa, where breast cancer is rare, to Israel, a nation with high incidence. Initially, their breast cancer risk is one-half that of their Israeli counterparts. But risk rises rapidly with duration of stay: within thirty years, African-born and Israeli-born Jews show identical breast cancer rates. Jewish women from the Middle East and Asia also increase their risk of breast cancer upon arrival in Israel, although the pace at which they do so is considerably slower.

Likewise, in the United States, the breast cancer rates of European, Chinese, and Japanese women immigrants all eventually rise to conform to the U.S. rate, but they do so at different speeds. Polish women assume

U.S. rates of breast cancer quickly. Japanese women migrating to the U.S. mainland require two generations to achieve our breast cancer rate. First-generation Japanese immigrants show a rate intermediate between that for Japan and the United States; their daughters, however, reflect the U.S. rates completely. Immigrant Hispanic women have lower rates of breast cancer than their U.S.-born counterparts. However, the longer they stay in the United States, the greater their risk for breast cancer.

Happily, the reverse is also true. Women moving to a new country with lower breast cancer rates experience a decline in their chances of contracting the disease—as when, for example, an English woman immigrates to Australia.

These results lead us back to the Möbius strip of lifestyle and environment. Both change simultaneously when someone moves from one part of the world to another. At present, no one understands precisely how these changes interact to create the patterns described above.

~~~

By 1991, half the homes on Karo Street had a cancer patient residing there. It also seemed to some residents that Normandale's children were unusually susceptible to eye and ear infections. In one neighborhood, fourteen residents were diagnosed with cancer over a ten-year period. These numbers were calculated by the people themselves and presented to the health department and the local newspaper. A citizens' group was organized and a letter dispatched to the Tazewell County Health Department requesting an investigation of cancer incidence in their community.

Those quoted in the newspaper mentioned neighbors who died of cancer, as well as those who had moved away out of fear of it.

"Oh, but we've lost so many," said one.

~~~

Because we have no nationwide cancer registry, we also have no definitive geography of cancer incidence in the United States. The National

Cancer Institute does provide an interactive atlas of cancer mortality—in essence, a collection of customizable cancer death maps. Death from cancer is not randomly distributed in the United States. Shades of red consistently light up the northeast coast, the Great Lakes area, and the mouth of the Mississippi River. For all cancers combined, these are the areas of highest mortality; they are also the areas of the most intense industrial activity. The trend maps show that *rates* of increase, on the other hand, are actually higher in the parts of the country with lower mortality, indicating that cancer deaths are tending to become more geographically uniform as time passes, possibly due to the growing urbanization of formerly rural areas, the increasing mobility of the population, and the rising use of pesticides. For two cancers, these maps reveal latitudinal patterns: Southern regions predominate in the maps of melanoma, a pattern consistent with exposure to sunlight. Deaths from breast cancer also follow a north-south gradient but with rates higher in the North, especially in the highly industrialized Northeast.

When studying cancer atlases, a reader must keep in mind that these maps display cancer *deaths*, not cancer diagnoses. Counties with higher levels of contamination may also have worse health care: cancer patients from more polluted, more pesticide-saturated counties may be dying at faster rates simply because they are receiving poorer treatment. On the other hand, the rates of death from other causes—such as cardiovascular or infectious diseases—are not as closely linked to environmental contamination as cancer is. Disparities in health care, then, cannot account for all of the differences in geographic distribution of cancer deaths.

Cancer atlases offer the opportunity to overlay maps of cancer with maps of industrial and agricultural sites to see what patterns exist. In one study, researchers found significant associations between agricultural chemical use and cancer mortality in 1,497 U.S. rural counties. In another, investigators found a close overlap between cancer mortality and environmental contamination: concentrations of industrial toxins were higher in the top-ranked cancer counties than in the rest of the country.

In England, where cancer mortality data have been collected and analyzed for over a century, geographic analysis can be highly sophisticated.

In 1997, a team of researchers mapped the home residences of all 22,458 children who had died of leukemia and other cancers in England, Wales, and Scotland between 1953 and 1980. They then created a second map that charted the locations of every potential hazardous site—ranging from power plants to neighborhood auto body shops. They then superimposed the two maps. Their findings reveal that children face an increased risk of cancer if they live within a few miles of certain kinds of industries—especially those involving large-scale use of petroleum or chemical solvents at high temperatures. These include oil refineries, airfields, paint makers, and foundries. The danger was greatest within a few hundred yards and tapered off with distance. Among children who had moved during their short lives, the relationship was stronger for their birth address than it was for their address at the time of their death. This result strongly suggests that very early—probably prenatal—exposures to environmental carcinogens create the threat of cancer in children.

Another way of mapping cancer is to examine how it distributes itself among people of various occupations. Just as cancer is not scattered uniformly across the physical landscape, neither does it afflict with an even hand the landscape of work. Understanding occupational cancers is important not only because people spend so many hours of their lives in the workplace but also because it yields critical clues about cancers beyond the factory wall and the office door. Released into air or water, hauled away as toxic waste, or mixed into consumer products, most cancer-causing agents in the workplace ultimately become part of the general environment in which we all live. Workplace carcinogens are largely identical to those agents that cause cancer in the general population. Indeed, the near half of the substances now classified as known human carcinogens by the International Agency for Research on Cancer were first identified in studies of workers. Let's look at farmers first.

Farmers from industrialized countries around the world exhibit consistently higher rates of many of the same cancers that are also on the rise among the general population. In other words, farmers die more often from the same types of tumors that are also afflicting, with increasing frequency the rest of us. These include multiple myeloma, melanoma, and prostate cancer. Farmers also suffer from rates of non-Hodgkin lymphoma and brain cancers higher than those of the general population—although these excesses are more modest. In spite of lower overall mortality and lower rates of heart disease (and lower overall cancer occurrence), farmers also die significantly more often than the general public from Hodgkin disease, leukemia, and cancers of the lip and stomach. Likewise, migrant farmworkers suffer excess rates of multiple myeloma, as well as of stomach, prostate, and testicular cancers. These results are consistent with the geographic patterns revealed within cancer atlases: death rates from multiple myeloma are highest in rural farming areas. Found in the central Corn and Wheat Belt region of the United States are high rates of leukemia and lymphoma.

Additional clues about farm chemicals and cancer are emerging from the ongoing Agricultural Health Study. Begun in 1993 and sponsored by the National Cancer Institute, this investigation has been following a cohort of fifty-seven thousand farmers in Iowa and North Carolina. Spouses and children of these farmers are also enrolled in the study. Among the findings to date is the good news that the overall cancer rate among the study's farmers appears significantly lower than that of nonfarmers. Farmers use less tobacco and enjoy higher levels of physical activity and lower levels of diabetes than their off-farm counterparts. And yet, certain cancers nevertheless stalk these farmers and their families with greater frequency than the general public. Prostate cancer is one.

The Agricultural Health Study is revealing other patterns as well: One pesticide (permethrin) shows associations with bone marrow cancer (multiple myeloma). Two different weed killers show associations with pancreatic cancer. Parental pesticide application is linked to lymphoma in children. Children whose fathers do not use gloves when handling pesticides are also at higher risk for leukemia. Farm children whose fathers used atrazine recently had higher concentrations of this herbicide in their urine than farm children living where atrazine had not been recently applied. Women who use pesticides have longer menstrual cycles and later age at menopause. Yet pesticide use does not associate with breast cancer risk. On the other hand, [illegible] were [illegible] of pesticide application do suffer modestly elevated rates of breast cancer.

Occupational studies of other professions reveal still more associations. Elevated cancer rates are found among painters, welders, asbestos workers, plastics manufacturers, dye and fabric makers, miners, printers, and radiation workers. Workers exposed to formaldehyde are more likely to contract leukemia. Firefighters are twice as likely to develop testicular cancer and suffer elevated rates of non-Hodgkin lymphoma, prostate cancer, and cancer of the bone marrow. Barbers and hairdressers have elevated risks for bladder cancer. Finnish women exposed to solvents and gasoline on the job also show increased risks of bladder cancer, as well as liver cancer. Finnish women workers exposed to diesel exhaust have higher rates of ovarian cancer—and risk of disease rose as exposure increased. Taiwanese women electronics workers exposed to chlorinated solvents in one particularly contaminated factory had increased incidence of breast cancer. There are ongoing concerns about the breast cancer risks of those who work in beauty parlors or nail salons.

People who work in a number of white-collar jobs are also at higher risk: for example, chemists, chemical engineers, dentists and dental assistants, and—perhaps most ironically—chemotherapy nurses. (Many of the chemicals used to treat cancer are themselves carcinogenic, as the high rate of adult cancers among childhood leukemia survivors attests.)

As we saw with farmworkers, the children of adults who work in specific occupations also have higher rates of cancer. Childhood brain cancers and leukemias are consistently associated with parental exposure to paint, petroleum products, solvents, and pesticides. Some exposures may occur before birth. Children can also be exposed when these materials are carried into the home on their parents' clothes and shoes, through breast milk (which can be contaminated directly or through maternal contact with the father's clothing), or even through exhaled air: because solvents are, in part, cleared by the lungs, parents can expose their children to carcinogens simply by breathing on them. In this way, a father's home-coming kiss and work-clothed embrace can contaminate his child.

In response to the questions raised by the people of Normandale, two health studies were quickly conducted—one by the state health department and one by the county. Neither involved mapping disease patterns, identifying pollution sources, estimating actual exposures, locating those who had moved away, or, for those who had died, interviewing their next-of-kin. No blood, urine, or fat samples were collected to test for the presence of contaminants. In fact, the study design did not require public health officials even to set foot in Normandale.

In the first study, the Illinois Department of Public Health pulled up from its computerized cancer registry banks all the cancers diagnosed in Pekin's ZIP code area—as reported to the Illinois Cancer Registry between 1986 and 1989. From these data, researchers calculated an *actual* cancer rate for the whole town. Based on the statewide rates, researchers then generated an *expected* number of cancer cases for a hypothetical town the size of Pekin. Cancers were categorized by location in the body (colon, ovaries, breast, and so forth), the actual numbers were compared to the expected numbers, and . . . no statistically significant differences were found.

On December 19, 1991, the headline in the *Pekin Daily Times* read, STUDY: AREA CANCER RATES NORMAL.

∿∿∿∿∿

If cancer-causing chemicals in the environment play a significant role in actually causing cancer, then we should expect to find high rates of the disease in areas where carcinogens are highly concentrated. The industrial workplace, where such chemicals are manufactured or used, is one such area. Hazardous waste sites, where such chemicals are dumped, are another.

Quite a few of us are included in the population of the potentially exposed. By 1980, the EPA had tallied up 32,645 sites of past chemical waste dumping in need of cleanup. Some of these are actual hazardous waste landfills, but many are former manufacturing sites where drums full of chemicals have simply been abandoned. The names of the most notorious appear on the EPA's National Priorities List. These are the so-called Superfund sites, named for the super fund of money put together by Congress in 1980 to clean them up. In 2009, the Superfund list contained 1,331 sites. One-quarter of the U.S. population lives within four miles of one of them. Among those living within one mile are an estimated 1.1 million children under the age of six. Currently, Illinois is home to fifty Superfund sites.

Most of these sites did not exist before the end of World War II, when most plastics, solvents, detergents, pesticides—and all the unwanted byproducts of petrochemical manufacturing—made their debut on the planet. Poor and dispossessed children have lived cheek-by-jowl with carcinogenic waste ever since soot-encrusted chimney sweeps in eighteenth-century England were discovered to be at high risk for scrotal cancer. But those of us born after World War II are the first generation to grow up in such large numbers near such large amounts and diverse assortments of manufactured chemical refuse. Between the late 1950s and the late 1980s, more than 750 million tons of toxic chemical wastes were discarded.

Several large studies have detected elevated cancer rates around hazardous waste sites. One of them was conducted in New Jersey, a petite state with an astonishing 133 Superfund sites. Researchers asked whether cancer mortality was associated with environmental factors of various kinds, including the location of toxic waste dumps. Their results showed that communities near toxic waste sites had significantly elevated mortality from stomach and colon cancers. Additionally, in twenty-one different New Jersey counties, breast cancer mortality among white women rose as the distance from residence to dump site shrank. However, many of the clusters of excess cancer occurred in heavily industrialized counties so that air pollution from these sources confounded the results. Thus, a woman with breast cancer in northeastern New Jersey cannot know with certainty whether she is dying because of the air wafting down from the factory stacks or because of the water contaminated by the dump site.

In another large study, researchers scoured the United States for counties that met two criteria: first, their hazardous waste sites had contaminated the groundwater, and second, this groundwater served as the sole source of the drinking water for the residents. Meeting these qualifications were 593 waste sites in 339 counties in forty-nine states. Next, researchers obtained for each of these 339 counties ten years' worth of cancer mortality data and compared them to cancer mortality data from counties without hazardous waste sites.

Here are the results: men living in hazardous waste counties suffered significantly higher mortality from cancers of the lung, bladder, esophagus, colon, and stomach than did their contemporaries residing in counties without such sites. Women living in hazardous waste counties suffered significantly higher mortality from lung, breast, bladder, colon, and stomach cancers. Indeed, counties with hazardous waste sites were 6.5 times more likely to have elevated breast cancer rates than counties without such sites.

Other studies corroborated these results. Looking specifically at breast cancer, researchers found that mortality rates at the county level were significantly correlated with Superfund sites. Counties with the highest breast cancer mortality had four times as many facilities that treated and stored toxic waste than the national average.

Studies such as these two are considered preliminary rather than definitive because possible confounding factors could not be controlled. These include the possibility that residents living in counties with hazardous waste facilities are getting more cancers not because of the dumps but because they work for the companies that create the waste or because they smoke more and drink harder.

Among other things, the term *ecological fallacy* refers to the temptation of assuming that all associations are causative when one examines statistical patterns. My statistics professor was fond of telling the story of the boy and the department store escalator: the boy wondered what caused the escalator to move. After hours of observation, he concluded the escalator ran on the energy generated by the revolving door, because when the door ceased turning at the close of the day, the escalator stopped.

Ecological fallacy became a real issue for me when I started work as a field biologist. In Minnesota, I wanted to know why pine trees were failing to reproduce. Absence of new seedlings was correlated with high population levels of deer—but also with low frequency of forest fires and high populations of hazel shrubs. Which, if any, was the root cause of the problem and which were the confounders? Or, if fire, hazel, and deer all conspired to contribute to the demise of the pines, how exactly did they do so? Once I had established the pattern, I needed to design experiments that would uncover causal mechanisms. I found this work very exciting.

But as a woman with cancer who grew up in a county with numerous hazardous waste sites, several carcinogen-emitting industries, and public water wells that, from time to time, show detectable levels of toxic chemicals, I am less concerned about whether the cancer in my community is more directly connected to the dump sites, the air emissions, the occupational exposures, or the drinking water. I am more concerned that the uncertainty over details is being used to call into doubt the fact that profound connections do exist between human health and the environment. I am more concerned that uncertainty is too often parlayed into an excuse to do nothing until more research can be conducted.

~~~

By 1991, I am living a long way from Normandale. My sister still lives nearby.

"What's the latest?" I ask into the phone.

"People are worried about their dogs over there. They say there's a problem with cancer in pets. One man has a German shepherd with breast cancer."

I call an old high school teacher who has served a long term on the city council. The questions raised in Normandale have him thinking
~~~

Epidemiologists investigate patterns of disease in human populations. They look at the world through a wide-angle lens. While the focus of medicine is the treatment and prevention of diseases in individuals, epidemiology attempts to explain and prevent the occurrence of disease in large groups.

One type of investigation is what epidemiologists call ecological studies. In these, investigators compare the frequency of a given disease (e.g., cancer) in populations that differ in some factor of interest (e.g., the presence or absence of a leaking hazardous waste site). Statistics are then used to determine whether the frequency of disease is significantly different in the two types of communities. Researchers can often complete ecological studies without ever talking directly to any of the human subjects or assessing their exposure levels to the contaminants in question. The studies in Pekin and Normandale were ecological studies. As strange as it sounds to ecologists, the word *ecological* is used by epidemiologists simply to mean a descriptive, rather than an analytical, approach. Ecological studies, like circumstantial evidence, provide the weakest demonstration of proof.

Included in epidemiology's analytical category are two basic study designs. One is the case-control study. Here, a group of diseased people are identified (the cases) and compared to a group of people drawn from the larger population (the controls). The point of comparison is their exposure to possible disease-causing agents. Mary Wolff's study of DDT and breast cancer, discussed in Chapter One, is an example. Her cases were women with breast cancer; her controls were women without breast cancer (matched for age, menopausal status, and other variables of personal history); exposures were assessed by measuring blood levels of DDT and PCBs. Her results showed that women with breast cancer had significantly higher DDT levels than women without breast cancer.

Closely related to the case-control study is the cohort study, in which people are classified as exposed or unexposed and are followed through time until disease or death occurs. (The farmers and their families monitored in the ongoing Agricultural Health Study form one such cohort.) In this way, we compare the rate of disease in people known to be exposed to a possible carcinogen to disease rates in unexposed persons. The ratio of the two is known as relative risk.

One needs to understand a bit about the inner workings of cancer epidemiology in order to understand why the topic of individual cancer clusters is such a vexing one. Determining from ecological studies that communities near hazardous waste sites tend to suffer from an excessive risk of cancer is one kind of investigation. Determining that any one particular community has an elevated cancer risk due to any one particular waste site is a very different kind of project. The second kind is the one most people are interested in. We live in particular communities, not general ones, and our concerns are about the health of the particular people in our families and neighborhoods. Indeed, almost all cancer cluster studies are initiated by alert citizens contacting their health departments to request such investigations. Their phone calls and letters often tell of "cancer streets," along which the prevalence of cancer seems extraordinarily high, or of growing numbers of neighborhood children afflicted with disease. This is exactly what happened in Normandale.

In spite of public concern, many public health officials become downright apoplectic when the subject of community-level cancer clusters is raised. Some consider the investigation of alleged clusters a disparaged practice and lament the inability of common people to grasp the statistical concept of randomness. Within the medical literature, publications advise health authorities on how best to deal with citizen requests for cluster studies. They are often overtly dismissive. Change a noun or two and some of them could double as guidelines for how to deal with people who wish to report a U.F.O. sighting. Typically, the message relayed back to these vigilant citizens seeking explanations is that their questions are misguided. Too rarely are they told that the tools of epidemiology are just too blunt to provide answers.

One problem with cancer cluster studies is that investigations of individual communities have limited power to identify existing problems. In this context, the word *power* refers to the ability to detect a significantly increased cancer rate if indeed the increase really exists. The word *significantly* also has a particular meaning. Significance is a statistical standard that limits a finding to only those increases in cancer rates we are reasonably sure did not occur by chance. *Reasonably sure* is traditionally defined as 95 percent sure, so 95 percent is accepted as the conventional cutoff for significance. If I roll a pair of dice six times and they always come up sixes, I can be more than 95 percent certain that this event is not due to chance. The finding is statistically significant. I conclude the dice are loaded. However, if I roll only one die one time and I get a six, the finding is not considered significant. By chance alone, the odds of this outcome are 16.7 percent. The dice may indeed be fraudulent, but my test does not have the power to say so.

Looking for a cancer cluster in a single, small community is like rolling the dice only once. Before the possibility that the cluster has occurred by chance can be ruled out, cancer rates in small communities must reach extraordinarily high levels—sometimes as high as eight to twenty times higher than levels for the surrounding areas. Because of the small sample size, lesser increases will not attain sufficient power for the study to be conclusive.

The second problem with cancer cluster studies is that there often remain no unexposed populations to use as a comparison group. In cluster studies, epidemiologists look for an increase over and above some background level, but if the people in the background are also becoming increasingly contaminated, the researchers are paddling a boat in a moving stream. Differences are harder to see.

Suppose, for example, we want to know whether people living near a particular hazardous waste dump are getting cancer because of it. Suppose the chemicals wafting into the air and trickling into the groundwater include several pesticides, some vinyl chloride, and an industrial solvent called trichloroethylene (TCE), classified by the EPA as a probable human carcinogen. With such contents, this dump would be utterly typical: TCE is the most frequently reported substance at Superfund sites, vinyl chloride is not far behind, and half of all hazardous waste sites contain pesticides. We have already seen that almost all of us experience chronic incremental exposure to vinyl chloride and pesticides from our air, food, and water.

Most of us are also exposed regularly to molecules of TCE. Used by industry to degrease metal parts, TCE is now estimated to be in 34 percent of the nation's drinking water. Most processed foods contain traces as well. TCE is also found in paint removers, spot removers, cosmetics, and rug cleaners. An estimated 3.5 million workers are exposed to TCE on the job. Not so long ago, TCE was also used as an obstetrical anesthetic, a fumigant for grain, an ingredient in typewriter correction fluid, and a coffee decaffeinater. These uses have been phased out, but there is still sufficient release of TCE into the general environment to ensure that traces of this vaporized metal degreaser persist in the ambient air that we all breathe—including detectable amounts in the air above the Arctic Circle. Therefore, if we design a study that compares cancer rates between people living near this hypothetical dump site and a control group of people drawn from the general population, our results might reveal little about why either group is getting cancer. The cluster group and comparison group share exposures, and there is no unexposed control. As one nurse has observed, "To the public, it is no consolation that we are all exposed to environmental pollutants equally, and therefore not at increased risk of cancer because of it when compared with anybody else."

At least two additional problems with cluster studies exist, and they both have to do with the nature of cancer. First, cancer usually requires a long period of time to develop after exposure occurs. This lag time makes exposure assessment very difficult. Researchers must rely on old, incomplete records—which may not exist at all—or people's memories, which may be imprecise. Second, the origins of cancer are multiple, often resulting from exposures to combinations of substances—vinyl chloride plus heavy drinking, for example. These two features of the disease dilute and confuse epidemiologists' finest efforts to understand its causes in any given community. In a case-control study, some of our cases may have contracted cancer from prenatal exposures, some from the dump site, some from their jobs, some from pesticide residues, and some from a combination of these. Furthermore, unexposed people migrate into the community, and exposed people move out. Epidemiologists cannot ask people living near carcinogens to stay put for ten years so that they may conduct a decent cohort study.

Consider epidemiology's most dramatic success in cracking a disease cluster not involving cancer: the Case of the Eleven Blue Men.

In 1953, New York City police reported to the health department that eleven homeless men in one neighborhood had all been discovered to be very ill and that all of them had turned sky blue. This particular skin color is the hallmark symptom of a disease called methemoglobinemia. Eleven cases in one neighborhood is thousands of times above background level. Knowing that this disease is associated with the ingestion of sodium nitrite, epidemiologists interviewed the men about their eating habits and discovered all had frequented a particular neighborhood diner and all had used the saltshaker there. The said shaker was impounded, laboratory tests were run, and the discovery was made that sodium nitrite had indeed been substituted for sodium chloride. The cook had made a mistake. Mystery solved.

Now imagine that cancer made people turn blue. And further imagine a skid row saltshaker containing a powerful chemical carcinogen that eleven customers unwittingly sprinkle over their food and that eventually causes them to develop cancer. In spite of their telltale color, the reason for their disease would probably never be uncovered. Because of the delay between exposure and onset of disease, at least ten years would pass before any of the eleven turned blue, and some of them would undoubtedly move away during this time. Because cancer is a disease with multiple causes, other dwellers with blue complexions, who contracted cancer for unrelated reasons, would move into the area. The saltshaker itself would be long gone. Thus, although a cluster of people did indeed contract cancer from a single, identifiable source, a study of all blue-faced people in the neighborhood would not likely be able to establish the fact.

To overcome the limitations of basic epidemiology, state-of-the art cluster studies now incorporate geographic information systems (GIS) mapping and exposure assessment. GIS tools can create visually compelling pictures of potential cancer clusters. These spatial patterns can then be statistically tested for randomness. Exposure assessment can take the form of biomonitoring, which consists of collecting samples—urine or blood, for example—from bodies of the potentially exposed people themselves and testing them for the presence of particular contaminants. (Or it can involve sending the household dust bunnies off to the chemistry lab.) Analytic methods now allow for the direct testing of at least three hundred chemicals. Results can then be compared with baseline human exposure data, collected by the Centers for Disease Control and Prevention.

Even with all these methodological upgrades, the work of investigating cancer clusters suffers bedeviling problems. GIS mapping was designed to provide snapshots of locational information for businesses, not track chronic disease patterns. It lacks a temporal dimension. Cancer data are often aggregated by ZIP code—a geographic unit intended to speed mail delivery, not serve to analyze disease statistics by geography. It lacks standardization. And rarely can biomonitoring detect exposures from years earlier that may be influencing cancer risk now. But the biggest obstacle standing in the way of a good faith, due-diligence cluster investigation is not technological. As identified by the Pew Environmental Health Commission and reiterated in a recent issue of *American Journal of Public Health*, the biggest hindrance to cancer cluster inquiry is ignorance. Our nation lacks basic knowledge about the toxic properties

al chemicals in commerce and, until 2002, lacked any system for environmental health tracking at all. At the state level, there is no consistent [illegible] agency with responsibility for following up on cancer cluster reports by citizens. There are no standard protocols for doing so, no rapid response teams at the ready, nor any systematic record keeping on [illegible] investigations. We track pizza deliveries and overnight packages more closely than we track toxic chemicals or cancer diagnoses.

Sometimes cluster studies go forward in spite of this paralyzing incapacity. Sometimes they yield instructive results. An investigation of twenty-five bladder cancer clusters in Florida revealed that advanced bladder cancer aggregated near arsenic-contaminated drinking water wells. Bladder cancer mortality was found elevated among men living in Clinton County, Pennsylvania, near a 46-acre chemical dump containing benzene and aromatic amines. An investigation in Sugar Creek, Missouri, found significantly elevated rates of Hodgkin lymphoma in a benzene-contaminated town where an oil refinery with a history of leaks now sits abandoned. In Endicott, New York, researchers verified elevated rates of lymphoma among former workers in a computer manufacturing plant. The plumes of TCE-contaminated water near the plant are now receiving close scrutiny as are the cancer rates of community members. In Ohio, state officials pinpointed a cluster of childhood cancers in Sandusky County. No one has an explanation yet, but the inquiry continues. Says the chief of Ohio's cancer control program, "We owe it to these kids, parents, and future generations of kids to try to find out."

~~~

The kind of investigation conducted in Normandale was standard epidemiological fare. The statistical standards used in the analysis, however, were out of the ordinary.

Recall that statistical significance is traditionally defined as a less than 5 percent probability that any observed differences are attributable to chance. Curiously, state officials conducting this study chose 1 percent, rather than 5 percent, as their cutoff level for significance. This is
~~~

an unusually strict measure, which, not surprisingly, causes differences to disappear. Two excesses in the Pekin area did in fact attain statistical significance at the usual 5 percent level: ovarian cancer and lymphoma.

No mention of the study's statistical methodology was made in the newspaper account headlined AREA CANCER RATES NORMAL. Nor was any discussion devoted to what *normal* means in this context. Tazewell County's toxic emissions are high, but in comparison to those in the rest of the state, they are not off the chart. Thus, statistics aside, a discovery that Pekin's cancer rates are rising in tandem with the rest of the state's says nothing about whether our cancers, or anyone else's in Illinois, are or are not attributable to environmental exposures.

A rising tide raises all ships. Is this situation normal?

~~~

The investigations that have successfully documented cancer clusters and traced them to possible sources have all involved brilliant environmental sleuthing and heroic perseverance by ordinary citizens working together with researchers.

One of these places is Long Island, New York. In 1994, the state health department released the results of a case-control study of Long Island women that showed that women with breast cancer were more likely to live near a chemical plant than women without the disease. Breast cancer risk rose with number of facilities: the more chemical plants in the community, the higher the incidence of breast cancer. And the closer a woman lived to one of these plants, the greater her chance of developing breast cancer.

This study was the first to show that breast cancer has links to air pollution. It was undertaken as a reaction to an earlier study that had dismissed environmental links to breast cancer in Long Island, concluding instead that breast cancer incidence in Long Island correlated with affluent lifestyles. The Centers for Disease Control reviewed these findings and, in 1992, recommended no further follow-up. Nevertheless, when women found various flaws in the study's design, they took matters into
~~~

their own hands. Some began creating their own maps and others began petitioning Congress to order a federal investigation. In the fall of [illegible], a group of advocates hosted their own scientific conference, the first in the nation to bring together scientists and women with cancer to design a program of study. It was within this atmosphere that the 1994 study linking breast cancer and proximity to chemical plants emerged.

In the meantime and after considerable pressure, the U.S. Congress directed two federal agencies to begin a multi-million-dollar study: the Long Island Breast Cancer Study Project. This project is actually a collection of ten different study projects, the results of which are still coming in. Teams of researchers investigated such environmental features as aircraft emissions, pesticide practices, and plumes of contaminants in groundwater. Exposures were measured directly. Blood from thousands of Long Island women with and without breast cancer was analyzed for organochlorine pesticide residues and industrial chemicals.

At this writing, here is what we know from the Long Island Breast Cancer Study Project. First, there is no evidence for a link between adult exposure to individual organochlorine chemicals and breast cancer risk. Women with breast cancer had body burdens of DDT that were no higher, on average, than in women without the disease. With the benefit of hindsight, we now know that this is not a surprising result: as discussed in Chapter One, coherent evidence now indicates that the operative risk factor associated with DDT is early-life exposure when breast tissue is developing rapidly. The Long Island study measured levels in adult women after the time of diagnosis and missed this period of developmental sensitivity. It also looked for associations with individual chemicals rather than chemical mixtures. The importance of developmental exposures and real-life mixtures were not fully appreciated at the time the Long Island Study was designed. By the time the results were released in 2002, awareness of the study's design limitations were widely acknowledged. And yet, these negative results were overgeneralized in the media as proof that pesticides of all kinds are not linked to breast cancer . . . and then still later as proof that the very act of searching for environmental links to cancer was a fool's errand and a waste of tax dol-

lore. Like the story of the fish that got away, the fish got larger with every retelling.

Meanwhile, positive findings have quietly emerged from the Long Island study. Most notably, breast cancer risk is significantly higher among women who report using pesticides in their lawns and gardens. And it is also higher among women whose blood cells show signs of DNA damage caused by the inhalation of polycyclic aromatic hydrocarbons. Also known as soot.

Across Long Island Sound lie the shores of Connecticut and Rhode Island. Just beyond, past Buzzards Bay, Cape Cod emerges from the Massachusetts coastline like a girl's arm, curled and slender. As a midwestern adolescent, I became enchanted with Henry David Thoreau's account of walking the length of this narrowest of peninsulas as it unfurled into the Atlantic. Cape Cod seemed to me a place of danger, beauty, and wild escape.

In the 1980s, year-round residents of the Upper Cape began agitating for an investigation into the relationship between environmental hazards and cancer rates. It seemed to them that cancer in their isolated communities was unusually common, and they were also aware of many environmental hazards, including pesticide use in cranberry bogs and golf courses as well as groundwater and air contamination from a nearby military reservation. Many could recall how the entire Cape had been drenched with DDT for several years during the 1950s in a failed campaign to eradicate the gypsy moth.

Residents were correct about the cancer rates. Records from the state cancer registry revealed excesses in prostate, colon, and lung cancers, along with elevations in pancreas, kidney, and bladder cancers. Of the ten towns with the highest breast cancer incidence in Massachusetts, seven are located on the Cape, and nearly all the Cape's towns have higher-than-average rates of breast cancer.

Persistent citizen pressure led to two studies: an Upper Cape study completed in 1991 by two Boston University epidemiologists and another, the Cape Cod Breast Cancer and the Environment Study, which began in 1994 with a $1.2 million appropriation from the state legislature and continues to this day under the direction of scientists at the Silent Spring Institute in Newton. Let's look first at the ongoing study.

Of particular interest to Silent Spring investigators is the underground aquifer that supplies nearly all the area's drinking water. Covered with sandy soil, the aquifer is vulnerable to all manner of contamination—from pesticides to septic tank effluent to jet fuel and solvents spilled at the military base. Ironically, environmental regulations that protect the Cape's coastal marine sanctuary mean that all waste water is discharged onto land, where it trickles through the sand and into the groundwater. Many of the chemicals contained in this waste are believed to play a role in breast cancer. Plumes of contaminated groundwater and areas where pesticides were sprayed were mapped and compared to maps of breast cancer incidence on the Cape. Residential history was factored in, so distinguishing between newcomers and old-timers. Using GIS techniques combined with models that allow for spatial data to be analyzed together with temporal data, researchers have created an animated picture of breast cancer's shadow as it moved across the Cape over a forty-seven-year period.

Here is what Silent Spring researchers have learned so far: the space-time maps revealed an association between breast cancer risk and living in some areas near the military reservation from 1947 to 1956, but most of the elevated incidence was far from the military base. While the groundwater aquifer used for drinking water is contaminated with wastewater that contains hormonally active agents, limited evidence showed no connection between breast cancer and contaminated drinking water. Nevertheless, risk of breast cancer rises with duration of residency on the Cape, leaving open the question of causes in the ongoing study.

While researchers in the ongoing study are focusing specifically on breast cancer, those in the 1991 Cape study, by contrast, were looking at nine different cancers. Organized in case-control fashion, the study's cases comprised Upper Cape residents diagnosed with cancer between 1983 and 1986, and the controls were a random sample drawn from the entire population of Upper Cape residents. Exposures were assessed through interviews. In this way, potential confounding factors such as smoking and other lifestyle habits could be uncovered and corrected for. The study was impressively thorough: when dealing with people who had already died from their disease, researchers matched them with nonliving controls—people who had died from other diseases and whose names were selected randomly from death certificate registries. To gather exposure information on cases and controls no longer alive, researchers interviewed their next-of-kin.

After three years of research, the study's chief investigators reached the following conclusion:

> In summary, this inquiry was begun because of concern about the generally increased cancer rates in the Upper Cape region along with the presence of known or suspected environmental hazards. After an extensive review of environmental factors it is clear that there was ample cause for concern.

While low statistical power prevented researchers from explaining all of the cancer increases, several interesting results emerged. The rates of both lung and breast cancer were elevated among residents living near the gun and mortar positions on the reservation. One possible explanation is airborne exposure to the military's chemical propellant, dinitrotoluene, which is used for firing artillery. Classified as a probable human carcinogen, dinitrotoluene has been shown to cause breast cancer in laboratory animals. The study also yielded evidence for an increase in brain cancer among people living close to cranberry bogs, and it revealed elevations in leukemia and bladder cancer among those whose homes were fed by a particular type of water distribution pipe.

These water pipes had long been under suspicion. In the late 1960s, a new innovation in cement water pipes was introduced into New England: pipes with plastic liners that improved the water's taste. A large number were laid in the Upper Cape, which was then undergoing rapid development. In manufacturing these pipes, workers applied vinyl paste to the inside surface, using a solvent called tetrachloroethylene. For reasons known only to organic chemists, tetrachloroethylene is more commonly referred to as perchloroethylene, PCE, or simply perc. Like its chemical cousin [illegible] trichloroethylene, perchloroethylene is classified by the International Agency for Research on Cancer as a probable human carcinogen.

The assumption by the manufacturers of these waters pipes was that all the solvent would evaporate during the curing process. It did not. In fact, substantial quantities remained and slowly leached into the drinking water. Thus, the drinking water of the Upper Cape is contaminated not only by chemicals leaking from the land's surface into the sole-source aquifer from which public water supplies are drawn, but also, in some areas, by the pipes carrying this water into individuals' residences. The knowledge that Upper Cape water pipes were shedding perc into the drinking water was not a new discovery. This phenomenon had been known since the 1970s, but perc was not a substance regulated in drinking water during that decade. In 1980, plastic-lined water pipes were finally banned for use.

The link between perchloroethylene and bladder cancer was also not a new discovery. Perc is a familiar substance to almost all of us. Since the 1930s, it has been the chemical of choice for dry-cleaning clothes. Compared to the general population, dry cleaners have twice the rate of esophageal cancer and twice the rate of bladder cancer. Thus, a discovery of a bladder cancer cluster among the folk of the Upper Cape should come as no surprise. Further studies of the Upper Cape's water pipes, published in 1993, showed that people's actual exposure to perc varied widely, depending on the length, shape, size, and age of the water pipe, the pattern of water flow, and the person's length of residence in that house. For those people with highest exposure, bladder cancer risk was four times higher and leukemia nearly twice as high when compared to people without such pipes.

The *Journal of the American Water Works Association* first reported on the problem of perchloroethylene leaching from drinking water pipes in 1983. The following words, written by scientists studying the Upper Cape, were published exactly ten years later:

In conclusion, we have found evidence for an association between PCE contaminated public drinking water and leukemia and bladder cancer. In some EPA surveys, 14–[illegible] percent of groundwater and [illegible] percent of surface water sources have some degree of PCE contamination. Thus, its carcinogenic potential is a matter of significant public health concern.

The second study in Normandale supported the first one. Because it could not provide cancer incidence data on any scale smaller than ZIP code level, the state health department turned the remainder of the investigation over to the county. County officials promised to conduct a door-to-door survey with the goal of determining whether "the cancer cases in Normandale are out of sync with the rest of the ZIP code." They did not. Instead, questionnaires, which recipients were asked to fill out and mail back, were sent to 184 Normandale residences. Sixty-seven completed forms came back—a 37.5 percent response rate—and among these, eight cases of cancer were described.

The headline on March 6, 1992, announced, STUDY: NO CANCER CLUSTER, and the accompanying story read, in part:

> The Tazewell County Health Department found no significant cancer problem [in Normandale], officials said Thursday in announcing results of the department's cancer survey of the 40-acre subdivision. . . . The findings put an end to five months of investigation by state and county health officials into some residents' fears that they were living amidst a cancer cluster.

One does not need to be an epidemiologist to see the glaring problems with this study. First, the numbers are too small to draw conclusions one way or the other. Second, there is no way of knowing whether respondents represent a random sample of the community. Perhaps responding households were, on average, healthier or better educated than nonresponding households. Perhaps households providing care to a family member with cancer were more likely to misplace the mailing or were more likely to be too grief stricken or too overwhelmed to sit down and fill out answers to lengthy questions. Perhaps they were too busy fighting with insurance companies or planning funerals to write up a detailed family history and remember to mail it. Perhaps families with cancer were more likely to be out of town. Perhaps illiteracy prevented some from responding. Perhaps those angriest about the county's broken promises chose to boycott the questionnaire. Or perhaps, conversely, families with cancer patients paid more attention to the questionnaire. In short, without direct human contact, no one can know why nonresponding households, the majority, remained silent.

Furthermore, anyone who had lived alone and died from cancer had no chance to be counted at all. The local newspaper obtained county death certificates showing that at least five cancer deaths in the community were never reported to the county's survey. These included one case of liver cancer, two cases of breast cancer, one case of leukemia, and one of ovarian cancer.

How can silence be statistically evaluated? How can such a flawed, limited response to a questionnaire lead to an assertion that there is no problem?

These questions were not lost on the people of Normandale, many of whom expressed doubts about the study's validity. Still the inhabitants of Normandale are not the residents of Cape Cod nor the women of Long Island. They are not positioned to reject the results of a county investigation and insist on a multi-million-dollar federal study. They have no friends in Congress. They are unlikely to invite world-renowned scientists to convene proceedings in the parking lot of the A & W root-beer stand.

The citizens of Cape Cod and Long Island have struggled mightily to bring scientific attention to the link between cancers and environmental contamination in their communities. Still, the resources they command are starkly different from those among Normandale's residents. My meetings with the breast cancer activists of Long Island have taken place on college campuses and convention hotels. I have spoken with the cancer activists of Cape Cod in a beachfront conference center. When I met with a community leader in my own hometown, we held our discussion in the back room of an auto repair shop and towing company.

The Massachusetts report concerning the alleged cancer clusters on Cape Cod is more than five hundred pages long. The two reports detailing the state and county investigations into cancer rates in Pekin and the Normandale subdivision together total eight pages.

Said a man from Normandale who lost his wife to ovarian cancer, "I think the state has a way of putting things to the side or overlooking what's the real truth."

five

war

When my father, at age sixty-nine, wrote his memoirs on a manual typewriter and sent copies to all surviving members of his family, he did so to commemorate the fiftieth anniversary of the Allies' victory in the Mediterranean theater. The significance of this event is emphasized throughout the text. It was his defining moment.

I have often imagined my father as a soldier in Italy. His two desires: to stay alive and to avenge the capture of his brother, my Uncle LeRoy, held as a prisoner of war in Germany. His one fear, which the Allied victory in Europe very nearly realized, was to be sent to the other theater—the blood-soaked Pacific.

My father firmly believes his life was saved by excellent typing skills. This was not a lesson to be lost on his daughters. The ninth child of a poor Chicago family, he moved a dozen times before finishing school and enlisting. How exactly he learned to type a hundred words per minute *with no errors* I do not know. It is part of my father's mystique. Throughout my childhood, the sounds of rapid, flawless typing filled my parents' bedroom. According to legend, his remarkable talent with the typewriter saved him for two reasons: first, because he was selected to work in correspondence at a U.S. Army office safely away from the front and, second, because he was therefore privy to orders about upcoming troop deployments. Thus forewarned, he deftly reenlisted in the right unit at the right moment and kept himself out of harm's way. His skills as a tank destroyer motto: Seek, Strike, and Destroy), for which he was trained, would go untested.

With these stories, I was encouraged to spend time practicing penmanship, dictation, and typing [illegible]. Like him, I am near-sighted and left handed. Neither were allowable excuses for sloppy work. But if I became more attracted to the sounds of the words than to the speed with which I could produce them, it was both for my plain lack of clerical talent and for the irrelevance introduced into the whole endeavor by computers. Still, until I read his error-free autobiography while sitting at my own big desk, I did not realize how deeply my father's stories had influenced me or how much I am like that nineteen-year-old army clerk furiously typing up casualty reports. My own work as a writer is a legacy of a war ended years before I was born.

World War II is mentioned throughout the chapters of *Silent Spring*. Carson's references are casual, and they seem designed to remind already-aware readers that the technologies developed for wartime purposes had changed chemistry and physics forever. The atomic bomb was only the most arresting example. More intimate aspects of the human economy were also changed. The multitude of new synthetic products made available after the war altered how food was grown and packaged, homes constructed and furnished, bathrooms disinfected, children deloused, and pets de-flea'd. Carson described this transformation almost offhandedly, as though the connection between lawn-care practices and warfare was perfectly obvious.

Carson made at least two other points about World War II. First, because many of these new chemicals were developed under emergency conditions of wartime, they had not been fully tested for safety. After the war, private markets were quickly developed for these products, and yet their long-term effects on humans or the environment were not known. Second, because wartime attitudes accompanied these products onto the market, the goals of conquest and annihilation were transferred from the battlefield to our kitchens and farm fields. The Seek, Strike, and Destroy maxim of my father's antitank unit was brought home and turned against the natural world. This attitude, Carson believed, would be our undoing. All life was caught in the crossfire.

When *Silent Spring* was published, the victory days of the Second World War had not yet reached their twentieth anniversary. Compared to Carson's generation, those of us born after World War II are not as aware of the domestic changes wrought by this war. We have inherited its many inventions—as well as the waste produced in their manufacture—but we do not have a keen sense of their origin. In seeking explanations for the unprecedented cancer rates among our ranks, we need to examine them.

Taped above my desk are graphs showing the U.S. annual production of synthetic chemicals. I keep them here to make visible a phenomenon I was born in the midst of but am too young to recall firsthand. The first consists of several lines, each representing the manufacture of a single substance. One line is benzene, the human carcinogen known to cause leukemia and suspected of playing a role in multiple myeloma and non-Hodgkin lymphoma. Another is perchloroethylene, the probable carcinogen used to dry clean clothes. A third represents production of vinyl chloride, a known cause of angiosarcoma and a suspected breast carcinogen. They all look like ski slopes. After 1940, the lines begin to rise significantly and then shoot upward after 1960.

A second graph shows the annual production of all synthetic organic chemicals combined. It resembles a child's drawing of a cliff face. The line extending from 1920 to 1940 is essentially horizontal, hovering at a few billion pounds per year. After 1940, however, the line rockets skyward, becoming almost vertical after 1960. This kind of increase is exponential, and in the case of synthetic organic chemical production, the doubling time is every seven to eight years. By the end of the 1980s, total production had exceeded two hundred billion pounds per year. In other words, production of synthetic organic chemicals increased one-hundred-fold between the time my mother was born and the year I finished graduate school. Two human generations.

No category of synthetic chemicals has increased faster than plastics, the annual production of which increased from a half million tons in 1950 to over 260 million tons in 2009. Indeed, the quantity of plastics synthesized in the first decade of the twenty-first century approaches the total produced in the entire one hundred years that preceded it.

T[illegible] organic and synthetic are slippery ones. Organic has two definitions that very nearly contradict each other. In popular usage, *organic* describes that which is simple, healthful, and close to nature. Food certified as organic is supposed to be produced without the use of artificial, *synthetic* chemicals.

In the parlance of chemistry, *organic* simply refers to any chemical with carbon in it. The study of organic chemistry is the study of carbon compounds. The word *synthetic* means essentially the same as it does in everyday conversation: a synthetic chemical is one that has been formulated in a laboratory, usually by combining smaller substances into larger ones. Most often, these substances contain carbon. Indeed, many organic chemicals now in daily use are synthetic—they do not exist in nature.

Of course, not all organic substances are synthetic. Wood, leather, crude oil, sugar, blood, coal—these are all carbon-based, organic substances found in the natural world. But, insofar as they have carbon atoms in their structures somewhere, the vast majority of synthesized chemicals are also organic. Plastic, detergent, nylon, trichloroethylene, DDT, PCBs, and CFCs are all synthetic organic compounds. The close alignment between organic and synthetic leads to the absurd but truthful concept that organic farmers are those who shun the use of (synthetic) organic chemicals.

Synthetic organic compounds are derived from either coal, or, far more commonly, petroleum, which is a generic term for the mixture of oil and natural gas that is trapped within the fractures and pores of some sedimentary rock formations. When petroleum is brought to the surface, crude oil can be refined into gasoline, heating oil, and asphalt. Or the oil can serve, along with natural gas, as a starting point for making new chemicals. Petroleum is the mother of synthetic chemical manufacturing. Synthetic chemicals are, by and large, petrochemicals. Like coal, petroleum started out, millions of years ago, as organisms buried in mud, which is why both coal and oil are called fossil fuels. Whereas coal is dead plants, petroleum is dead animals. (Animals are oily.)

These facts bring the widely divergent definitions of the word *organic* together. Organic substances are those that come from organisms living or dead. Long chains of carbon atoms compose the chemical infrastructure of all life forms, including the petrified organisms and liquefied organisms who lived on the planet eons ago and who have since been extracted from their burial grounds in the form of a boxcar of coal or a barrel of crude oil. A molecule of pesticide is made up of rearranged carbon atoms distilled from some creature's once-living body. So is a plastic shower curtain.

And here lies the problem. Synthetic organic molecules are chemically similar enough to substances naturally found in the bodies of living organisms that, as a group, they tend to be biologically active. Our blood, lungs, liver, kidneys, colon—with the help of an elaborate enzyme system—are all designed to shuttle around, break apart, recycle, and reconstruct carbon-containing molecules. Thus, synthetic organics easily interact with the various naturally occurring biochemicals that constitute our anatomy and participate in the various physiological processes that keep us alive. By design, petroleum-derived pesticides have the power to kill because they chemically interfere with one or another of these processes. DDT wrecks the conduction of nerve impulses. The weed killer atrazine poisons photosynthesis. The phenoxy herbicides act like growth hormones.

Plenty of synthetic organics are inert in their finished forms. Indeed, this is why they are not biodegradable: their molecules are so large or otherwise so complex that they do not decay. They are thus exempt from the global carbon cycle that is constantly building up and breaking down organic molecules. And, of course, this exemption is what you want in a window frame or roof gutter.

But undegradable substances like vinyl—that is, polyvinyl chloride—are only inert during the middle of their life spans. In the beginning, they are created out of synthetic chemicals that are highly reactive. By accident or on purpose, these industrial feedstocks are routinely released, dumped, or spilled in the general environment. For example, while PVC plastic is, biochemically speaking, quite lethargic, the vinyl chloride from which it is manufactured is a powerful liver carcinogen. It is also violently explosive. In the presence of moisture, it turns into hydrochloric acid and so will liquefy your lungs on inhalation. At the end of its life, PVC turns toxic again as it degrades in landfills or burns in incinerators. When perfectly benign piles of vinyl siding are shoveled into a garbage incinerator, poisonous dioxin rises from the stack. The incinerator itself, in this case, acts as a de facto chemical laboratory synthesizing new organic compounds from feedstocks of discarded consumer products.

Through all of these routes, we find ourselves facing a rising tide of biologically active, synthetic organic chemicals. Some tinker with our hormones. Some attach themselves to our chromosomes and trigger mutations. Some cripple the immune system. Some light up our genes and so enhance the production of certain enzymes. If we could metabolize these chemicals into completely benign breakdown products and excrete them, they would pose less of a hazard. Instead, a good many of them accumulate.

In short, synthetic organic chemicals confront us with the worst of both worlds. They are similar enough to naturally occurring chemicals to react with us but different enough not to go away easily.

A number of these chemicals are soluble in fat and so collect in tissues high in fat content. Synthetic organic solvents, such as perchloroethylene and trichloroethylene, are an example. They are specifically designed to dissolve other oil- and fat-soluble chemicals. In paint, they carry oil-based pigments. As degreasing agents, they sweep oils from lubricated metal parts. As dry-cleaning fluids, they whisk away greasy stains. They also all work splendidly to dissolve oils on our skin and can thus easily enter our bodies upon touch. In addition, they are readily absorbed across the membranes of our lungs. Once inside of us, they take up residence in fat-containing tissues.

Many such tissues exist. Breasts are famous for their high fat content and serve as storage bins for synthetic organic chemicals. But organs less renowned for fat content also collect these chemicals. The liver is surprisingly high in fat. So is bone marrow, the target organ for benzene. And, because nerve cells are sheathed in a fatty coating that serves as electrical insulation, so are our brains. Many synthetic solvents have been used as anesthetic gases due to their ability to alter consciousness. Chloroform is one.

Its medical uses long since discontinued, chloroform is now used to make refrigerants for home air conditioners and supermarket freezers. It's also feedstock for the synthesis of pesticides and dyes. At last count, U.S. annual production of chloroform is was about 565 million pounds. And it's not just a commodity and an ingredient for the manufacture of other commodities; it's also a waste product, unintentionally created when other chemicals are synthesized. So, it leaks from facilities that make chloroform, as well as from those that use chloroform to make other synthetic chemicals, as well as from those in which chloroform is an unwanted waste product. Chloroform is also formed when drinking water is chlorinated. Its vaporous fumes rise from half of the hazardous waste sites on the Superfund list. It is detectable in air from all parts of the United States and in almost all public drinking water. Chloroform causes cancer in animals and is classified as a probable human carcinogen. With a half-life of a mere eight hours, its residence time in the body is actually quite brief. The problem with chloroform is not longevity but the fact that we are continuously exposed through multiple routes. Through food, water, and inhalation, all of us, according to the U.S. Agency for Toxic Substances and Disease Registry, receive daily, low levels of exposure to chloroform.

~~~~~~

First synthesized in 1874, DDT languished without purpose until drafted into World War II, and it proved its mettle by halting a typhus epidemic in Naples. My father arrived in this occupied city not long after. According to his wartime account, Naples lay in ruins, its people hungry, dirty, and in great despair. Little wonder they were also vulnerable to typhus. DDT's ability to annihilate the insect carriers of this disease—fleas, lice,
~~~~~~

and milk—must have seemed miraculous. Shortly thereafter, DDT was loaded onto American bombers and sprayed over the Pacific Islands to control mosquitoes. War production of DDT soon exceeded military requirements, and by 1945 the U.S. government allowed the surplus to be released for general civilian use.

As documented by historians, this decision marked a profound change in purpose. It is one thing to fumigate war refugees falling ill from insect-borne epidemics and quite another to douse the food supply of an entire nation not at risk for such diseases. It is one thing to rain insecticide over war zones ravaged by malaria and quite another to drench suburban Long Island. The skillful advertising that accompanied this transformation advocated a whole new approach to the insect world. Various insect species—some, mere nuisances—were recast in the public's imagination as deadly fiends to be rooted out at all cost. Cohabitation was no longer acceptable. In demonizing the home front's new enemy, one cartoon ad even went so far as to place Adolf Hitler's head on the body of a beetle.

Synthetic pesticide use thus began in the United States in the 1940s. Two other chemicals participated in this debut: parathion and the phenoxy herbicides 2,4-D and 2,4,5-T. Parathion—and its sibling malathion—belong to a group of synthetic chemicals called organophosphates, which are created by surrounding phosphate molecules with various carbon chains and rings. Like the chlorinated pesticides, they attack an insect's nervous system, but they do so by interfering with the chemical receptor molecules between the nerve cells rather than by affecting the conduction of electricity, which is DDT's mode of action. Like the chlorinated pesticides, organophosphate poisons played a starring role during the war—but as villain rather than hero. Developed by a German company as a nerve gas, members of the first generation of organophosphate poisons were tested on prisoners in the concentration camps of Auschwitz.

By contrast, the phenoxy herbicides were an Allied weapon. As we have already seen in Chapter Three, they were mobilized in the 1940s with the goal of destroying enemy crops. Another American invention—

the atomic bomb ended that war before field testing could yield to full scale chemical warfare. Twenty more years would pass before 2,4-D and 2,4,5-T would enter combat—this time in Vietnam's rainforests under the nom de guerre Agent Orange. In the meantime, they were introduced into U.S. agriculture for weed control and into forestry for shrub control. By 1960, 2,4-D accounted for half of all U.S. herbicide production. The hoe was fast on its way to becoming obsolete.

The historical picture of pesticide use in the United States closely resembles the graphs of synthetic chemical production: a long, gentle rise between 1850 and 1945 and then, like the side of a mesa rising from the desert, the lines shoot up. Insecticide use begins ascending first; herbicide use closely follows. The line for fungicide use rises more gradually. All together, within ten years of their introduction in 1945, synthetic organic chemicals captured 90 percent of the agricultural pest-control market and had almost completely routed the pest-control methods of the prewar years. In 1939, there were 32 pesticidal active ingredients registered with the federal government. At present, 1,290 active ingredients are so registered and are formulated into many thousands of products. Contemporary trends in pesticide usage are hard to gauge. In 2001, the last year for which estimates were made, U.S. pesticide use exceeded 1.2 billion pounds.

Agriculture consumes the lion's share of this total, with only about 5 percent used by private households. Nevertheless, three-quarters of U.S. households use pesticides of some kind—from yard and garden weed killers to pet flea collars. These uses place us in intimate contact with pesticide residues, which can easily find their way into bedding, clothing, and food. Pesticides persist longer indoors, where no sunlight, flowing water, or soil microbes can help break them down or carry them away. Yard chemicals tracked indoors on the bottoms of shoes can remain impregnated in carpet fibers for years.

As part of a 2009 study conducted by the U.S. Department of Housing and Urban Development, researchers swiped dust samples from kitchen floors in 500 homes across the United States and analyzed them for the presence of twenty-four different pesticides, some of which had been banned for [illegible] years earlier. The results: each of the twenty-four pesticides tested for turned up in somebody's house. Almost everyone's kitchen floor had [illegible] insecticide, and many long-banned pesticides were among the frequent detections. The authors concluded that pesticides are essentially ubiquitous in our living areas whether or not directly applied by the householder.

Many studies have found associations between home pesticide use and childhood cancers. The strongest and most consistent connections are between household insecticides and child leukemia. The earlier in the life the exposure, the greater the risk, with exposures during pregnancy representing the highest danger. In one study, children whose yards were treated with pesticides were far more likely to have soft tissue cancers than children living in households that did not use yard chemicals. Other studies have found links with brain tumors and lymphomas. A 2007 review of the evidence concluded, "one can confidently state that there is at least some association between pesticide exposure and childhood cancer." A 2008 review of the evidence concluded, "studies published in the last decade have reported positive associations with home use of insecticides, mostly before the child's birth." In spite of all these conclusions, little has changed. A 2009 study found higher levels of household pesticides in urine samples collected from children with leukemia *and from their mothers* than in the urine of mother-child pairs living in households unaffected by leukemia. Not all of the mothers of child cancer patients used pesticides themselves. In fact, most did not. (See paragraph above.)

Of course, the postwar boom in synthetic organics was not limited to pesticides. Other products derived from petroleum also exploded onto the scene. In this case, World War II simply accelerated a process set in motion years earlier.

Historians of chemistry trace the twentieth-century rise of the petrochemical industry to the near extermination of whales in the nineteenth century: lack of whale oil for lamps created a market for kerosene, one of the lighter fractions of petroleum. Another petroleum derivative, gasoline, found purpose with the advent of the automobile. With the blockades against imported materials during World War I, the chemical industries of all warring nations were stimulated to invent new products. Germany, for example, developed artificial fertilizers when its supplies of Chilean saltpeter were cut. The same manufacturing process proved quite useful for producing explosives—as the fertilizer-derived bomb that destroyed the Oklahoma City federal building in 1995 illustrates.

With a large supply on hand for making dyes, Germany turned to chlorine gas to serve as a wretched weapon of chemical warfare in the trenches of France. Chlorinated solvents were also introduced during this time. After the war ended, new chemical products in the United States were protected by high tariffs, the war's losing parties surrendered their chemical secrets to the victors, and considerable wealth and prestige accrued to the chemical industry. By the 1930s, petroleum began to outpace coal as the source of carbon for new chemical investigations.

The cliff face of exponential growth in synthetic organic chemicals, however, did not begin until the 1940s. The all-out assaults of World War II created instant demands for explosives, synthetic rubber, aviation fuel, metal parts, synthetic oils, solvents, and pharmaceuticals. The innovations in chemical processing developed in the wake of World War I—such as the cracking of large, heavy petroleum molecules to produce many lighter and smaller molecules—were perfected and tested in large-scale production. When the war ended, the resulting economic, housing, and baby booms created unprecedented consumer demands as wartime chemicals, aided by skillful advertising, were transferred to civilian posts. Fearing a return to economic depression, national leaders encouraged the conversion of military products to civilian use. "In the United States," the historian Aaron Ihde has wryly noted, "peace did not prove catastrophic to an industry grown to monstrous proportions in response to the needs of war."

From an ecological point of view, World War II was a catalyst for the transformation from a carbohydrate-based economy to a petrochemical-based economy. For example, plastic—the manufacture of which now consumes 8 percent of world oil production—was formerly derived from plants. Invented in the 1870s, it was called celluloid. Clear plastic film made from wood pulp with adhesive on one side was introduced as cellophane tape. Plant-derived substances were also used to make steering wheels, instrument panels, and spray paint. Thus, while the carcinogen vinyl chloride was first synthesized in 1913, its production did not begin to jump until after World War II when research on botanical matter was replaced by research on petrochemicals. Automobile interiors would no longer come from cotton fibers or wood pulp, but from a barrel of oil.

Wood glue was formerly derived from soybeans. After the war, soybeans were replaced by formaldehyde for adhesive purposes. Classified as a probable human chemical carcinogen by the EPA (and as a known human carcinogen by the International Agency for Research on Cancer), formaldehyde is linked to leukemia and several other forms of cancers. It consistently ranks among the top fifty chemicals with the highest annual production volumes in the United States. In 2000 alone, 11.3 billion gallons were produced—a tenfold increase since 1960. We have an intimate relationship with this chemical. It is used as an antiseptic in personal care products—makeup, nail polish, mouthwash—and it is sprayed on fabric to create permanent press. But most of formaldehyde's annual production is used for synthetic resins in plywood and particle board. Its subsequent evaporation from construction materials and furniture makes this chemical a significant contributor to indoor air pollution. This property of formaldehyde made headlines in the aftermath of Hurricane Katrina when displaced families were so exposed while living in the government-issued trailers.

As with chloroform, the problem with formaldehyde is not that it accumulates in our tissues but that we are continuously exposed to small amounts of it from so many sources—from our subflooring to our wrinkle-free sheets—and throughout our life spans. Or beyond them. As embalming fluid, formaldehyde accompanies us into the grave. The National Toxicology Program notes that professional embalmers have excess rates of brain cancers, which may be attributable to formaldehyde exposures at the funeral home.

Before the war, plant oils of all kinds played leading roles in industry. Soybean oil was used in fire suppressant foam and wallpaper glue and as a base for paints, varnishes, and lacquers. Oils extracted from corn, olives, rice, and grape seeds were used to make paint, inks, soaps, emulsifiers, and even floor covering. The word *linoleum* shows the name of its original key ingredient: linseed oil. Castor oil, from the tropical castor bean tree, was used to lubricate machined parts during cutting and grinding. Replacing castor oil in machine shops after the war were synthetic cutting oils, which can expose workers to N-nitrosamines, a contaminant formed during their manufacture. By the 1970s, cancer among machine operators and its possible relationship to synthetic cutting fluids began receiving attention. In one study the researchers concluded:

> Until now, N-nitrosamines have not been directly associated with human cancers because no population groups had been identified that were inadvertently exposed. Cutting fluid users have the dubious honor of being the first such population group to be identified.

~~~

In the fall of 2009, the EPA's director, Lisa Jackson, conceded that the thirty-three-year-old federal law governing the manufacture of hazardous substances had failed to protect public health. She announced her intent to reform it. However this turns out, it is safe to say that, legally speaking, all of us born before 2010 entered this world under the protection of an environmental law that is now understood to be ineffective. It seems worth understanding the origins of this law, the Toxic Substances Control Act (TSCA), and the reasons for its impotency.

The rapid birthrate of petrochemicals began in 1945 and soon swamped the ability of government to oversee their production, use, and disposal. By 1976, sixty-two thousand synthetic chemicals were in
~~~

commercial use. How many of these were carcinogens? No one knew. No testing was required for chemicals to enter the marketplace. In that year, Congress passed the TSCA, which mandated the review of new chemicals and put the EPA in charge of their appraisal. Pre-existing chemicals were not covered by the new rules. Let me put a finer point on this: the entire standing inventory of sixty-two thousand was exempt from testing.

And, as of this writing, they still are. *They still are.* Of the eighty thousand or so chemicals now believed to be circulating (no one knows for sure), only 2 percent of them (this is the General Accounting Office's best guess) have been thoroughly assessed for toxicity. The only possible conclusion is that many chemical carcinogens remain unidentified, unmonitored, and at large. Here, then, is a deeply underacknowledged truth: most chemicals in commerce have never been vetted. We know nothing about them. Too often, this unknowingness is paraphrased as "there is no evidence for harm." And this in turn is sometimes translated as "the chemical is harmless." Lack of knowledge about safety becomes an implicit endorsement of safety.

TSCA has other down-the-rabbit-hole quirks. For one, the law does not precisely require the testing of new chemicals, either. Instead, it requires manufactures to divulge what they know about the risks of any new chemical they propose to commercialize. This allows for regulations based on comparisons: the structure of a newly synthesized molecule is compared to the structures of previously synthesized molecules, and then regulators make an estimate of risk. In other words, the assessment of new chemicals involves on-paper modeling and does not require independent, hands-on science at a lab bench somewhere. Second, concerning the regulation of pre-existing chemicals, TSCA compels the EPA to balance economic benefits of any chemical against its health risks. EPA may only regulate chemicals that present risks that are "unreasonable."

But with no baseline data on toxicity, it's impossible to know which chemicals are behaving unreasonably. And needless to say, petrochemical industry representatives and breast cancer advocates view the scales for weighing costs and benefits very differently. It's easy enough to load up the pan on the benefits side of the scales' fulcrum with numbers that have dollar signs in front of them. But without knowledge about inherent toxicity, what goes into the other pan? Much of the work that goes on under TSCA is the mediation of an unceasing argument about that very question. Essentially, TSCA has turned the EPA into a referee of a bickering match. Tellingly, since TSCA's inception, only five chemicals have been taken off the market. No chemicals have been banned within the past nineteen years, as U.S. chemicals policy sinks deeper into the quicksand of logical paralysis: TSCA says manufacturers need only provide toxicity data when there is demonstrable evidence of risk, yet risk cannot be demonstrated without toxicity data. Hence, the calls for reform.

Pesticides, which are subjected to somewhat more scrutiny, are regulated by twin laws: the Federal Food, Drug, and Cosmetic Act (FFDCA) and the Federal Insecticide Fungicide, and Rodenticide Act (FIFRA). FFDCA governs pesticide tolerances on agricultural commodities—that is, it sets legal limits for pesticide residues allowed in food and animal feed. FIFRA requires companies manufacturing pesticides to test their products for toxicity and submit the results to the federal government. Unlike TSCA, FIFRA does haul in for testing its backlog of old chemicals. Amendments to FIFRA require reevaluation of untested pesticides approved before the current requirements for scientific testing were put into place. Initially scheduled to be completed in 1976, this reregistration process is still under way. Until then, the old, untested pesticides can be sold and used. As one critic has noted, it is as if the bureau of motor vehicles issued everyone a driver's license but did not get around to giving us a road test until decades later.

~~~

In the 1970s and 1980s, various right-to-know laws began springing up as a response to this ever-expanding mosh pit of toxic chemicals. The first group of laws established employees' right to know about hazardous substances in their workplaces. A second group sanctioned citizens' right to
~~~

know about the presence of toxic chemicals in their communities and, finally, about the routine release of some of these chemicals into the environment. For nearly four decades after the widespread introduction of [illegible] were not ours. The identity of chemicals released by industry was considered privileged information—trade secrets. Those of us born during this time—the 1940s until the mid-1980s—will never know with certainty what we were exposed to as children and what carcinogenic risks we have assumed from such exposures. We can, however, obtain partial information about our current exposures.

Significantly, neither set of laws came about because legislators and manufacturers calmly agreed that citizens should be made aware of their chemical exposures. Rather, workplace right-to-know laws are rooted in a long history of labor struggle, and the community-based laws—codified as the Emergency Planning and Community Right-to-Know Act (EPCRA)—passed the U.S. Congress in 1986 over intense industry opposition.

The linchpin of EPCRA is the Toxics Release Inventory (TRI). As the SEER Program registry is to cancer incidence, TRI is to carcinogens and other toxins. It requires that certain manufacturers report to the government the total amount of each of some 650 toxic chemicals released each year into air, water, and land. The government then makes these data public information. As a pollution disclosure program, TRI has many deficiencies. It relies completely on self-reporting. It does not address the presence of carcinogens in consumer products. Small companies are exempt. And 650 is a small fraction of the total chemicals in use. Between 2001 and 2008, TRI was scaled back, and, with a homeland security rationale, public information on industrial accidents and locations of chemicals stored at chemical facilities was also removed from Web sites. The number of facilities required to report to TRI fell. A new curtain of secrecy descended. (In 2009, many of these changes were reversed.)

Furthermore, loopholes in reporting requirements allow industries to play a shell game with their wastes. Researchers tracking the flow of track chemicals through the economy point out that declines in toxic waste *releases* have not always been accompanied by parallel declines in toxic waste *production*: the generation of toxic waste by TRI-reporting facilities remains high. Where, then, is the waste going? Without thorough materials accounting, which is not currently required, no one is exactly sure.

Nevertheless, under EPCRA, any citizen can obtain a list of the reported toxic releases in his or her home county by typing their ZIP code into a government Web site (see Afterword for details). Access to this information is still a fundamental public right.

In some communities, the TRI has served as a powerful tool for pressuring factories to reduce pollution. Its most important function may be the implicit recognition that a so-called private industry is engaging in a very public act when it releases toxic chemicals into a community's air, water, and soil. Conceptually, we all know the industries in our communities pollute the environment. We may even be able to see and smell the results. But very often, the picture does not come into focus for us until we actually stare at the list of specifics, as when the names and the numbers are printed in our local newspapers: how many pounds of which known or suspected carcinogens were released by which companies into the air we breathe or into the rivers we fish and from which we draw our drinking water?

The TRI's first report, released by the EPA in 1989, had just such an effect. It revealed that billions of pounds of toxic chemicals were being routinely emitted each year into the nation's air, water, and land. Nearly all who read the report were amazed. This was the first attempt to gather together routine toxic releases, and the sum was an unquestionably staggering amount. Said a representative from Monsanto: "The law is having an incredible effect. . . . There's not a chief executive officer around who wants to be the biggest polluter in Iowa."

In the first year of reporting, only about 5 percent of toxic releases in the country were reported under TRI, and yet the effect of ending the silence about toxic releases was huge. According to the most recent TRI, 4.1

billion pounds of toxic chemicals were released into the environment in 2007. Of these, [illegible] million pounds were known or suspected carcinogens. Mexico and Canada now also both require reporting of industrial pollutants released into air and water, so it is possible, for the first time, to look at patterns of toxic emissions across all of North America. In 2007, Montreal's Commission for Environmental Cooperation combined the nation's three inventories, which all together represent pollutant releases from thirty-five thousand industrial facilities. The two big patterns are these: first, the petroleum industry is, all by itself, responsible for one-quarter of reported toxic releases on the continent. Second, as restrictions on toxic disposal increase the cost of dumping, toxic waste is increasingly transported to recycling facilities rather than landfills.

What the 2007 TRI reveals for central Illinois, however, is something else. Peoria County now ranks number fourteen in the nation for toxic emissions, and emissions and toxic transfers are going sharply up rather than trending down. This is partly attributable to all the industry that lines the river valley—Peoria County is home to a quarter of Illinois' manufacturing facilities—but it is mostly due to a single hazardous waste landfill that sits above the Sankoty Aquifer in a community near the edge of Peoria. Siting a toxic waste dump above the drinking water supply is no longer allowable practice, but this particular dump was open for business long before laws prohibiting such sitings went on the books, so it was grandfathered. It is one of only eighteen hazardous waste landfills in the nation that is permitted to accept the baddest of the bad: heavy metals, known human carcinogens, waste from Superfund cleanups. In 2005, the Peoria Disposal Company ranked number twenty-one in the list of facilities reporting the largest toxic releases in the United States. To be sure, Peoria Disposal's releases are to land—meaning that the toxic materials are released into a hole in the ground—and the operators claim the waste is all safely contained therein. For now.

When the Peoria Disposal Company applied for an expansion a few years ago, community members objected. Organized as Peoria Families Against Toxic Waste and with local physicians and hospital administrators as allies, they convinced the county board to deny a permit. In response, the landfill operators sought approval—and received it—to delist one of its materials as hazardous so that a treated form of it can be diverted to Hopedale's Indian Creek landfill on my side of Illinois River in Tazewell County. This leaves more room in the Peoria landfill for the truckloads of toxic waste that arrive here from all over the Midwest. As other hazardous waste landfills around the country close or restrict dumping, everyone else's carcinogens are coming right at us.

In a favorite photograph of myself as a child, I am hanging determinedly onto a tricycle, wearing a goofy expression and my father's army hat. The determination came from trying to salute my father, the photographer, while simultaneously pedaling. It is 1962. The setting is the concrete patio on the south side of our house. A construction worker before the GI Bill returned him to the typewriter as a college student, my father poured this patio himself and laid the brick walkway leading out into his 1.5 acres of former cow pasture.

After the war, my father married a farmer's daughter with a degree in biology and another in chemistry. He built his house on Pekin's east bluff and planted lines of silver maple and white pine in the sod. Before the trees grew up to form a wall around the borders of his property, the patio offered a spectacular view. To the east, cows grazed. Although afraid of them, I liked to stand at the fence and watch them eat—their purply tongues and black plumes of flies, the ripping sounds of the grass.

Just beyond, the bluff's pastures unrolled into what was once—and I am guessing here—hill prairie. Here lay vast fields of corn and soybeans. I liked the corn—each stalk a green man waving his arms. In September, the soybeans turned brilliant yellow and then deepened into an orange-brown far richer than my burnt-sienna crayon.

"What color would you call soybeans?" I inquired of Aunt Ann, who farmed two counties east from us.

She didn't miss a beat. "At six dollars a bushel, I would call them gold."

My father drove west to work every morning. Looking through the patio screen from my tricycle, I could see the smokestacks, cooling towers, and distillation chambers of the river valley's three dozen industries. I liked the steam clouds, trails of smoke, and mysterious shimmering vapors. I was especially fond of the pink-and-white-striped towers belonging to the ethanol distillery and the coal-burning power plant just upriver. At night, they became lighthouses—great blinking columns warning planes away.

~~~

Tazewell County, Illinois, is home to two distinct cultures, one emblemized by the lone figure on his tractor and the other by picket lines of striking plant workers. Our house was situated in the transitional zone between the two.

Among farmers, Tazewell is known as the birthplace of Reid's Yellow Dent, a famous strain of field corn that became the ancestor of many hybrid seed lines. Among industrialists, Tazewell County is known for its ethanol plant and as the proving grounds for Caterpillar tractors, backhoes, and bulldozers. Caterpillar's management offices are headquartered across the Illinois River in Peoria. A hydrologist's description of the area from 1950 is as good as any: "The Peoria-Pekin area is a highly industrialized district requiring an enormous volume of water. The industrial areas are surrounded by the fertile agricultural prairie lands of the corn belt."

Settled before the prairie was sod-busted, Pekin began as a military fort. War and manufacturing have frequently danced together here. Distilling and brewing began as a means of transforming grain into a nonperishable cash commodity that could easily be shipped east. Wartime needs for industrial alcohol then provided a huge new market and inspired new production technologies. One of Pekin's distilleries was founded in 1941 expressly to provide the U.S. military with ethanol. In 1916, the U.S. Army ordered the first Caterpillar tractors, which it used
~~~

to drag cannons, ammunition, and supplies to the front. During World War II, Caterpillar machines were used to bulldoze airstrips, grade roads, clear bomb wreckage, and topple palm trees.

Other photographs from the turn of the century show child laborers posing in the sugar-beet fields. The black smoke of the sugar factory forms a dramatic backdrop against the little white faces. When sugar beets proved harder to grow than corn, the sugar works became Corn Products, which produced Argo cornstarch and Karo syrup. In 1924, a starch explosion incinerated forty-two workers. In 1980, in a joint venture with Texaco, Corn Products became Pekin Energy, which became, by the mid-1990s, the nation's second-largest producer of ethanol. In 2003, Pekin Energy became Aventine and, together with its sister plant in Nebraska, produces each year 100 million gallons of fuel-grade ethanol. In 2007, this plant reported the release of 4,613 pounds of known and suspected carcinogens.

In September 1994, I drove along the Illinois River banks to pass by where those old beet fields must have been—less than two miles from the house I grew up in. The floodplains are now a landscape of docks, stacks, rail yards, conveyers, elevators, hopper bins, pits, lagoons, coal piles, tailings ponds, settling tanks, power lines, and scrap heaps—all that I had seen at a distance as a child. A union billboard announced, "You Are Now Entering a War Zone," a comment not on the environment but on labor's latest showdown with management at the nearby Caterpillar plant.

There are some places in this world that prompt one to ask, "Where did all this come from?" The fish, vegetable, and flower markets of New York City always bring me to this question. Tazewell County is another kind of place. Spend some time on the Pekin docks. Watch the barges of coal, grain, steel, chemicals, and petroleum products. "Where is all this going?"

There are partial answers. The grain elevators and the mills ship corn and palletized animal feed south to New Orleans and from there to Asia and Europe. The coal-fired power plant called Powerton sends electricity 165 miles north to Chicago via high-power lines. In 1943, its smoke prevented landings at an airport forty-one miles away. In 1974—the year I turned fifteen—this plant was named the worst polluter in the state of

Illinois Trace Chemicals formulates pesticides. I don't know where they end up. The brass foundry makes huge cylinders, called bushings, for draglines, drills and crushers used in strip mines. Caterpillars end up [illegible], I looked out the window of a bus heading south along the Nile River in eastern Sudan and found myself face to face with Caterpillar's familiar logo—a giant capital C—painted on a billboard near a military installation.

About the other industries lining the river valley I know less. I don't know what goes on at Airco Industrial Gases, the Sherex Chemical Company, the Agrico Chemical Company, or the aluminum foundry. I know that Keystone Steel and Wire makes nails and barbed wire out of scrap metal. In 1993, the company faced charges for polluting the sand aquifer below its facilities with TCE and another synthetic degreaser, 1,1,1-trichloroethane, a suspected carcinogen.

Like a film of gasoline on a pond's surface, an emotional blankness coats my words here. There are, of course, many ways of expressing the relevance of the historical past to the personal present. Surely there is one that could describe the private thoughts of an East Bluff girl returning home and passing by the hospital where, years before, she was diagnosed with a type of cancer known to be caused by exposure to environmental carcinogens. Surely there is a language able to explain why such a woman would again drive along Distillery Road, breathing the acrid air, searching for nineteenth-century sugar-beet fields and twentieth-century hazardous waste sites.

A silence spreads out. I cannot make her speak.

It is not the silence of resignation or paralysis. It is the fear that speaking intimately about this landscape—or myself as a native of this place—would make too exceptional what is common and ordinary. I feel protective of my hometown. Its citizens are not unusually ignorant or evil or shortsighted. Away from the river, the city itself is lovely. Between the fields and factories are nice, old neighborhoods, beautiful parks, the county fairgrounds, and reasonably good schools. There is nothing unique or even unusual about Tazewell County, Illinois. As is true everywhere else, its agricultural and industrial practices—from weed control to degreasing parts—were transformed by chemical technologies introduced after World War II. As is true everywhere else, these chemicals, many of them carcinogens, have found their way into the general environment. As is true almost everywhere else, no systematic investigation has been conducted to determine whether any connection exists between the release of these chemicals and the rates of cancer here.

"We know the emissions are present, and the cancer, but we don't know if the two are related," said a state toxicologist quoted in the local paper in March 1995. This article concluded:

> The impact of tons of toxic emissions on the health of industrial workers and the public never has been systematically studied and may be impossible to determine. . . . Health statistics in Peoria and Tazewell counties are troubling, but the connection between emissions and health problems is not clear.

I cried after I first read through the TRIs for Peoria and Tazewell Counties. Hundreds of pages itemize the toxic emissions for area industries during the years since 1987, when this information was first compiled. In 1991, for example, large manufacturers in Peoria and Tazewell Counties legally released 11.1 million pounds of toxic chemicals into the air, water, and land. Among the known and suspected carcinogens released were benzene, chromium, formaldehyde, nickel, ethylene, acrylonitrile, butyraldehyde, lindane, and captan. Captan is a carcinogenic fungicide prohibited for many domestic uses in 1989. In 1987, according to the TRI, 250 pounds of captan ended up in the Pekin sewer system. In 1992, 321 pounds were released into the air.

Tips of all kinds of icebergs are revealed in other right-to-know documents. For example, I have a partial record of pre-TRI toxic releases in Tazewell County dating back to 1972. The carcinogens catch my eye first—PCBs, vinyl chloride, benzene—but the list also includes other frightening and surreal items: printing ink, jet fuel, asphalt sealer, dynamite, scrubber sludge, fuel oil, antifreeze, fly ash, coal dust, herbicides, furnace oil, and "explosive vapors."

In addition, I possess a twenty-four-page list of facilities in Tazewell County with permits to discharge wastes into particular rivers and streams ("local receiving waters: Farm Creek . . . local receiving waters: Illinois River," etc.). I've also obtained a thirty-four-page list of each and every facility—from the local crematorium to the auto body shop—permitted to deal in any way with hazardous materials. Right-to-know legislation has given me access to a hefty off-site transfer report, a document particularly revealing because it shows the flow of toxic wastes coming into Tazewell County. I know, for example, that the Sun Chemical Corporation of Newark, New Jersey, sent 250 pounds of friable asbestos to the Pekin Metro landfill for disposal in 1987. Tazewell doubled the amount of hazardous waste it generated and shipped off-site between 1989 and 1992, but, as one of the state's top receiving counties, it still received four times more waste than it produced.

The spill report for Tazewell County details chemical accidents. Here is the first entry as it appears on the list:

DATE: 6/11/1988
STREET: RTE 24
MATERIAL SPILLED: METHYL CHLORIDE
AMOUNT SPILLED: 2,000 LBS
WATERWAY/OTHER: AIR RELEASE
EVENT DESCRIPTION: WEIGH TANK/WHILE PREPARING FOR INSPECTORS, VALVE INADVERTENTLY OPENED/EXACT CAUSE UNDER INVESTIGATION
ACTION TAKEN: TEMPORARILY EVACUATED AFFECTED BUILDING FOR TWO HOURS . . . SHUT VALVE TO STOP RELEASE

Methyl chloride causes kidney cancers in mice and degeneration of the sperm-carrying tubules in rat testicles. Methyl chloride is considered "not classifiable as a human carcinogen" because human data are too limited to so classify it.

Amid a flood of information, an absence of knowledge. Amid a thousand computer-generated words, a silence spreads out.

Seek. Strike. Destroy. Of all the unexpected consequences of World War II, perhaps the most ironic is the discovery that a remarkable number of the new chemicals it ushered in are estrogenic—that is, at low levels inside the human body, they mimic the female hormone estrogen. Many of the hypermasculine weapons of conquest and progress are, biologically speaking, emasculating.

This effect occurs through a variety of biochemical mechanisms. Some chemicals imitate the hormone directly, while others interfere with the various systems that regulate the body's production and metabolism of natural estrogens. Still others seem to work by blocking the receptor sites for male hormones, which are collectively called androgens. In 1995, fifty years after its triumphant return from the war and entry into civilian life, DDT again made headlines when new animal studies showed that DDT's main metabolic breakdown product, DDE, is an androgen-blocker. Since that discovery, evidence has accumulated to suggest that a whole raft of synthetic chemicals have the power to derange the male testicle if exposures occur at key moments in development. The consequences to men may include physical deformities such as undescended testicles, lowered sperm counts, and testicular cancer. This chemically induced complex of problems goes by a terrifying name: testicular dysgenesis syndrome.

Among the leading suspects in this ransacking of manhood is a ubiquitous class of petrochemicals called phthalates. Phthalates give PVC plastic flexibility and so are responsible for the signature smell of new vinyl shower curtains and new car interiors. Because they are oily, phthalates are also used to carry fragrance in perfumes and lotions. Phthalates have also turned up in foods, especially those with high fat content, such as eggs, milk, cheese, margarine, and seafood. And they have also turned up in us. In [illegible], the Centers for Disease Control released the results of its extensive body burden survey, which measured concentrations of common chemical contaminants in a representative sample of the U.S. population. The CDC survey found nearly ubiquitous exposure to phthalates among all ages, with the highest levels in children. These results were worrisome because, in studies of animals, damage to male testicles can occur during prenatal life at close to these levels.

Recent human research has focused on two different stages of male life: baby boys and men trying to become fathers. In one of the baby studies, researchers measured mothers' levels of phthalates before birth and anatomical features of the babies' genitals after birth. The results showed that prenatal exposure to phthalates was associated with altered development of the testicle and penis. In one of the adult studies, researchers looked at men who had visited a fertility clinic. The results here showed that increasing levels of phthalates in urine were associated with increasing levels of damaged sperm.

Many synthetic materials known to function as endocrine disruptors belong to a chemical group called organochlorines. Lindane, DDT, heptachlor, chlordane, PCBs, CFCs, TCE, perchloroethylene, 2,4-D, methyl chloride, PVC, dioxin, chloroform, and atrazine are all members of this group. (Formaldehyde and phthalates are not.)

Organochlorines involve a chemical marriage between chlorine and carbon atoms. A few are formed during volcanic eruptions and forest fires. For the most part, however, chlorine and carbon move in separate spheres in the natural world—and in the bodies of humans and other mammals. To force the two together, elemental chlorine gas is required.

Although it holds a rightful place in the periodic table of elements, pure chlorine is a human invention. It can be produced by passing electricity through brine in a procedure first undertaken on an industrial scale in 1893. A powerful poison, chlorine gas became known to the world during World War I, but its manufacture grew slowly until World War II, then rose exponentially. About 1 percent of this production is used for disinfecting water and about 10 percent for bleaching paper, and the majority is combined with hydrocarbons derived from petroleum to make organochlorines.

In its elemental form, chlorine is highly reactive with carbon, which is why so many combinations are possible. Like houses of different styles, some organochlorines are small and plain, and others sprawling and ornate. One of the simplest is chloroform, which consists of a single carbon atom with one hydrogen and three chlorine atoms attached to it like four spokes on a hub. Vinyl chloride, the dry-cleaning solvent perchloroethylene, and the industrial degreaser trichloroethylene are also relatively simple. On the more elaborate side are chlorinated phenols. These consist of a ring of six carbons with various chlorinated groups hanging off the corners. The pesticide lindane, the herbicide 2,4-D, and DDT are also ringed structures.

And then there are the PCBs. The elders of the group, PCBs are referred to in the plural for a reason. Polychlorinated biphenyls are two rings of carbon atoms welded directly together, around which are attached any number of chlorine atoms. In fact, there are 209 possible combinations and therefore 209 different PCBs. Some of these chemical combinations are estrogenic and some appear not to be, but no one has worked this out definitively. A study published in 2009 found strong evidence that PCBs block male hormones. Among members of the Mohawk Nation in Akwesasne Territory, men with higher PCB levels had lower testosterone levels. Akwesasne, along the St. Lawrence River, is contaminated with PCBs from upstream industries, and so are the local fish.

As a group, organochlorines tend to be persistent in air and water. When they evaporate and are swept into the wind currents, some fall back to the earth close to their origins, while others can circulate for thousands of miles before being redeposited into water, vegetation, and soil. From there, they enter the food chain. Diet is thus believed to be a major route of exposure for us.

Not all organochlorines are deliberately constructed. Whenever elemental chlorine is present, the natural environment will synthesize additional, unwanted organochlorine molecules. These reactions can take place when water containing decayed leaves is chlorinated. It can happen when chlorinated plastics are burned. It can happen during the manufacture of other organochlorines. The production of 2,4,5-T, the burning of plastics and certain methods of bleaching paper all contribute to an unwelcome organochlorine known as dioxin. A chemical of no usefulness and never manufactured on purpose, dioxin has been linked to a variety of cancers and is now believed to inhabit the body tissues of every person living in the United States. Dioxin is a beautifully symmetrical molecule, consisting of two chlorinated carbon rings held together by a double bridge of oxygen atoms.

~~~~~~

As the daughter of a World War II veteran, I am grateful that my father did not die in a typhus epidemic in Naples. But as a survivor of cancer, as a native of Tazewell County, and as a member of the most poisoned generation to come of adult age, I am sorry that cooler heads did not prevail in the calm prosperity of peacetime, when a longer view on public health was once again permissible and necessary. I am sorry that no one asked, "Is this the industrial path we want to continue along? Is this the most reasonable way to rid our dogs of fleas and our yards of crabgrass? Is this the safest material for a baby's pacifier or a tub of margarine?" Or that those who did ask such questions were not heard.

To these queries we now have a new one: Do we really want to keep creating vast amounts of synthetic materials from petroleum? Or might it be time to revoke petroleum's de facto title of Feedstock for Life? Petroleum is the same substance whose combustion is driving climate change. It is the same substance that, by accident of geology, we must import from distant lands, so distorting our economy and undermining national security. It is the same substance that verily defines the word *unrenewable* (as in, "God is not putting any more of it in the ground") and whose excavation from the earth has peaked, or is peaking, or will peak soon. It is this substance that drizzles cancer-causing and hormone-disrupting chemicals into our environment at every stage of its production, use, and disposal, and again after its conversion into synthetic, organic chemicals . . .
~~~~~~

and during the use of those chemicals . . . and after the disposal of those chemicals in a hole above Peoria's drinking water supply and elsewhere.

Here is where the nascent struggle to divert our economy from its crippling dependencies on petroleum meets the long-standing but equally fervent struggle to lessen the burden of cancer. These are not separate tasks. Shared by each is the need to draft a new chemicals policy. In this, I believe, we cancer survivors have insight and wisdom to offer. Our current environmental regulatory apparatus is paralyzed by uncertainty. It cannot regulate unless there is proof, but proof is unattainable. Making difficult decisions in the face of uncertainty is something cancer patients have experience with. We know how to find the courage to stop dithering. We look at our lab reports, we study the MRI images, we get a second opinion, and then, *based on the best evidence available to us in the present moment*, we choose a path forward that seems most likely to preserve life. Waiting for proof, we discover, is folly. As much as we might prefer to indulge in denial and hand over to others the terrible burden of deciding what to do, we somehow find it in ourselves to set a course of action. Along the way, we become resourceful, determined, and unafraid. We unparalyze ourselves. We get up and walk.

These are exactly the skills required to re-imagine our nation's ineffectual chemicals policy. Happily, we may not have to start from scratch, as alternative systems have already been devised and implemented in other places in the world. Let's look briefly at what other courses of action are available that might serve as models for us.

Recall the worrisome body of evidence linking childhood cancers to pesticides. Recall also the HUD survey that found pesticides in kitchens whether residents actively used them or not. Hence, choosing voluntarily, as individual homeowners, to forswear pesticides in and around our own homes is only part of the solution. We need policies that compel nonchemical methods of pest control. In Canada, these exist. Ontario and Quebec, as well as 152 cities across the rest of Canada, now prohibit the use of pesticides for cosmetic reasons. Within these provinces and municipalities, using synthetic pesticides to improve the appearance of lawns and gardens is now illegal.

Upheld by the Canadian Supreme Court, these bans were supported by the Canadian Cancer Society, a counterpart to our American Cancer Society. An endorsement by the Ontario College of Family Physicians was also influential in their passage. Reviewing the evidence for the cancer health effects of pesticides, Ontario's family physicians concluded,

> Studies looking at pesticide use and cancer have shown a positive relationship between exposure to pesticides and the development of some cancers, particularly in children. Because most studies assessed use of multiple pesticides, the authors recommend that exposure to all pesticides be reduced.

In other words, because it is not possible—and will never be possible—to prove which cancers are caused by which pesticides, they should be used as little as possible and not at all for nonessential purposes like growing weed-free lawns. Benefit of the doubt goes to children, not to chemicals. In joining this call to action, the Canadian Cancer Society asserted that the risks of a link between pesticides and cancer were too great to ignore. And those arguments won the day. And there are still beautiful French gardens in Montreal. They are just organic French gardens. And there are still parks and playing fields in Toronto. They are just organic parks and playing fields. Indeed, they have become a selling point for tourism. As the complimentary magazine in my downtown Toronto hotel recently boasted:

> Thanks to a recent bylaw, all . . . green spaces are pesticide-free. In 2004, Toronto became the largest municipality in the world to ban cosmetic use of lawn and garden pesticides. The Sierra Club of Canada reports a clear link between pesticide use and breast cancer; many other studies have shown the dangers to children from chemical exposure to pesticides. . . . So Torontonians can take comfort that homeowners, renters, lawn-care companies, golf courses, and property managers are all subject to a fine of up to $5000 for applying pesticides.

Meanwhile, across the Atlantic, the European Union has reconfigured its entire chemicals policy. Adopted by the EU Parliament in 2006, the new legislation is called REACH—Registration, Evaluation, Authorisation and Restriction of Chemicals—and its seeds were planted in 1998 when the governments of the EU member states recognized the previous policy as unable to protect people and the environment. REACH requires that producers and importers of chemicals disclose toxicity data in order for their products to enter or remain on the market. No data, no market. And the same rules apply for new and old chemicals alike, including the sixty-two thousand chemicals exempted under U.S. law. This provision rescues chemicals regulation from the inertias of the U.S. system. In the United States, the burden of demonstrating that a chemical is dangerous falls to the government. In Europe, the obligation is shifted to industry to demonstrate a chemical is safe, and the government is left with a freer hand to restrict chemicals and compel substitutions of toxic chemicals with safer ones. Over the next decade, thousands of chemicals will have to be registered in Europe. With so many substances requiring basic testing under REACH, the law also stimulates the development of new methods to rapidly screen chemicals for toxicity (more on these in Chapter Six). More critically, the swift identification of hazardous chemicals and their replacement with safer alternatives should lower human exposures to cancer-causing substances.

Also inspiring is the Stockholm Convention on Persistent Organic Pollutants, a United Nations treaty that became international law in 2004. This treaty aims to eliminate worldwide production and use of synthetic organic chemicals that are inherently toxic and remain intact in the environment for a long time. Within the family of persistent organic pollutants are many carcinogenic organochlorines. As we have seen, these chemicals travel globally, siphon their way up the food chain, and disrupt hormone systems. No nation can manage them alone. At this writing, twenty-one chemicals are listed under the Stockholm Convention. They include many of the pesticides Rachel Carson warned about—aldrin, dieldrin, chlordane, heptachlor, lindane, mirex, and toxaphene. DDT is targeted for eventual elimination, with recognition that some countries will continue to use it for malaria prevention. [illegible] PCBs are on the list, and provisions within the treaty support efforts around the world to identify and contain its remaining stockpiles.

All three of these initiatives—Canada's provincial bans on cosmetic pesticides, the new European chemicals policy, and the global Stockholm Convention—deserve the praise and support of cancer advocates. They also offer us ideas for policies we may wish to replicate or build on here at home.

Cancer survivors—and there are more than ten million of us in the United States—can also throw our mighty weight behind green chemistry. This back-to-the-drawing-board approach seeks to design chemicals in such a way as to reduce or eliminate the generation of pernicious substances in the first place, including carcinogens.

Design is the operative word here. Green chemistry is utterly uninterested in end-of-the-pipe solutions to the many problems that petrochemistry has created. It has no truck with filters, scrubbers, Superfund cleanups, and arguments about acceptable pesticides residues. Instead, green chemistry attempts to eliminate hazards by making chemicals—including synthetic organics and the feedstocks for them—that are inherently safe. For green chemists, the problems of persistence, bioaccumulation, endocrine disruption, and long-distance transport are considered at the creation stage rather than after their materials are already released into commerce and into our shared environment. From this perspective, a chemical that causes or contributes to human cancer has a design flaw. Green chemistry has twelve basic principles. One of them is *use renewable feedstocks* (instead of petroleum, start with, say, potato peels). Another is *attend to atom economy* (all atoms of starting materials should end up in the final product, eliminating byproducts in need of disposal).

For all of us nonchemists, the idea that synthetic molecules have designers at all—like cars and clothes—comes as new thought. And, in fact, green chemical design tends to be much more sophisticated and complicated than traditional petrochemistry protocols. Green chemistry looks to nature as a model for synthesizing chemicals, and nature typically employs several more steps in the creation of new molecules than most industrial chemists who start with petroleum. Nature designs elaborately. One of green chemistry's biggest [illegible] soy-based adhesive for use in plywood that can replace formaldehyde. Winning the 2007 Presidential Green Chemistry Challenge Award, this adhesive imitates, in its molecular behavior, a protein that mussels use to hang on to rocks.

With elegant examples like mollusk-mimicking glue, what is preventing green chemistry from becoming just plain chemistry, that is, the default approach for synthesizing all the materials we use? The interlocking infrastructure of the petroleum and chemical industries is surely one hang-up, but there are at least two others. First, because our chemical regulatory system under TSCA does not compel the generation and disclosure of knowledge about the toxic properties of chemicals already commercialized, manufacturers have little basis for claiming that products made of green chemicals are safer than those assembled from their nongreen counterparts. The ability of green chemistry to innovate runs aground on the rocks of institutionalized ignorance. Second, as long as it is cheap and socially acceptable to continue using air, water, and soil as a repository for all the noxious wastes of petrochemistry, economic investment in green chemistry remains trivial.

Here is where we ten million cancer survivors—and our allies—come in. We can argue, loudly and clearly, that the promotion of green chemistry and the comprehensive reform of TSCA are cancer prevention strategies as legitimate as smoking cessation efforts and exercise programs. Locally, we can make the routine releases of cancer-causing chemicals a big, expensive headache. We can help raise the costs of business as usual. Appeals to shame are permissible.

Maybe you are reading these words while receiving chemotherapy infusion. Maybe you are sitting in an oncology waiting room or at the bedside of someone recovering from cancer surgery. Maybe you recall that you flunked high school chemistry and have only the vaguest notion where carbon is positioned in the periodic chart of elements. (It's number 6.) No matter. You still have online access to the TRI. You can write letters to the editor. You can show up at county board meetings when decisions are being made about toxic chemicals. You'll know what to say. When the people of Peoria succeeded in halting the expansion of the toxic waste dump, one of the most influential public statements came from a local doctor. He said, "I have enough patients with cancer in my waiting room already."

I do not contend that all synthetic organic chemicals should be banned. Neither do I advocate a return to the days of celluloid and castor oil. From what I understand, celluloid was flammable and brittle, and I'm sure castor oil had its own problems. However, I am convinced that human inventiveness is not restricted to acts of war. The path that chemistry took in the last half of the twentieth century is only one path—and not even a particularly imaginative one.

It is time to start pursuing alternative paths. From the right to know and the duty to inquire flows the obligation to act.

six

animals

Bathed in a brilliant yellow-green light, they look like bats floating in a perfectly round pond. Never before have I stared at living cancer cells. Alive, they look like bats.

"Now compare this one to this one."

The first petri dish is removed and replaced by another, and I look again through the microscope. In this second watery landscape, they look more like fallen leaves—some drift together in large masses, others in smaller clusters.

"Okay, here's dish number three."

Now they are everywhere. A mosaic of islands and jutting peninsulas. Pieces of a crazy quilt tossed into a lake. A raft of vines tangled with shards of crockery. There is no one way to describe them. Collectively or alone, cancer cells are more chaotically arranged than the scurrying animals from which the disease—as well as the zodiac constellation—derives its name. Cancer, carcinogen, carcinoma, from the Greek *karkinos*: the crab.

I am visiting the laboratory of cell biologists Ana Soto and Carlos Sonnenschein in downtown Boston. The three petri dishes I've been asked to compare contain estrogen-sensitive breast cancer cells derived from a human cell line called MCF-7. The first dish is the control. Its culture medium, the broth that nourishes the growing cells, contains no estrogen. The third dish is a control of the opposite sort. Its medium was inoculated with the most potent known form of human estrogen, which is called estradiol. It's also the dish with the most luxuriant growth, th[illegible] tumors grow faster with estrogen, and MCF-7 cells are well-known exemplifiers of this principle.

It is the second dish, the one with the intermediate growth rate, that reveals the significant finding. Its culture medium has been laced with trace amounts of endosulfan, an organochlorine pesticide. These three dishes are part of a series of experiments showing that endosulfan is estrogenic. Like the hormone it mimics, endosulfan stimulates breast cancer cells to divide and multiply.

In this ability, endosulfan is much less effective than a woman's own estradiol. However, endosulfan can act in concert with other xenoestrogens, that is, chemicals foreign to the body that, directly or indirectly, act like estrogens. When a mixture of ten different synthetic chemicals, all estrogen mimics, are added to the culture medium at one-tenth the minimal dose required for proliferation of MCF-7 cells, proliferation ensues. Like raindrops eroding a boulder, quantities of weakly estrogenic chemicals too small to exert effects on their own have an impact when combined.

Endosulfan was first marketed in 1954 and was widely used on cotton and salad crops. Like DDT, endosulfan blocks testosterone. It is also highly toxic, bioaccumulative, and capable of long-distance transport: Endosulfan from southern cotton fields can end up in the fat of Arctic seals and fish. All these features qualify this pesticide as a persistent organic pollutant, and so, in 2008, it was recommended for induction into the Stockholm Convention list of chemicals prohibited from use. If its nomination is approved, endosulfan will be entirely banished from worldwide use and production. It is already banned for use in nation states of the European Union and in twenty other countries. By contrast, the United States uses about 1.4 million pounds each year—on tobacco, tomatoes, and a variety of fruits and vegetables—and this fact has ignited a firestorm of fury from scientists, health advocates, and leaders of Arctic tribal governments. Endosulfan is no longer produced domestically, but it is imported, leading some to wonder if our permissive approach to regulation has not turned the United States into a dumping ground for chemicals no one else wants. Californians have the most at stake. Half of all endosulfan used each year is used in California, and it is a common contaminant in the Imperial Valley's Alamo River (found in 64 percent of all samples collected). Under pressure, the EPA cancelled home and garden uses of endosulfan and then, two years later, made a peculiar decision. It determined that endosulfan residues in food and water posed unacceptable risks but allowed it to stay on the market anyway. In 2008, more than fifty-five scientists petitioned the EPA to revisit that decision. And so it has. As I am writing these words, the EPA is deliberating.

Soto and Sonnenschein's discovery that endosulfan can speed the growth of human breast cancer cells is part of the damning evidence against this insecticide. It's an important finding because the animal studies that tested endosulfan's potential to cause cancer are problematic. In one study, the animals all died before tumors could form. Another had too small a sample size. It is "not possible to assess the validity of these conclusions," complained the Agency for Toxic Substances and Disease Registry in its toxicological profile of endosulfan.

~~~

Let's go back for a moment to the microscope and look once more at the cells named MCF-7. Whose breasts did they come from, and what was her fate?

Finding answers to such questions isn't easy. Medical researchers maintain a comfortable distance between themselves and the cancer patients who provide the human tissues used in their experiments. The results of research involving MCF-7 cells are reported in numerous published articles. Even as the cells' various properties are described in depth, these papers mention almost nothing about their human origins.

Here is what I do know. All successfully established cancer cell lines, including MCF-7, are immortal, meaning that they will reproduce endlessly in covered dishes so long as they are provided with the proper nutrients.
~~~

Under such conditions, most human cells—even most cancer cells—tend to die out after a finite number of cell divisions. No one knows why some cancer cells can attain immortality while others cannot. Because they can be shipped all over the world, immortal cell lines allow many laboratories to conduct research on cells from the same tumor over long periods of time. Immortal cells are to cancer researchers what sourdough starter is to bread bakers.

MCF-7 is among the oldest of breast cancer cell lines and is also considered the most reliable—the coin of the realm, according to one researcher. Its name reveals a few interesting clues. *MCF* stands for Michigan Cancer Foundation, the Detroit institution that makes this cell line available to laboratories around the world. The trailing seven refers to the number of attempts that were required to establish a self-perpetuating stock of cells from the body of the particular woman patient who consented to this effort. Immortality was finally achieved on the seventh try.

"Does this mean cancerous cells were withdrawn multiple times?" I ask into the phone, trying to imagine the procedure, wondering if it was painful, wondering how many attempts she was willing to submit to.

"Yes, that's right," says Joe Michaels of the Michigan Cancer Foundation.

I learn that her birth name was Frances Mallon. At the time of her diagnosis, she was a nun—Sister Catherine Frances—at the Immaculate Heart of Mary Convent in Monroe, Michigan. Strangely enough, I have been there. The Immaculate Heart of Mary was the setting for a conference I once attended on organochlorine contamination of the Great Lakes. So, not only have I looked at the cells of her breasts, but I have walked through the corridors of her home and eaten in her dining room.

Sister Catherine Frances died of her disease in 1970. An old newspaper clipping reports that "she was a slightly built woman of medium height, with auburn hair, gray eyes and hands that were remarkable for their delicate beauty." Before entering Immaculate Heart in 1945, she had worked for twenty-five years as a stenographer at the Mueller Brass Company in Port Huron. Both her mother and sister died of cancer before her. The cancer cells that ultimately begat the MCF-7 line were extracted from fluid trapped in her chest cavity.

This is all I know about her. If endosulfan is banned, I will light a candle for Sister Catherine Frances.

~~~

In science, an assay is an evaluation of a biological or chemical substance. Estrogens, for example, are defined as substances that stimulate proliferation of uterine and vaginal cells. Thus, the traditional assay for estrogenicity involves injecting the substance to be evaluated into female rats or mice, letting a period of time go by, killing the animals, and then noting whether or not their genital tracts have gained weight, in comparison to the tracts of a control group.

These assays are complex, messy, and expensive. For these and other reasons, screening of environmental chemicals for possible hormone-mimicking effects is not routinely done. This raises the question of whether human breast cancer cells growing in petri dishes can serve as an alternative to rodents for an assay of endocrine disruption. So far, concordance between animal assays and breast cancer cell line assays has been high.

Cell line assays, like those used by Soto and Sonnenschein, may eventually replace animal studies altogether for toxicity testing. As the European Union prepares to evaluate thousands of never-before-scrutinized chemicals under REACH, cancer bioassays using animals look more time-consuming and expensive than ever.

Consider that eight hundred animals are typically required for a single carcinogenicity assay. The first step is to assign animals of two different species—usually rats and mice—and both sexes to one of four groups. Groups one through three represent high-, medium-, and low-dose exposures to whatever substance is being tested. The fourth serves as the unexposed control. Each group thus contains two hundred animals, approximately fifty of each sex and species. Next, the test substance is administered by inhalation, ingestion, or skin application on a regular basis throughout the animals' life spans. At the end of the experiment, which lasts two years, researchers compare tumor patterns in
~~~

exposed and unexposed animals and determine whether differences exist among the four groups. Pathologists then stare and review the tumor data for several more years. This is the drill for testing one chemical. Now recall that about eighty thousand chemicals are on the market, with about seven hundred new ones added every year. There is not mice, money, or time to test them all—let alone examine their effects when administered as mixtures or compare the effects of early-life exposures to adult exposures.

The International Agency for Research on Cancer conducts two-year rodent bioassays for the World Health Organization and has been at this work since 1971. At last count, it had assayed about nine hundred chemicals and identified four hundred of these as human carcinogens or potential carcinogens. In the United States, the National Toxicology Program has the job of conducting these tests and informing the nation of the results (in the venerable *Report on Carcinogens*, now in its eleventh edition). To date, it has completed rodent cancer bioassays on six hundred-odd chemicals. About half have showed some evidence of carcinogenicity. This does not mean, by extrapolation, that half of all chemicals in commerce cause cancer. The chemicals selected for the costly undertaking of a two-year bioassay are not randomly chosen; they all have prior criminal records of some kind. Based on the overall results of all the tests conducted so far, toxicologists estimate that somewhat fewer than 5 to 10 percent of all chemicals in commercial use might reasonably be considered human carcinogens. Nevertheless, 5 to 10 percent means four thousand to eight thousand different chemicals. The difference between those totals and the three hundred to four hundred chemicals identified as carcinogens by our government's toxicology program and its international equivalent presumably represents the number of unidentified carcinogens circulating among us.

Faced with the huge black hole called Lack of Data on Basic Toxicity of Common Chemicals, many scientists have asked if it might not be time to abandon the time-honored, two-year rodent assay and move on to the development of new chemical screening tools. Ideally, these could evaluate chemicals quickly and cheaply and survey the whole range of mechanisms by which chemicals could potentially contribute to cancer. In 2007, the National Research Council began this conversation in earnest by calling for a new approach to toxicity testing that incorporates advances in genomics and computational science. Front and center within this vision are high-throughput screening assays. These are automated (think robots), rapidly performed tests that can evaluate many thousands of chemicals over a wide range of concentration. Instead of whole animals, the new tests use human cells or cellular components.

Focusing on the biological processes within cells that are known to be disrupted by chemical exposures, these tests screen for changes in, say, gene activity or cellular communication. In so doing, they rely on emerging knowledge about the interconnected network of genes, proteins, and receptor molecules that work together to maintain normal cell functioning. Chemicals that are known carcinogens disrupt certain pathways within that network. Thus, any chemical that perturbs those same pathways in the same way can be presumed also to be a carcinogen. That's the theory, anyway. We have discussed one of these pathways already: estrogen signaling results in the proliferation of breast cells. Therefore, any chemical that has the same effect can be rightly called estrogenic. No need to inject a rat with the chemical in question, sacrifice it, and weigh its uterus.

The danger of a rush to high-throughput testing and other biotechnology-inspired methods (of which there are many, with awful names like toxicogenomics and bioinformatics) is the real possibility of generating voluminous data on tens of thousands of chemicals that we wouldn't know how to interpret. If indeed a welcome new day in toxicity testing has dawned, the first task of the morning is to validate the new assays using chemicals about which we already have good knowledge. Can the new assays accurately identify as carcinogens chemicals that have already been shown to cause cancer in animals? At this writing, that work is just beginning. And then the second question is, what do we do until toxicologists figure this out?

At a national breast cancer meeting, I am introduced to a well known researcher whose work I admire. Over dinner we discuss his current experiments, and I ask him [illegible]

[illegible]

"Did you know that she was a nun?"

There is a long pause. I watch him grope toward this unexpected bit of information. He blinks several times and takes a few swallows from his glass of ice water.

"Then, MCF is her name, her initials?" His voice is low and gentle.

"Actually, no. . . ."

I propose a rechristening of MCF-7. Let them be called IBFM-7: the Immortal Breasts of Frances Mallon, attempt number seven. Let them be known as a sacrament: *This is my body, which is broken for you. This do in remembrance of me.*

~~~

We are so overwhelmed with unexamined chemicals that our animal friends can no longer help us make sense of them. Animals live, grow tumors, and die too slowly. Even though a mouse's natural lifespan is only 2.5 years and a rat lives for only 3, our situation is such that we need methods for testing tens of thousands of chemical combinations, and we need the results today. I am nevertheless grateful for the knowledge that animals have provided us so far and am sorry, given how many animals were sacrificed for the data, that we largely failed to act on the information they provided. As a bladder cancer patient, I owe a particular debt to dogs.

The history of carcinogenicity testing in animals is intimately linked to the history of organized labor. In 1938, in a series of now-classic experiments, exposure to synthetic dyes derived from coal and belonging to a class of chemicals called aromatic amines were shown to cause bladder cancer in dogs. These results helped explain why bladder cancers
~~~

had become so prevalent among dyestuffs workers. With the invention of mauve in 1854, synthetic dyes began replacing natural plant-based dyes in the coloring of cloth and leather. By the beginning of the twentieth century, bladder cancer rates among this group of workers had skyrocketed, and the dog experiments helped unravel this mystery. Decades later, these dogs provided a lead in understanding why tire-industry workers, as well as machinists and metal workers, also began falling victim to bladder cancer: aromatic amines had been added to rubbers and cutting oils to serve as accelerants and antirust agents.

In their ordinary life as our pets and companions, dogs are still providing us clues to the link between environment and cancer. A study of more than eight thousand dogs showed that bladder cancer in these animals is significantly associated with residence in industrialized counties, a pattern that mirrors the geographic distribution of bladder cancer among humans. Bladder cancer among pet dogs is also significantly associated with direct exposure to insecticidal flea and tick dips, especially if dogs are obese or live near another potential source of pesticides. In Italy, dogs are more likely to have lymphoma if they live northeast of Naples, where illegal waste disposal is a rampant practice. (People living in this community, known as "the triangle of death," also have elevated cancer mortality.) Military dogs in Vietnam exposed to the herbicide Agent Orange suffered from high rates of testicular cancer. Scottish terriers in Indiana have higher rates of bladder cancer if their owners use lawn chemicals. Scotties are particularly prone to bladder cancer and so serve as a sensitive sentinel for bladder carcinogens in the environment, according to veterinarians who are tracking cancer incidence in this breed.

Whatever plan prevails in the ongoing debate about how best to screen chemicals, human cell lines growing in petri dishes are likely to have limitations. Until more sensitive in vitro assays are developed, lab animals will continue to provide us important clues about how carcinogens ply their trade. For instance, how might chemical exposures in early life alter the course of development in ways that predispose for cancer in later life? Nowhere is this question more important than in breast cancer research. And to answer it, you need whole organisms. And you need time.

The breast is that gland that defines all mammals. While rodent breasts and human breasts would seem to have little in common—after all, mice have ten of them and we have only two—the similarities in development and anatomy are remarkable. Early in life, both female rodents and female humans possess immature breast tissue that consists of a bundle of slender tubes called ducts. The mammary ducts open into the nipple, which is essentially a sieve. At the other end, in the back of the chest wall, are the duct's tear-shaped terminal end buds, which give each duct the shape of a canoe paddle. During puberty, under the direction of female hormones, the ducts begin to branch and elongate, with the terminal end buds leading the way, like little snowplows. During maturation, the buds disappear, metamorphosing into clusters of lobules designed to produce milk. A cushion of fat surrounds both the ducts and their lobules, and the whole structure comes to resemble an orchard of fruiting trees. This process unfolds in all female mammals according to the same basic blueprint, although the timing of events is different. Across all species, the periods of life during which the terminal end buds are present are before, during, and just after puberty.

The cells with the terminal end buds are extremely susceptible to chemical carcinogens. The fewer of them, the better, and the sooner they complete their transformation during puberty, the better. Studies with rats show us that anything—including any environmental chemical—that increases the number of cells in the terminal end buds during early life or that delays maturation of the buds can raise the risk for breast cancer, even without genetic damage. The lingering presence of terminal end buds in the gland after puberty raises the susceptibility of the breast to carcinogenic damage. As we have seen in Chapter One, the weed killer atrazine has such effects: in laboratory rodents, atrazine exposure before birth retards the maturation of the mammary gland in puberty and increases the number of end buds. In one strain of rat, low levels of atrazine metabolite mixtures administered before birth perturbed glandular development in the female pups in ways that persisted into adulthood. That is, the breasts of rats that had been prenatally exposed to atrazine had a different interior anatomy than the breasts of rats not so exposed. When those rats went on to reproduce, the nursing pups of the atrazine-exposed mothers gained less weight. Their mothers had less-developed mammary glands, and therefore quite possibly insufficient milk production. Needless to say, these kinds of findings would not be possible using cell lines alone.

But atrazine is also a lesson in how animal studies do not offer perfect concordance: atrazine exposure increases breast cancer in one strain of laboratory rat, but the pathway by which this rat develops breast cancer is believed not relevant to humans. Specifically, atrazine triggers premature reproductive aging, which, in this particular rat, is thought to be associated with elevated estrogen levels. By contrast, reproductive aging lowers estrogen levels in humans. The 2003 decision of the EPA to allow the ongoing use of atrazine pivots on the evidence—some say presumption—of a biological difference between this particular rat and us.

~~~

I had just turned twenty when I was diagnosed with bladder cancer of a type called transitional cell carcinoma. It is something I have in common with at least one beluga whale in the St. Lawrence River.

Proceeding northeast from Lake Ontario, the St. Lawrence River slants through the Canadian province of Québec and flares open like a trumpet as it pours itself into the North Atlantic. Where the river's current meets the ocean's tide is one of the world's deepest, longest estuaries. About 650 beluga whales, a remnant of the thousands that once lived here, inhabit this transition zone. This estuary also receives tributarial waters that have traversed some of the most industrialized landscapes of southern Canada and the northeastern United States.

Belugas are small, toothed whales. Their skin is white.

Transitional cell carcinoma among the belugas was first discovered during an autopsy of a carcass that had washed ashore in 1985. It was a
~~~

particularly provocative finding because workers in nearby aluminum smelters, which release their wastes into the St. Lawrence, had also been found to have an elevated incidence of this type of bladder cancer.

Gross hematuria, or noticeable blood in the urine, is the usual way bladder cancer presents itself. I do not know how a whale would experience this—perhaps through sense of smell. As for myself, gross hematuria arrived as I was finishing up a morning shift at a truck-stop diner. After making my final rounds with the ketchup bottles and syrup dispensers, I stopped in the restroom. Turning to flush, I froze. My urine looked like cherry Kool-Aid. I stood there a long time.

And then I remembered the beets—sliced red beets, which the cook had prepared for the lunch special and which I had eaten in great quantity during my break. Could beets make urine turn pink? What other explanation could there be?

I swore off beets. Three weeks later, I returned home from a night shift at a pancake house, tore off my waitress uniform, went to the bathroom, turned to flush, and . . . the toilet was full of blood. Brilliant and thick. I drove to the emergency room.

I was wrong about the beets.

Bladder cancer is one of several cancers striking the beluga population of the St. Lawrence. In 1988, a team of veterinarians found tumors in the bodies of four dead whales from a group of thirteen that had washed up over a period of ten months along a polluted stretch of the river. In addition, the immature breast ducts of one young female showed abnormal proliferation. Called ductal hyperplasia, this condition is considered a strong risk factor for breast cancer in women. (Whales are mammals and so have breasts; in belugas, they are located on either side of the vagina, with only the nipples visible and the mammary glands themselves hidden beneath a layer of blubber.)

Autopsy reports on twenty-four other stranded carcasses were published in 1994. Twenty-one tumors were found in twelve carcasses. Among these tumors, six were malignant. By 2002, 129 stranded carcasses had been autopsied, and cancer was found in 27 percent of whales, a percentage similar to that found in humans living in the area. To date, cancers identified in the beluga include bladder, stomach, intestinal, salivary gland, breast, and ovarian. The prevalence of intestinal cancer is especially high. No cases of cancer have been reported in belugas inhabiting the less contaminated Arctic Ocean. Indeed, the St. Lawrence belugas die at an earlier age than their counterparts in northern Alaska, in part because of their high rates of cancer. The researchers concluded, "These observations suggest that a human population and a population of long-lived, highly evolved mammals may be affected by specific types of cancer because they share the same habitat and are exposed to the same environmental contaminants."

The beluga whales of the St. Lawrence estuary have more wrong with them than cancer. They also have trouble reproducing. When chemical analyses of their blubber were conducted to illuminate possible causes of both problems, a mixture of persistent organic pollutants—at some of the highest levels ever recorded in a living organism—were all found dissolved in the whales' fat. Free-swimming live whales carry these same pollutants in their blubber but at lower concentrations, according to biopsies obtained by darts and crossbows. Two of these contaminants, PCBs and DDT, have a history of use in the St. Lawrence basin, but three of them, the pesticides chlordane, toxaphene, and mirex, do not. (As of 2004, all of these chemicals are now banned under the international Stockholm Convention.) Toxaphene and chlordane are both found in the waters and sediments of the estuary, presumably because they are carried into the seaway by winds blowing up from the southern United States, where both were once used heavily. But mirex, long ago used to control fire ants in the South, is found in neither the water nor the silt of the estuary. It is not found in the bodies of any other marine mammal living in the estuary. So how did it come to lodge in the fat of the belugas who live here?

The answer may be eels. Beluga whales love to eat them. Eels run through the icy, deep Laurentian channel on their autumn migration from Lake Ontario to the warm waters of the Sargasso Sea. There is mirex in the tissues of St. Lawrence eels. And there are two sources of mirex in the Lake Ontario basin where the eels originate: a pesticide-manufacturing plant near Niagara and a river called the Oswego, where mirex was once accidentally spilled. The eels are the apparent courier between these contaminated sites and the beluga whales living six hundred miles away.

Eels are strange. Like salmon, they migrate thousands of miles to spawn, crossing between fresh and salt water to do so. However, eels make their journey in reverse: they spend twelve to twenty-four years living in lakes and rivers, and then they head out to the ocean to lay their eggs. Baby eels, each the size and shape of a willow leaf, spend their first year of life trying to swim back.

Less is known about what toxins eels might bring back from their birthplace, which is also a contaminated site. Elliptical and still, the Sargasso Sea lies within the clockwise current of the Gulf Stream. The islands of Bermuda rise from its center. Eels from freshwater rivers in North America, Europe, and Africa all converge here to spawn. The Sargasso sits at the center of a whirling gyre of currents, and so accumulates seaweed and debris from all over the Atlantic—but especially from the U.S. and Caribbean coasts. Along with the ocean's other detritus, chemical pollutants—such as DDT and balls of tar—also slowly drift in and accumulate here to join what the poet Ezra Pound once called "this seahoard of deciduous things."

I tried to be kind to my hospital roommate. No one else was. We were both recovering from surgery, but her situation was more typical of what happened to girls in Pekin: A fast car. Drunk boys. She was the only one pulled out alive, and the story made the front page. When the nurses refused to tell her what had happened, I read aloud to her from the newspaper account. Mostly, she slept and watched TV. I spent a lot of time staring at her.

Outside this room, our lives were on two different tracks—in my view, at least—and I was trying to figure out how I had ended up here with her. I was a high-achieving biology major with a penchant for modernist poetry, who viewed drink, drugs, TV, and junk food as tickets to nowhere, who was back in this town only for the summer, now college had resumed and I was still here. Some malevolent current had deposited us together in this hospital. But unlike my partner in the next bed, no one had any explanations for my situation. The newspaper said she was expected to survive. Was I?

I examined the outline of my legs under the thin blanket, the shadow my hand cast on the sheet. Between the sheet and the blanket snaked the wretched catheter tube. I felt flattened down, like an animal wounded by something cruel and meaningless. My roommate looked over at me and touched her hands to her discolored face. Her boyfriend and brother were both dead.

"I think I'm going to stop partying for a while."

It was the kind of moment where laughing and crying were synonymous. We started laughing.

"I think I'm going to start."

The belugas have local problems as well as global ones. Aluminum smelters and other industries lining the river basin have contaminated their waters with benzo[a]pyrene, a potent and well-known carcinogen.

A type of polycyclic aromatic hydrocarbon (PAH), benzo[a]pyrene is seldom manufactured on purpose. With molecules consisting of twenty carbon atoms arranged as two hexagonal rings nestled on top of three hexagonal rings, it is created during the combustion of all kinds of organic materials from wood to gasoline to tobacco. It also occurs in coal tar, which is distilled to make a couple of familiar products. One of these is creosote, used to preserve wood (think of the smell of telephone poles on a hot summer day), and another is pitch, used in roofing and aluminum smelting.

Benzo[a]pyrene causes cancer in a simple, direct way. Nearly all living things have in common a group of cellular enzymes responsible for detoxifying and metabolizing possibly harmful chemical invaders. When this enzyme group encounters benzo[a]pyrene, it inserts oxygen into the foreign molecule, the first step toward breaking it down. However, ironically, this addition activates benzo[a]pyrene rather than detoxifies it. The altered molecule now has the ability to bond tightly to a strand of DNA. A chemical invader so attached is called a DNA adduct, and it has the power to produce genetic mutation. If uncorrected, this type of damage can become a crucial step leading to the formation of cancer. (Some PCBs have the power to activate the suite of enzymes that converts benzo[a]pyrene into a mutating carcinogen, leaving open the possibility of a deadly interaction between these two co-occurring contaminants.)

The number of adducts attached to an organism's DNA is considered a useful measure of benzo[a]pyrene exposure. DNA from the brain tissue of stranded St. Lawrence belugas bore impressively high numbers of adducts. They approached values found in laboratory animals exposed to levels of benzo[a]pyrene sufficient to cause a response in bioassays. In contrast, DNA adducts were not detectable in beluga whales inhabiting Canada's more pristine estuaries.

Discharged at last from the hospital, I opened the door to my dormitory room and saw that bare mattress. I became secretive and territorial. I staked out a favorite stall in the women's bathroom. In my return every third month to the hospital for cystoscopic checkups, cytologies, and other forms of medical surveillance, I told no one where I was going. The interval between checkups approximated one semester's worth of time, one season. These seasons went by. I waited tables and pursued perfect grades. I finished college and began graduate school. I stopped studying grasses and started studying trees. I compulsively checked the toilet bowl for the presence of blood. I still do.

After five years, my checkups became annual, and I was no longer tethered so tightly to the medical system. This change was almost unnerving—as though it were normal to think of the interior landscape of one's body as a study site that required constant data collection. I immediately accepted a fellowship in Costa Rica, where I became involved in a field study of ghost crabs—delicate creatures that occupy burrows along the Pacific beaches at the edge of the rainforest. At the study's conclusion, the night before we were to fly out, I had a vivid dream:

I am walking by the ocean and discover a pale orange crab, as large as a whale, washed up on the beach. It is dying. I lie down next to it, and slowly it wraps a great, clawed arm around me. Reaching my arm over its carapaced body, I return the embrace. I am not afraid. As if in the final frame of a movie scene, giant letters appear in the sky above us, spelling out a single word—G R A C E.

Among those of us who had spent days out in the tropical sun trying to monitor the movements of these reclusive, lightning-fast animals, the dream was hugely funny. Not until I returned home did I connect the dream to the end of five intense years of monitoring the possible movements of cancer. Cancer, carcinogen, carcinoma, from the Greek *karkinos*.

Ghost crabs.

At the International Forum for the Future of the Beluga, conservationist Leone Pippard asked the following questions:

> Tell me, does the St. Lawrence beluga drink too much alcohol and does the St. Lawrence beluga smoke too much and does the St. Lawrence beluga have a bad diet . . . is that why the beluga whales are ill? . . . Do you think you are somehow immune and that it is only the beluga whale that is being affected?

~~~~~~

In 1964, the pathologist and physician Clyde Dawe discovered white suckers with liver cancer in Deep Creek Lake, Maryland. This was the first time this disease had been found in a wild population of fish, and Dawe was worried. While isolated incidents of fish tumors had been previously described, liver cancer in a large number of individuals had never been seen before. What might be going on with other populations of fish or with other species?

The following year, at Dawe's instigation, the National Cancer Institute opened the Registry of Tumors in Lower Animals to facilitate
~~~~~~

the study of cancer in cold-blooded vertebrates (fish, amphibians, and reptiles), as well as various invertebrates (such creatures as corals, crabs, clams, snails, and oysters). For the next forty years, field biologists who discovered animals with tumors could send them—live, frozen, or preserved—to the registry. [illegible] they tested, a number of the registry's 7,000 accessions, representing more than a thousand species were submitted by concerned citizens. The NCI ended its sponsorship of the registry in 2007.

The evidence compiled by this registry, together with other research on cancer in wildlife, indicates that cancer in at least some species of lower animals, but especially liver cancer in fish, is intimately linked to environmental contamination. In these patterns is writ an urgent message to us higher animals. First, the preponderance of cold-blooded animals with cancer are aquatic bottom feeders, such as the brown bullhead catfish and the English sole. And the dark beds of rivers, lakes, and marine estuaries are precisely where the highest concentrations of contaminants are found. What's more, when extracts of these sediments are painted onto healthy fish, injected into their eggs, or added to clean aquarium tanks in the laboratory, these fish contract cancer in significant numbers. Second, fish populations with liver cancer are clustered around areas of environmental contamination—the Great Lakes tributaries, Puget Sound harbors, and bays along the East coast—while cancer among members of the same species who inhabit nonpolluted waters is virtually nonexistent. Similarly, liver cancer in marine fish from around the world also correlates with the presence of chemical pollutants, as compared to a virtually zero background rate in nonpolluted waters. Finally, aquarium studies in the laboratory show that the same carcinogens known to cause cancer in humans and rodents also cause cancer in fish and mollusks—and they are often metabolized in the same way.

~~~

I have developed an idea for a pilgrimage that involves people with cancer traveling to various bodies of water known to be inhabited by animals
~~~

with cancer—from Cobscook Bay in Maine to the mouth of the Duwamish River in the Puget Sound. It involves an assembly on the banks and shores of these waters and a collective consideration of our intertwined lives. We could start with the question posed by Leone Pippard at the beluga conference: Do you think you are somehow immune? I would like us to end up at the Fox River, which flows south from Wisconsin and joins the Illinois River near the river town of Ottawa, about seventy-five miles upstream from my hometown. Tumors from an assemblage of Fox River fish—walleyes, pickerel, bullheads, carp, and hog suckers—were among some of the first identified. But I would not take us to the old industrial sites thought to be responsible. Instead, I would bring us up to Buffalo Rock, a ninety-foot-tall bluff that abuts the Illinois River just a mile or two downstream from its confluence with the Fox. From here we could see the entire river valley spread out before us.

The ecology of Buffalo Rock was ravaged in the 1930s by a strip mine that tossed highly toxic shale and pyrite onto the topsoil. All life forms, animal and plant, were killed. For decades, it remained a landscape of jagged escarpments that funneled acidic runoff into the river.

Then in 1983, the artist Michael Heizer was commissioned to help reclaim Buffalo Rock. Inspired by ancient Native American earth mounds, Heizer used bulldozers to sculpt the thirty-foot furrows into the shapes of five river animals: water strider, frog, catfish, snapping turtle, and snake. Each of these earth sculptures is hundreds of feet long. Rye grass now grows on top of them. You can climb the catfish's whiskers, walk along the strider's legs, lie down on the snake's head.

I would choose a September afternoon with a sky as blue as cornflowers. Cancer survivors would gather on the backs of these monumental animals and, in this place of damage and reclamation, bear witness.

seven

earth

"All flesh is grass."
—Isaiah 40:6

To hear my mother tell it, butchering day was a festive occasion. Preparation started weeks in advance when all five girls got new outfits, each hand-sewn by my grandmother. Butchering dresses, they were called. These were not work clothes, my mother stresses here, but real dress-up dresses. They honored the company that would be arriving to assist with the slaughter and share in its spoils.

To hear my Aunt Lucy tell it, butchering day was a time of somber anticipation. Before dawn, the yard was filled with fire and cauldrons of boiling water, the kitchen with knives and large strangers. She recalls tiptoeing through the house, listening for and finally hearing . . . the gunshot. Aunt Lucy firmly disputes the existence of butchering dresses.

Girls do not appear in my uncles' accounts. In their rendition, butchering day was a ceremony of manhood. Sons accompanied their fathers. They shot pigs. They shared secret knowledge. And they partook of hog testicles.

The story of Hog Butchering on the Farm—featuring gleefully indelicate food items, special costumes, and gunfire—easily withstood repeated retellings to the next generation of children, which included me and my various cousins. I always preferred my mother's version because she could provide the most vivid details of sausage making. She starts by describing the scraping of bristles from the skin of a scalded hog, spends considerable time on the part where the women clean out the intestines (soon to be- [illegible] [illegible] and how to twist a pork-engorged casing to form a chain of perfect links. Coiled into five-gallon crocks and covered with melted lard, these sausages could safely pass the winter in the crypt of a farmhouse cellar.

Somehow, all these images explain why great jars of beans, flour, grain, rice, cereal, and pasta line my shelves. Something in my kitchen is always boiling or soaking, as though an enormous family dined here. I like the purchase of commodities in bulk, the laying up of food, the heft and volume of it all. I like fruit in the fruit bowl, greens in the crisper, onions in the cupboard, potatoes under the sink, a wreath of garlic bulbs nailed to the wall. The more dirt on the carrots the better. And although I store no crocks of sausage, I like to imagine old Uncle Sander bent over the extruding press, with a lengthening garland of pork looped about his shoulder like a string of Christmas lights.

~~~~~~

American agriculture has changed dramatically since *Silent Spring* was published. For one thing, the number of farms has declined sharply. Illinois now has fewer than half the number of farms it had in 1960 and in 2008, my home county could lay claim to only 998 of them. For another thing, land ownership has become separate from farming: over half the farmland in the United States is now farmed by persons who do not own it. My cousin John, for example, leases most of the thirteen hundred acres he farms near Saybrook. The agricultural landscape has also become more uniform. Farm animals have vanished from the barnyards as the raising of livestock has become an enterprise separate from the growing of crops. In 2007, 3 percent of Illinois farms had milk cows and 8 percent had chickens. One hundred years ago, those figures were 80 and 97 per-
~~~~~~

cent, respectively. And the diversity of crops grown on any one farm has itself declined. Over the past five decades in Illinois, harvests of fruits, vegetables, hay, wheat, and oats have all fallen off. Orchards, pastures, vegetable plots, and woodlots have been plowed into ever-larger fields of corn and soybeans. In [illegible], field corn and soy, which no one but livestock eats directly, accounted for more than 90 percent of the total cash receipts for crops in Illinois.

All of these changes conspire to make farming an increasingly remote activity. We know less and less about how and where our food is grown and by whom. And the number of folks around who can explain such things to us is diminishing. The disappearance of farm talk from our communities is not included in compilations of agricultural statistics, but it is a very real change.

I feel lucky to have grown up around such talk—not only because my mother descends from a line of farmers but because I waitressed in roadside restaurants. Rainy mornings brought farmers in from the fields to sit and visit for a while. One had only to pour coffee and listen. Another group came in very early before heading out to work. Indeed, a magical hour occurred between 4:00 and 5:00 A.M. when the last of the shift workers still occupied the booths and the first wave of farmers began lining up at the counter. Talk of union contracts mingled with discussions of weather and grain prices. Outside the windows, corn and bean fields slowly took shape as the darkness faded into gray light.

When I take my nephews to the cafés where I once carried platefuls of pancakes and chili-mac, we don't see many farmers there. Among the idle talk, there is silence.

By other measures, farming has changed remarkably little since Rachel Carson described what she saw as a dangerous trend toward an increasing reliance on pesticides. Indeed, many of the changes outlined thus far have come about, directly or indirectly, because agriculture has traveled even farther down the path of chemical dependency depicted in *Silent Spring*.

Synthetic chemicals introduced into agriculture after World War II reduced the need for labor. At about the same time, profits per acre plummeted. Both these changes forced farmers into managing more acres to earn a living for their families. Average farm size increased. (Or as my Uncle Roy says, "These days, you have to plow half the state or forget it.") Federal farm programs that encouraged farmers to specialize in a single crop further increased the need for chemicals to control pests. And the use of these chemicals themselves set the stage for additional ecological changes that only more chemicals could offset.

The increasing use of herbicides to control weeds, for example, has discouraged the practice of crop rotation, a kind of repertory theater in which a sequence of crops is cycled through a field—corn, oats, hay, alfalfa, corn—each one altering the chemistry and physics of the soil in slightly different ways. Because their residues can overwinter in the soil, herbicides prevent the sowing of chemically sensitive crops the following season. Herbicide-insensitive corn, therefore, may end up planted in the same field year after year. (Farmers call this sequence "corn on corn.") Lack of crop rotation, in turn, encourages insect pests—whose reproductive cycles are no longer disrupted by changing vegetational patterns—and these outbreaks invite the use of insecticides. Alfalfa has natural weed-suppressing effects when rotated in with corn and other grain crops, but without farm animals, there is little local market for it. Simple plowing helps control many kinds of weeds and pests, but the enormous size of most fields locks farmers in to low-till and no-till practices in order to prevent topsoil from blowing away.

A cornfield and a soybean field are very different places.

Bean fields are humble; they start out that way and stay that way. For reasons I can't explain, they are also a little bit sad. Walking through a soybean field, I feel like myself, only sadder. A soybean is a delicate plant. Like all other legumes—clover, peas, alfalfa—the soybean plant has a softness in its leaves. Fully grown, it is mostly shaped like a little bush that never extends much above the thighs, but, late in the season, an inconspicuous twining reveals its origin as an Asian vine. In spite of their modesty, the high yielding varieties of soybeans are given brawny names—Jack, [illegible]son, Pharoah—that sound like brands of condoms.

Illinois soybeans bloom in midsummer and produce somewhere between sixty and eighty pods per plant. The pods are fuzzy—almost bristly—and bulgy. Each holds three (on average) round, pale-yellow beans. At the end of the season, the plants seem to lose all sense of individuality. Turning from deep green to brilliant yellow to an indescribable shade of red-brown, they all blur into each other and sink back toward the earth, forming an upholstered surface that invites one to walk in and lie down.

Not so with cornfields. There is something animate about corn, which starts out as a merged, green expanse and then sharpens and stiffens into lines of individual, human-shaped forms. Corn is proud; it seems to stand in judgment somehow. It has conspicuous body parts—ears, tassels, silks, stalks—and it has the power to alter the landscape. By summer's end, county roads that had once offered unbroken views of the horizon become chutes between solid walls of corn. But this view offers only a one-dimensional perspective. To really understand how corn occupies space, you have to push past this screen and step into the deep interior of the field.

Walking through a field of corn can feel very sheltering. It can also make you have crazy thoughts and bring on panic, as when swimming in a large lake and suddenly realizing you are too far from shore. Much of corn's power to soothe and disorient surely has to do with the fact that it becomes taller than we are. By the end of July, you cannot see over it or walk through it without being touched along the length of your body by long, fibrous leaves. Moreover, even when green, corn makes a continual noise. It is a kind of soft hiss, not unlike the sound of snow falling. *Listening to the corn grow* is a phrase of respect, not an expression of boredom.

Corn is a grass and, as such, owes its existence to the wind. The silk-shrouded ear is the female flower of the corn, and each stalk has between one and three ears. The entire harvest depends on the ability of pollen grains to blow loose from the [illegible] at the top of the stalk, float as a cloud over the field, and alight on the ends of the silks as they wave from the ears. Each strand leads back to a single kernel. Like following a thread through a maze, a grain of pollen sends its chromosomes down the entire length of a silk to reach one kernel and fertilize it. As each ear holds about six hundred kernels and each acre about thirty thousand individual cornstalks, this process must happen independently and successfully innumerable times.

More than soybeans, corn in central Illinois defines certain experiences. "Corn-growing weather" means hot, sunny, and humid. Such conditions are probably also good bean-growing weather, but no one ever calls it that. Soybeans, introduced from China in the late 1800s, are relative newcomers to Illinois farming. We have not yet completely incorporated them into our myths about ourselves. Corn parables, on the other hand, abound. My mother is an especially skillful narrator. Consider, for example, the Story of the Scarred Knee.

As a young girl, my mother was running through a newly harvested cornfield when she tripped and tore open her knee on the sharp end of a broken-off stalk. My grandmother, while tending to the wound, noticed a bit of stringy debris stuck to the flesh. She gently pulled on it. Out from inside my mother's leg came a wedge of cornstalk *as long as a pinky finger.*

As a means of encouraging caution in wild children, this proved a successful tale. My sister and I could be stopped in our tracks by the mere mention of it. In fact, I am still spooked at the thought of running through a cornfield.

Many of my urban friends are surprised to learn that field corn is not harvested while green, as is sweet corn. Neither are soybeans. Both crops first need to turn brown under blue September skies, allowing the moisture content within the seeds to fall. The dry cornstalks that decorate Illinois porches for Halloween are a symbol of bounty, not death. Corn and soybeans are harvested at about the same time with modified versions of the same, basic, all-in-one machine called a combine. Again, however, the two activities have a very different feel to them. Combining corn seems almost violent—[illegible] from the cab of the combine. As the long tines of the combine's corn head [illegible] down the rows, the stalks first begin to tremble and wave their brittle leaves wildly, then jerk, snap back, and disappear. Somewhere out of sight, beneath the driver's feet, a series of augers, chains, cylinders, rasp bars, sieves, screens, and fans disassemble the ears, strip the corn from the cobs, and shoot the ground-up trash back into the field. In the meantime, out the rear window, the hopper fills with a thick stream of gold kernels.

From a distance, harvesting beans looks peaceful. Delicate spinning sickle bars sweep the pods, stems, and leaves into the hidden chambers of the combine (where different settings allow the same machinery to dissociate beans from pods as kernels from cobs). Showers of red, glittering chaff fall gently back to the ground in its wake. Beneath this apparent serenity, however, lies high-stakes anxiety. Soybeans nestle close to the ground. One good-sized rock pulled into the grain table can completely disable a 25-foot combine. Moreover, bean pods cannot withstand repeated rounds of soaking and drying out once they are ready for harvest. As my cousin puts it, "A little rainfall on the wrong day in October and you're done."

A bushel of corn weighs about 56 pounds, and a bushel of soybeans about 60. An average acre of corn in Illinois yields 180 bushels, and an average acre of soybeans around 43. In July 2009, corn was trading for $3.57 a bushel, soybeans for $10.06.

~~~

Central Illinois is the beginning of a human food chain that ends in meat and snack food. From a systems biology point of view, the landscape of central Illinois is a protein production machine. Whether it is sold domestically or exported—and almost half of the grain produced
~~~

in Illinois is sold for export—most of the corn and soybeans grown in the United States is fed to livestock. Illinois corn and soybeans are the raw materials for meat, eggs, and dairy. This fact is not news. Corn and beans were fed to animals when my grandfather farmed here, and my great-grandfather before him. The difference is that the animals so fed are no longer on or near the farm. The number of milk cows in Illinois peaked in 1935, eggs in 1955, cattle in 1957, chickens in 1927. Because of cheap oil and a subsidized infrastructure, hogs, cattle, and poultry can now be raised time zones away from the farms that feed them—even in places where not enough rain falls to grow grain. So when we go to the supermarket in central Illinois for hamburger, cheese, and a dozen eggs, we may be bringing home our own local crops, but only after they have traveled a half continent away to be transformed into animals . . . and then transported a half continent back to us. Consuming God knows how many gallons of diesel fuel along the way. Meanwhile, the corn syrup and soy oils from our signature crops find their way into soft drinks and cookies, which are also assembled in distant lands, packaged up and brought back to us. With low price tags attached.

This is very different circuit than the one my grandfather kept in motion in the 1930s when the corn in his field was fed to the cows in his barn, who produced cream to sell at the local creamery and skim milk to feed both his children and his hogs. Who lived next door to the cows. Who provide manure for his field. Furthermore, he fed the cobs, shorn of their kernels, to the stove in his house for heat. Corn, then, made my mother's family energy independent as well as food sufficient. The behemoth corn-soybean-livestock pipeline of today bears little resemblance to the tightly woven loops of my grandfather's farm, nor to the larger web of ecological relationships of which corn, beans, and livestock were, eighty years ago, but one part.

The Corn Belt's industrial food system, which glides along on subsidies, price supports, carbon, and credit and is overseen by a handful of corporations that control seed, grain, and protein sectors, has its loud critics. Some of them are eloquent as well as vocal, authoring bestselling books and inspiring public discussions about how to improve access to local and organically grown foods. Even among corn's fiercest defenders in the heart of Illinois, these conversations are swirling. What seemed to crystallize the situation here was the spinach scare in fall 2006. Bagged fresh spinach, contaminated with *E. coli*, sickened two hundred, killed three, and sent thirty-one people into kidney failure. With many victims living in Corn Belt states, public service announcements assured a terrified populace that the outbreak was limited to *California* spinach; local spinach was not affected. And so, being September, Illinoisans began to look for Illinois spinach. There was, essentially, none. (Indeed, spinach does not appear on the Illinois Farm Bureau's list of the state's specialty crops. Illinois does grow mustard greens, however. Thirty-two acres of them.) As central Illinois food writer Terra Brockman went on to explain, gently, to the readers of the *Chicago Tribune*: there is precious little food in farm country. Our state's cultural heritage of food security and self-reliance disappeared years ago, along with the pastures and orchards.

Many Chicagoans who imagined that they lived in an agricultural state were surprised by this news. Some rural folks were surprised, too. The long-cherished notion that our state's farmers fed the world needed some revision. It was, in fact, California growers who fed Illinois.

Joining a vanguard of writers calling for a change in the food environment are members of the public health community. Their battle cry is obesity, and their argument goes like this: a food system that makes unhealthful, calorie-dense food cheaper than fruits and vegetables is driving up the rate of obesity. Obesity in turn is driving up the cost of health care. Fanning the flames further is the economic contraction that began in 2008 and that has placed carrots and apples beyond the budgets of many families. Economic adversity increases consumption of commodity-based foods, which increases obesity, which increases health care costs, which contributes to economic adversity.

Commodity-based foods are those that Illinois farmers help produce.

The facts behind this line of reasoning are unassailable: Obesity rates among U.S. adults have doubled over the past three decades and tripled among children. (Childhood obesity may have peaked in 2005.) In 1991, no state had an obesity rate over 20 percent. In 2009, two thirds of all states had an obesity rate exceeding 25 percent. Over the past three decades, the price of soda and snacks has changed little, and the consumption of soda and snacks has nearly doubled. Over the same time period the price of fresh fruits and vegetables has risen markedly, and the consumption of fresh fruits and vegetables has declined.

Acknowledging the web of causation leading to weight gain, a 2009 report by the Robert W. Johnson Foundation and the Trust for America's Health provides an overarching national strategy to combat obesity that goes far beyond telling individuals to eat less and exercise more. Included in their recommendations are eliminating the marketing of junk food to kids, resurrecting daily gym class in schools, supporting farmers' markets, and taxing snack food. Making unhealthy food socially unappealing, as was done with cigarettes, is part of the strategy. Increasing access to fresh food in "the food deserts of urban and rural communities" is also part of the strategy.

I would like to add another factor to the obesity equation overlooked by this worthy report: land use. It is all very good to promote the consumption of fresh produce in the food deserts of rural communities, but, without also changing farming practices, it is hard to see how transformational that message would be in a state where 21,350,000 acres are planted in the commodity crops called corn and soy and 32 acres are planted in mustard greens. (We also have 50 acres of broccoli.) Where are the five to nine servings a day of fruits and vegetables for the entire population of Illinois going to come from and how will they arrive? (In care packages from California's Central Valley via diesel trucks?) Shouldn't we at least ask why, of all places, *rural* communities have turned into food deserts?

Thirty states now tax food of low nutritional value. Illinois is one. A snack tax is not a bad idea, but given that the corn that makes up the snacks is federally subsidized, it begins to feel like an example of what one hand giveth the other taketh away. Paying Illinois farmers to grow corn and then taxing consumers for buying the food that results is hardly a coherent strategy to fight obesity.

Since this is a book about cancer, let me connect a few more dots here. Obesity and weight gain are risk factors for several cancers, including esophageal cancer, pancreatic cancer, uterine cancer, colon cancer, and postmenopausal breast cancer. The mechanism by which excess weight contributes to cancer is not well understood, but it likely involves changes in circulating hormone levels—[illegible] estrogen, for example—and inflammatory processes. Childhood obesity may contribute to breast cancer risk by hastening the onset of puberty: as a group, chubbier girls develop breasts at younger ages than leaner girls; early sexual maturation is a known risk factor for adult breast cancers. Diets heavy in red and processed meats are also linked to increased risks for several cancers. The American Cancer Society emphasizes this evidence in its public educational literature, which attributes fully one-third of all cancer diagnoses to obesity, weight gain, sedentary lifestyles, or inadequate intake of fruits and vegetables. Given this level of certitude on cancer risk and food, does it not follow that we should examine our systems for making that food? Should we not look into the fields and see what is growing there, what we put on it, to what purpose it is turned, and how it is priced? Should we not ask, how does the landscape of Illinois influence the landscape of cancer?

I pose these questions with the greatest of respect for corn, for soybeans, and for the farmers who grow them. I hope that the plant kingdom may soon replace petroleum as feedstock for making things. I hope that soy-based adhesives for plywood can replace carcinogenic formaldehyde. And although I'm unconvinced by the argument that corn-based ethanol is a pathway out of petroleum dependency, I see that corn might play a role in the creation of new materials. (Ethanol's net energy production is too puny and its water requirements too huge to justify pulling acreage out of food production.)

But as an Illinois native and a cancer survivor, I worry that, over the past half century, a focus on producing two commodity crops in gargantuan quantities has resulted in a drastic simplification of Illinois' agricultural system. Consequently, the foods we should all be eating to prevent cancer and avoid obesity are no longer grown in my home state in amounts greater than negligible, and the long-distance transport of such food into Illinois depends on the availability of cheap, abundant petroleum, which soon will be neither cheap nor abundant. The American Cancer Society's recommended 5–9 servings of fruits and vegetables a day—sound advice—[illegible] 140–188 servings of produce per week for a family of four. This is an increasingly unaffordable goal for many. Whatever we decide to do with all the corn and soybeans we grow—send them to Asia or feedlots in Kansas, or turn them into soft drinks, snack foods, ethanol, glue, or biodiesel—the 12,901,563 people of Illinois still need to eat. Who will feed them? Who will pay?

It is time for an agricultural redesign.

Corn alone consumes half the total herbicides (weed killers) used in this country. Together, corn and soybeans are responsible for nearly three-quarters of all herbicides used. Indeed, the herbicides used on corn and soybeans account for 40 percent of all pesticides used in the United States for any farming purpose. (Pesticides include fungicides and insecticides as well as herbicides.) In short, Corn Belt weeds have become the number-one target of agrichemical warfare.

These are the names of the enemies: velvetleaf, foxtail, cocklebur, pigweed, smartweed, ragweed, morning glory, lambs-quarters, jimsonweed, dogbane, milkweed, nightshade, fall panicum, shattercane, nutsedge, Canada thistle. Each weed has its own manner of surviving and reproducing in Illinois corn and bean fields. Canada thistle is a perennial with creeping underground stems that give rise to new shoots. Cockleburs propagate the usual way but hedge their reproductive bets: of the two seeds housed inside each prickly bur, one germinates in the spring, while the other waits patiently in the soil and sprouts a year later. Lambs-quarters, foxtail, and pigweed are now-or-never annuals that pelt the landscape with a yearly seed rain, as weed scientists refer to their prodigious reproductive output. The seeds of these three species are among those most commonly encountered in Corn Belt fields, where the soil contains between 600 and 160,000 living weed seeds per square meter.

Cultivation by plowing, hoeing, or disking was once the dominant strategy for coping with weeds of all kinds in corn and bean fields. This was done by hooking various implements—the spring tooth harrow was a popular one—to the back of a tractor and dragging them through the soil. For perennial weeds, cultivation worked to repeatedly disturb the above-ground parts and eventually starve the roots. The aim for annual weeds was to cut them apart before their seeds formed. The get-big-or-get-out policies of the 1960s and 1970s, however, created fields so huge that mechanical weed control meant serious erosion problems—leaving farmers the terrible choice of watching their soil blow away or spraying it with herbicides. For conventional grain agriculture, then, weed control means chemical control. Since high levels of herbicide can also hurt the corn, herbicide-resistant varieties were developed through genetic engineering. By 2004, one third of Illinois' corn was made up of genetically modified strains, thus allowing for even more intensive applications of chemical weed killers.

Lest anyone assume that the habitudes of warfare are no longer part of chemical pest control, here are the trade names of some of them: Arsenal, Assert, Bicep, Brawl, Bullet, Burndown, Chopper, Dagger, Firestorm, Prowl, and Shotgun. Or if you prefer a cowboy theme for the poisoning of plants, you can choose among Cinch, Harness, Lariat, Lasso, and Roundup. Or for weed control that puts you in mind of a conquering frontiersman, there is Northstar and Yukon.

These herbicides kill by a variety of different poisoning mechanisms. Some interfere with plant hormones. The original synthetic herbicide, 2,4-D, for example, mimics a growth hormone, causing its target to grow too fast to absorb nutrients. A weed exposed to 2,4-D takes on a peculiar, twisted posture. Its stems become swollen and bent. As tissues burst, disease-causing agents invade and deliver the coup de grâce. Other herbicides halt the production of amino acids, from which proteins are made.

Still others strike directly at the process by which plants use sunlight to transform water and carbon dioxide into sugar and oxygen. This is the M.O. of atrazine: it interrupts photosynthesis.

Herbicide-resistant weeds are not mentioned in *Silent Spring*, as they did not yet exist. Today, weed scientists have identified 330 different types. The first one to emerge in Illinois was atrazine-resistant lambsquarters, making its first appearance in 1985. Illinois farmers now struggle with 18 different types of herbicide-resistant weeds, and the Weed Science Society of America makes available a website for tech support. In tracing the explosion of herbicide resistance among weed species that began in the late 1980s, researchers were forced to conclude that the "short-term triumphs of new pest control technologies have carried with them the seeds of long-term failure."

Atrazine has been in use since 1959, but the actual mechanism behind its ability to kill broad-leafed plants was not discovered until years later. We now know that the triazine herbicides, of which atrazine is one, poison a chain reaction that takes place inside the leaves' chloroplasts. (The chloroplasts of grass species, such as corn, rely on a different chain reaction and so are far less susceptible to the poisoning powers of atrazine.) Scattered across the leaf's surface like tiny Quonset huts, chloroplasts provide housing for the pigment chlorophyll, as well as the rest of the cellular machinery that runs the light-driven reactions of photosynthesis.

At its heart, photosynthesis depends on the handing off of electrons from one acceptor molecule to another, like a bucket of water passed down a fire brigade. These electrons are cleaved from molecules of water, and their liberation is intimately linked to the creation of oxygen. For photosynthesis to work, the electrons must reach a central reaction center. This is the link poisoned by atrazine. By binding to a protein in the reaction center, atrazine effectively blocks the bucket brigade's action. Without the transfer of these crucial electrons, the entire interlocking process grinds to a halt. "Excessive radiative excitement" builds up in the plants' green pigments, along with the toxic products of oxidation. The chloroplast swells and bursts apart.

The wisdom of broadcasting over the landscape a chemical that extinguishes the miraculous fact of photosynthesis—which furnishes us our sole supply of oxygen—is, in and of itself, questionable. Applied directly to the soil, atrazine is absorbed by the roots of plants and transported to the leaves. It poisons from within. Atrazine is thus water soluble. And because of its solubility, it tends to migrate to many other places.

Look at any map of atrazine usage in the United States, and Illinois will be the bull's-eye. It's the state completely shaded in red. Look at any map of atrazine in groundwater and surface water, and Illinois occupies the same central location. Illinois rivers and streams are universally contaminated with atrazine as are many of its groundwater supplies. Atrazine-sprayed fields are leaky fields. And atrazine's capacity to inhibit photosynthesis does not stop once it leaves the farm. It has demonstrated a remarkable capacity to poison plankton, algae, aquatic plants, and other chloroplast-bearing organisms that form the basis of the whole freshwater food chain. Atrazine also inhibits the growth of native prairie species. Once it enters the water cycle, atrazine becomes a component of precipitation, so that raindrops themselves are now laced with a chemical that possesses the wily ability to blow up chloroplasts. The inability to corral atrazine in the fields and prevent water contamination—even under the best of management practices—was the rationale for banning it in the European Union.

If photosynthetic inhibition were the only problem, the widespread use of atrazine in the Corn Belt would be worrisome enough. But as we have seen in Chapters One and Five, atrazine has other powers once it gains entry into the bodies of animals, including us. It is a proven endocrine disruptor. In laboratory rats, prenatal exposure alters the development of the mammary gland. It can also disrupt hormonal messages from the pituitary gland that play a role in ovulation. Several recent studies show atrazine causes sexual malformations in frogs. Results from human studies have been mixed. Although most human studies have not found an association between adult exposure to atrazine and breast cancer, no studies have yet looked at fetal or childhood exposure. There is a worrisome suggestion of a trend with non-Hodgkin lymphoma. A study of women farmers in Italy found an increased risk of ovarian cancer among those who worked with atrazine, but results from a California study were less persuasive. We do know that Iowa farmers have higher atrazine levels in their urine during spring planting and that their urinary levels are related to the amount of atrazine they apply. With these studies as the back story, we can now look more closely at atrazine's impact on the interior landscape of cells.

A cell that is sensitive to estrogen—like the cell of a breast—has estrogen receptors stationed along its nuclear membrane. Estrogen, or chemicals that act like estrogen, can bind to these receptors, which act as a kind of escort service to the VIP chamber of the nucleus. Here the genes hold court, receiving selected guests from the outside world and responding to their messages. Among them are the genes that respond to estrogen by, for example, ordering the cell to divide and proliferate. (That's the mechanism by which all of us women grew our breasts in the first place.) As a security detail, however, the estrogen receptor is notoriously gullible and will bind with an impressive number of chemicals, shuttling these hormone imposters along into the nucleus. Apparently, not much of a disguise is required to gain entry into the inner sanctum of the nucleus via the estrogen receptor. The password has not been changed in eons.

But atrazine does not look like or act like estrogen. It is no mimic. Ignoring the estrogen receptor entirely, atrazine involves itself with other elements within the cell—possibly including transcription factors, regulators, promoters, a receptor named NR5A, and a chemical compound called cyclic adenosine monophosphate that helps regulate genetic activity. At least, that's how it looks at this writing. The details of atrazine's subcellular activities and associations are still being worked out. What has been demonstrated is that atrazine, through a shadowy pathway, ramps up the production of an enzyme called aromatase. This enzyme is responsible for converting male androgen hormones into estrogen. In other words, atrazine tricks certain types of cells into making more estrogen, which then circulates around the organism, binds to estrogen receptors, and provokes estrogenic effects. No need to rely on estrogen mimicry. What is also becoming increasingly clear is that atrazine can exert these effects at very low concentrations. One experiment showed observable effects at two parts per billion, which is lower than the legal limit for atrazine in [illegible] water.

Stimulating aromatase may not be the only game that atrazine plays. In a 2008 study, atrazine stimulated ovarian cancer cells to proliferate but did so by binding to an entirely different receptor called GPR30. Meanwhile, yet another possible outcome of atrazine exposure is being examined by a team of researchers in Korea. Impressed by the remarkable similarity in the subcellular engineering of chloroplasts in plant cells (which atrazine is known to poison) and mitochondria in animal cells (which generate energy for the cell and so govern metabolism), they are investigating atrazine's possible effects on metabolic rate and risk of obesity. In a 2009 pilot study of rats, the researchers found that atrazine did indeed damage mitochondrial function. It also affected insulin signaling and induced insulin resistance. And the rats got fat.

~~~~~

In the waters of the Gulf of Mexico that lie beyond the mouth of the Mississippi River is a dead zone the size of Maryland. It has a simple cause: synthetic fertilizers. Through the provision of nitrogen, fertilizer makes corn grow faster, and, when it runs with the rain from fields into gullies and from gullies into streams, and streams into rivers and rivers to oceans, it has the same effect on marine algae. Their luxuriant growth sucks the oxygen from the room of the sea, and all life dies.

Synthetic fertilizers have also profoundly altered the agrarian landscape. More than any other factor, synthetic fertilizers are responsible for the segregation of the animal kingdom from the plant kingdom on U.S. farms. Commercial farms that grow grain are now, for the most part, strictly vegetarian. Manure from livestock is no longer needed. The
~~~~~

story of synthetic fertilizers, then, is the Biblical story of Noah run in reverse: the animals that once played a key role in maintaining soil fertility [illegible] and confinement facilities.

As we saw in Chapter Five, the Germans learned to make synthetic fertilizers in large amounts during World War I. Their widespread use in agriculture came only after the next world war, however, and their use has increased by fivefold since the 1950s. The process by which synthetic fertilizers are manufactured has changed little during this time. Almost all synthetic fertilizers are created using natural gas or other petroleum gasses as a feedstock, and more than 80 percent of the ammonia so created is used for fertilizer. Whereas household ammonia is a water-based solution, the signature white tankers that are pulled along Illinois' country roads contain anhydrous ammonia. *Anhydrous* means without water, and, because the boiling point of anhydrous ammonia is below zero, it must be stored under high pressure. The white paint on the pressurized tanks helps keep the temperature inside low. The price of anhydrous ammonia, as farmers know too well, rises and falls with the price of natural gas. Its ability to serve as a feedstock for the creation of methamphetamine has brought organized crime, with all its attendant miseries, into rural areas. As farmers also know too well.

Anhydrous ammonia and manure are not the only two ways to bring nitrogen to crops. Legumes, such as soybeans and alfalfa, have the ability to pull elemental nitrogen out of the air and fix it into nitrates that plants can use. More accurately, the fixing occurs within nodules of legumaceous roots, and the fixers are bacteria that live symbiotically within those nodules. Some of the nitrogen so fixed is used by the plants, and some remains in the soil as a gift for subsequent crops, like nitrogen-greedy corn. The heroic story of nitrogen-fixing soybean nodules is a lesson that all Illinois schoolchildren learn. Lightning, that mad scientist, also fixes atmospheric nitrogen, which rains fertility into the fields. My cousin John considers himself a friend of lightning and looks for it in the night sky when storm fronts wake him from sleep.

Like atrazine, nitrogen-fed fields are inherently leaky. Our primary route of exposure is through drinking water. Nitrate-contaminated drinking water is associated with cancer but, according to Mary Ward, senior investigator at the National Cancer Institute, there has been "no comprehensive research initiative to evaluate the heath effects due to nitrate ingestion and to determine whether the current regulatory limit is adequate." In the body, bacteria convert nitrate to nitrite and then to N-nitroso compounds, which are potent carcinogens in animal studies. The more nitrates you ingest in your drinking water, the more N-nitroso compounds you excrete in your urine. For these reasons, the International Agency for Research on Cancer concluded in 2006 that ingested nitrate is "probably carcinogenic to humans." Human studies have yielded mixed results. In Iowa, women with exposure to nitrates in public drinking water had higher rates of bladder and ovarian cancers. But other studies do not find such associations. One of the problems in designing studies to better understand the contribution of synthetic fertilizer to the burden of human cancers is that pesticides and nitrates often co-occur in drinking water so it is difficult to know which is the culprit or how they might be conspiring together.

~~~~~~

By now you are imagining that I'm going to make a pitch for organic agriculture. I am. But I want to do so by examining the two main arguments in favor of conventional agriculture. The first is that farming with pesticides produces higher yields than organic farming. The second is that pesticides allow farmers to bring to market fruits and vegetables at a lower price.

Organic farming, which is now regulated by the U.S. Department of Agriculture, uses ecological methods to control pests and enhance plant nutrition rather than petrochemical pesticides and fertilizers. Organic farmers rely on a sophisticated set of techniques—intercropping, cover cropping, trap cropping, rotational grazing, flaming, mulching, green
~~~~~~

manuring, and so on—to repel problems and direct nitrogen through the roots of their crops. Compared to soils on conventional farms, soils under organic cultivation have more fungi to help plants absorb water, more [illegible] to help with nutrient absorption, and more earthworms to make nutrients. And organic farming improves soil fertility for the next generation. That being said, yields on organic farms are—overall and in most cases, although there are notable exceptions—lower than yields on conventional farms. But here's the trade-off: organic farms burn up considerably less energy, and they tend to outperform conventional farms in extreme weather.

The longest view we have on the relative performance of organic and conventional agriculture comes from the Rodale Institute Farming Systems Trial, which began in 1981 and, for twenty-two years, compared a conventionally cropped field with two kinds of organically cropped fields. As analyzed by agroecologist David Pimentel at Cornell University, here are the results: For the first five years, corn in the conventional field system, like the proverbial hare, shot ahead with significantly higher yields. But, as the race went, corn in the two organic systems, playing the role of tortoise, eventually pulled even. And during a year of extreme drought, organic corn yields beat conventional by almost 50 percent. Organic soybeans, after twenty-two years, beat out conventional soybeans—but in only one of the two cropping systems. Soybeans from the other organic system came in dead last. Thus, the conventional beans still beat the average of the two organic beans by a nose. These results echo those of other field studies in the United States and Europe: conventional crops usually enjoy higher yields under normal circumstances or in wetter climates, whereas under drought conditions or drier climates, organic crops win. (Organic soils are better at retaining moisture.) Like municipal bonds, organic farms have what's called "long-term yield stability."

The bigger differences in the Rodale experiment are found in other metrics: after two decades, soil organic matter and nitrogen were significantly higher in the organic farming systems. The organic systems required 15 percent more human labor, but the work was more evenly distributed throughout the calendar year. The recycling of livestock waste turned a pollutant into a nutrient and solved a disposal problem. And this: after twenty-two years, total fossil fuel inputs were 30 percent lower in organic fields.

The possibility of diminishing the carbon footprint of farming by a third is, I think, impressive. I suspect, too, that we could bring organic yields up to par with conventional if research and development investments were directed toward improving organic agricultural practice. Currently, our agricultural policies provide little economic support for organic farmers, and our land grant colleges are notably uncurious about the complex ecological underpinnings of agricultural systems and how they might be recruited to ward off pests and encourage growth. That could change.

In regard to cost, there is no question that organic food is considerably more expensive than conventionally produced food. The principal reason for its premium price is that organic farming relies more on labor to control pests and less on chemicals, and, in the United States, labor costs more than chemicals. However, the higher retail price of organic food reflects the full cost of its creation, whereas the price of chemically grown food does not. Pesticides pass along some of their costs to others in society—the cost of testing food for pesticide residues, the cost of removing them from drinking water, and so on. These costs don't show up on anyone's expense statements. Economists call these costs externalities, and the bad thing about them, so they explain, is that they lead to market outcomes that are privately profitable but costly to society. The externalized costs of conventional agriculture have been calculated—no small task—by Dr. Pimentel at Cornell. His 2005 estimate is $10 billion a year. Included in this figure were the costs associated with the acute poisonings of farmworkers, loss of honeybees and other pollinators, loss of fish in rivers and streams, and cost of cleaning up water sources on their way to becoming tap water.

Another line item that appears in Pimentel's table of estimated economic costs of pesticides is medical treatments for pesticide-induced cancers.

We've already looked at the evidence for an association between childhood cancers and residential use of pesticides (Chapter Five) and for elevated rates of particular kinds of cancers within farmers using particular pesticides (Chapter Four). Let's put a face on this particular externality by looking more closely at California, where one-quarter of all pesticides used in the United States are deployed and where so many of the nation's fresh fruits and vegetables, our allies in the war on cancer, are produced.

Among the pesticides used each year in California are twelve million pounds of probable carcinogens and ten million pounds of possible carcinogens. Living in areas of the state where use of these carcinogenic pesticides is the most intense are 368,000 children under the age of fifteen. We suspect that children living here are exposed to the chemicals used in the nearby fields because researchers have tracked drifting pesticides from the winds that blow across those fields into homes located a half-mile or more from sites of application. We know with certainty that the children of farmworkers in the Salinas Valley are exposed. A 2006 study revealed elevated levels of recently applied pesticides in the household dust where farmworker families live and on the clothes and in the urine of farmworker children.

What we don't yet know is the number of cancers resulting from these known and potential exposures. There have been some human studies, but no clear pattern has yet emerged. California is the only state with a comprehensive pesticide registry—in place since 1991—that legally requires all growers to report all agricultural pesticide use. When maps of the state's agricultural pesticide use are overlain with maps of cancer incidence drawn from the state's cancer registry, nothing obvious jumps out. There is as yet no evidence that California women living in areas of recent high, agricultural pesticide use experience higher rates of breast cancer, for example. (By contrast, the national data do show some patterns: An exploratory study found significantly increased risk of childhood cancers in counties with intense agricultural activity, especially if at least 60 percent of the county's acreage was devoted to farming).

On the other hand, a California study that focused just on farmworker women did uncover some trends. Women farmworkers begin their work in the fields as children and teenagers and are potentially exposed to multiple pesticides from a young age onward. A study of Hispanic women farmworkers found increased breast cancer risk among younger women and those with early-onset breast cancer. Women with the highest pesticide use had a 40 percent higher risk of breast cancer than women with the lowest.

The Federal Insecticide, Fungicide and Rodenticide Act (FIFRA) is the law that protects people from pesticides and ensures their registration. However, this law has a cost/benefit clause that allows risks to humans, such as farmworkers, to be weighed against the economic benefits of the pesticide in question. Joan Flocks of the University of Florida investigates the cancer threat posed by pesticides to U.S. farmworkers. In her testimony before the President's Cancer Panel in October 2008, she said about FIFRA: "The fact that such limited regulation exists is misleading if it causes the public to believe that farm workers are protected from potentially carcinogenic substances when they are not."

~~~

Agriculture is 6,000 years old. For 5,940 of those years, it was practiced organically. But 60 years—two generations—is still a long time in human terms. It is long enough for us, as a culture, to have lost entire skill sets. It is long enough for the barns that once held fertilizer-providing animals to collapse. It is long enough for a county's entire fleet of hay balers to rust out, for local canneries to close up shop, and for knowledge about the weed-controlling abilities of sheep to slip away. It is long enough that the very idea of using chemicals linked to cancer to grow fruits and vegetables that we eat to prevent cancer no longer seems as bizarre as it might have
~~~

to our ancestors. Rachel Carson, who wrote near the beginning of that 60-year expanse, once remarked how strange it was to live in an age where carcinogens were a basic element in our system of food production. It is [illegible] In fact, partial answers and outright solutions exist all around us. Here are but two.

In central Illinois, a small organization called The Land Connection is attempting to rebuild the state's food security and tradition of self-reliance by creating a network of small, independent, diversified organic farms that grow food for people to eat. It does so by connecting young, organic farmers in need of land with parcels of farm land—often acquired when an elderly farmer retires. At the same time, it works to find markets for the crops so produced and rebuild infrastructure (like meat lockers) for storage and distribution of regionally produced food. It also facilitates year-long trainings for new recruits as well as a mentoring program that brings together organic farmers who are just starting out with veteran farmers of long standing.

In Black Hawk County, Iowa, a ten-year effort to foster a relationship between local farmers and institutional food buyers—retirement homes, grocers, restaurants, and college dining halls—has resulted in $2.2 million in sales for 2008. These were food dollars that had been flowing out of the county and the state. By capturing and investing them locally, the rural economy has been strengthened. In fact, every dollar invested in the work of promoting local food in the county has helped $14.6 circulate in the local economy. Part of the key to success was creating a local foods coordinator and making it a salaried position. The king of local food in Black Hawk County, by the way, is Rudy's Tacos, which, in both 2007 and 2008, purchased 71 percent of its ingredients—from the sour cream to the tomatoes—from local sources. I've eaten at Rudy's. The tacos are delicious. The lines to get to the tacos are too long.

With organic farming having enjoyed a huge boom in popularity from 1998 to 2008, there are many more small, satisfying stories I could cite here. The real issue, though, is how to replicate these narratives on a much larger scale. (For all the rebirth of organic agriculture, there were still only 26,367 acres under organic cultivation in Illinois in 2008. That represents 0.14 percent of all Illinois cropland.) Weaning U.S. agriculture from chemical dependency will not be easy and will require diligent effort on the part of many. The ten million of us who are cancer survivors surely have a stake in the outcome—whether we are motivated by a desire to reduce our exposure to possibly carcinogenic pesticides or ensure our access to affordable fruits and vegetables that do not require a 1,500-mile, petroleum-fueled journey to reach our plate. As we weigh how to redesign our agricultural system, here are some thoughts to consider.

First, Corn Belt farmers cannot be expected to change practices on their own. The infrastructure for transporting, storing, and marketing crops other than corn and soybeans is just not there. There are no grain elevators for flax or amaranth—even if they are ideally suited to the climate and soil and would make a great crop rotation. And, as agroecologist Laura Jackson points out, farmers who grow food for people to eat reduce the amount of their land eligible for corn subsidies, reducing its market value . . . and so reduce their collateral for loans. It is too much to expect farmers to transform farming by themselves.

Second, calling for bans on possibly carcinogenic agricultural chemicals needs to be part of a larger vision of a new food environment. Simply taking one herbicide off the market without making other changes may well accelerate herbicide resistance as farmers use more of the fewer chemical weapons remaining in their arsenal. A 2007 study—funded by the makers of atrazine—forecast dire consequences for farmers and the entire downstate Illinois economy if the use of atrazine were ever discontinued. Recalling that the proponents of chemical-based agriculture made similar claims about DDT in the 1960s helps to put predictions like this in perspective. Nevertheless, I asked myself, as I read through this report's analysis, is it truly the case that my home state's economy depends so completely on the growing of a single commodity (corn) and that this commodity in turn is entirely dependent on a single fifty-year-old chemical (atrazine)? So that no matter what the scientific evidence shows us

Out in the bean fields, my cousin John hollers over to me that Emily picked up a rock in her last pass on the combine. All its many working parts still seem to be working, however—the deafening decibel level of the engine certainly sounds the same. So after a brief inspection we climb into the cab together and swing the machine around. Usually, Emily runs the combine and her husband ferries the wagons back and forth to the storage bins, but they've switched places for a while.

The quiet of the cab gives us a chance to catch up with each other, gossip about the family, and laugh at our parents' various eccentricities. This kind of ride-along is a discouraged practice—guests in the cab are distracting and can compromise safety. John asks me to keep an eye out for rocks.

We're quiet for a while. The movement of the sickle bars is mesmerizing, like the churning paddlewheel of a large boat. If the weather holds off, Emily thinks they can get all their beans in by midnight. I look out at the many acres that remain—and the sea of corn beyond—and wonder how they will do it. A call comes in on the cellular phone from a friend near Bloomington who got rained out today. The price of beans has dropped a penny. The weather looks like it will pass to the north.

The combine's head lowers a bit as we descend into wetter ground. John points to the plants that rise above the beans—bright foxtail, glossy lambs-quarters, and the reedy flowering stalks of velvetleaf. He has a lot of respect for weeds. Since they're here to stay, he says, somebody ought to figure out a use for them. Besides, a few weeds means he's not using too many chemicals. John and Emily have five young kids. Most of their information on pest control comes from agrichemical distributors, but, John emphasizes, they want to become more environmental in their approach to farming.

An alarm sounds, indicating the hopper is full. John shakes his head. "Emily's going to ask me why I didn't dump this load on my last pass." We'll have to drive back the length of the field without harvesting any beans, which in Emily's thinking is not efficient farming.

All the crops they grow are for export. They sell to the river, as Emily puts it. Their entire corn harvest, for example, will be trucked to the Pekin docks. From Pekin it goes by barge to New Orleans; from there, who knows. Farmers have become as dissociated from those who eat their food as consumers have from those who grow it.

After John empties the hopper, Emily walks over to meet us. Instead of delivering the anticipated scolding, she smiles and hands me a plastic bag.

"These are tofu beans," she shouts over the machines. "We combined them yesterday. Right now, they're on their way to Japan."

I reach in and pull out a handful. They're a little bit bigger, rounder, and paler than the beans we just dumped in the wagon. John adds that he and Emily have never eaten tofu. Have I?

In fact, I have several vacuum-packed slabs in my pantry. "What's it like?" John shouts. Standing in a rented field that's full of them, I try, at the top of my lungs, to describe the taste of soybeans.

air

"Earth and the Sea feed air;
the Air those Fires Ethereal."

—JOHN MILTON,
PARADISE LOST, BOOK V

The thing about air in Illinois is that there is so much of it. Air is a more conspicuous element here than in any other place I've lived. It seems deeper, wider, more present.

I first learned to see air when an art teacher began coming to our grade school. Of all the ideas she introduced, none made more sense to me than the concept of the vanishing point—an invisible place on the horizon where parallel lines mysteriously converge. Simply by choosing a point and drawing in perspective a house or a road, I could draw air. This was a great discovery. And the Illinois landscape was full of such points through which all objects seemed to be striving to disappear. Grain elevators along a railroad track. A plowed field. The silver towers and looping cables of high-power lines. Everything eventually vanished into air.

What I never figured out was how to represent air's transformative properties. In rural Illinois, how objects appear depends on how much air they are viewed through. A brown chip in the sky turns, a half-mile closer, into a circling hawk. Fluttering black specks turn into black handkerchiefs and then into a flock of crows. The dead body on the side of the road eventually becomes a scrap of carpet.

In the fifth century B.C., the Greek philosopher and physiologist Empedocles declared that the atmosphere was not a void but a living substance. A thousand years later, his theory was animated by Paracelsus himself (of "dose makes the poison" fame), who professed that air was inhabited by elemental beings he called sylphs. I am thinking of these two scholars while driving through central Illinois during a period of record-breaking cold. It is many degrees below zero—dangerously cold—equipment malfunctioning, cars seizing up, and hourly warnings against venturing outdoors issuing from the radio. I should not be out myself, but it seems important to see the landscape at its most unmoving: Its constituent molecules vibrating at their slowest recorded speed. The earth with all its seeds frozen down to many feet. Water only a memory.

Leaving the engine running, I step from the car onto the stone floor of a cornfield. Only the atmosphere seems alive. Every inhalation brings pain, every exhalation a puff of crystals. Whatever sylphs are, they instantly find the cracks between scarf and collar, glove and sleeve. In seconds, I am exposed, although still standing fully clothed. And yet never is this element more invisible. Even objects at the vanishing point seem distinct and permanent, untransformed by miles of air.

~~~

Far from Illinois, the White Mountains lie across upper New Hampshire like a crooked crown. Their westernmost peak, Mount Moosilauke, rises above a tract of trees familiar to every student of ecology, the Hubbard Brook Experimental Forest. Here, researchers have used large-scale, long-term field studies to trace the slow cycling of nutrients through an entire living community. Much of what we know about the ecological pathways of nitrogen, phosphorus, and calcium, for example, derives from work carried out in these woods. Some of the first investigations of acid rain were conducted here as well.

A research team led by the biologist William H. Smith of Yale University discovered something else about the forest floor of the famed Hubbard Brook. Its newly fallen leaves and needles, as well as the composting earth
~~~

beneath, contained detectable amounts of both DDT (0.8 pounds per acre) and PCBs (2.3 pounds per acre). Even more remarkably, neither of these long-banned chemicals had ever been used, distributed, or produced in the immediate area where the samples were collected. Soil and leaf litter sampled from nearby Mount Moosilauke were found to be likewise contaminated—all the way up to its tundra-covered summit. Tellingly, the level of contamination rose with elevation and was greater on west-facing slopes. Such patterns are consistent with atmospheric deposition.

Smith and his colleagues thus postulated that molecules of DDT and PCBs were being carried into Hubbard Brook by prevailing winds. Their origin, however, remains obscure. Storm tracks that pass over major centers of agriculture and industry in the United States also pass frequently over New England. Quite possibly, regional air masses are spiriting these semivolatile, long-lived molecules from landfills, dump sites, and farm fields to this remote, pristine forest. It is also possible that global air currents are the transporting medium. Like undocumented aliens, these chemicals may have been swept in from other countries, even other hemispheres. Studies conducted in rain-fed bogs across eastern North America support this possibility. These peculiar habitats receive all their input of pollutants from the atmosphere, and not from ground or surface water. Therefore, they function as a living map, revealing in detail the historical and geographical contours of atmospheric deposition. And because peat preserves organic compounds almost completely, they also function as living archives.

More evidence for the role of global circulation comes from the world's trees. A survey found traces of twenty-two different organochlorine pesticides—including DDT, chlordane, and endosulfan—in tree bark gathered from ninety different sites around the globe. Bark is a remarkably oily tissue and therefore readily absorbs oil-soluble pollutants from the air. That trees growing in the agricultural regions of the midwestern and eastern United States should bear residues of pesticides in common use there is hardly surprising. However, researchers also found insecticides in trees growing in the Arctic. And they also found them in the bodies of the people who live there. Indeed, the highest concentrations of the most villainous chemicals of the twentieth century are inside the bark, blood, and tissue of the organisms living the farthest from the sources of those chemicals. This phenomenon is known as the Arctic Paradox.

As described by journalist Marla Cone in her 2005 book *Silent Snow*, this paradox is explained by a form of chemical nomadism called global distillation. When persistent organic pollutants are released in warmer climates, they evaporate and are carried by winds to cooler areas, where they condense and descend back to earth. These trespassers overwinter in soil, snow, or water. When the summer sun revaporizes them, air currents blow them further toward the pole. They then drift downward once again. During this process, the various chemical contaminants are spatially partitioned: those substances that evaporate at the lowest temperatures keep revaporizing and are thus continuously drawn to northern latitudes and higher altitudes. Finally, in the Arctic, they can ascend to the skies no more, and this is how the most pristine corner of the earth has become the most chemically contaminated. The rising and falling movements of global distillation explain not only why chemicals used in rice paddies eventually end up in the skin of Arctic trees but also why fish in Yukon Territory's Lake Laberge became so full of carcinogenic toxaphene—a pesticide used in cotton fields—that the Canadian government was forced to ban angling there. Global distillation also explains why, during negotiations for the United Nation's Stockholm Convention on Persistent Organic Pollutants, the most powerful testimony came from a delegation of Inuit mothers. They spoke about ecological genocide.

The phenomenon of global distillation shows that not all of the dangers from carcinogens in our air supply come from breathing. Some also come from eating. Poisons dumped and plowed into the earth are released, molecule by molecule, into the air, where they redistribute themselves back to the earth and into our food supply. In short, because of air, we each consume suspected carcinogens released into the environment by people far removed from us in space and time. Some of the chemical contaminants we carry in our bodies are pesticides sprayed by farmers we have never met, whose language we may not speak, in countries whose agricultural practices may be completely unfamiliar to us.

Some of the chemical contaminants we carry with us come from long-defunct products of industry—objects manufactured, used, and discarded by people of a previous generation. When we sit down to eat a meal of, for example, freshwater fish, we are linked to all these people through the medium of air.

Conversely, chemicals dumped and sprayed in our own neighborhoods, fields, and landfills have drifted to distant territories and found their way into the diets of the people who live there. I sometimes think of this multitude of connections while walking through Illinois corn and bean fields. I wonder where the chemicals sprayed in these fields when I was growing up here now reside. On what mountainside, in what forest or lake bottom, in whose bodies do they lodge now?

~~~~~~

Respiration is an ecological act. We inhale a pint of atmosphere with every breath. And yet, of all the component aspects of our environment, air remains mysterious. Air is the element most diffuse, most shared, most invisible, least controllable, least understood.

Of all the toxic chemicals released by industry into the nation's environment in 2007, more than one-third was released into air. These emissions include ninety-one million pounds of known or suspected carcinogens. When vehicle exhaust is added to the mix, air's burden of carcinogens grows larger. According to the International Agency for Research on Cancer, ambient air in cities and industrial areas typically contains a hundred different chemicals known to cause cancer or genetic mutations in experimental animals. And while air pollution in the United States has improved since the Clean Air Act of 1970, 60 percent of Americans—186 million of us—live in areas with unhealthful levels of air pollutants.

These are facts not in dispute. How much airborne carcinogens actually contribute to human cancer, however, remains an elusive question. Air can evade the rigors of scientific analysis through at least three means. First, its fluidity makes exposure very difficult to quantify. Wind speed and direction, as well as wind flow along river valleys, over hills, and around
~~~~~~

buildings, significantly alter the transporting path of airborne carcinogens. Residents of a single metropolitan area may all drink water from the same river and buy their food from the same supermarkets, but they may not all breathe the same air. Those who live downwind from the local industrial park may live in a very different atmosphere than those who live upwind. A centrally located air-monitoring system cannot account for differences in microclimate.

Secondly, there exist whole classes of air pollutants that we do not even attempt to monitor at all. Chief among these are fine (less than 2.5 micrometers) and ultrafine (less than 100 nanometers) particles. Some are so small that an electron microscope is required for viewing them. Fine and ultrafine particles are nevertheless visible: they are the haze that blurs the sunlight of a summer day. Well, not exactly. What we are seeing are not the specks themselves but the scattering of light waves they cause. Far more dangerous when inhaled than gritty generic particles, fine and ultrafine particles can slip past the defenses of the lung's cilia, glide straight through lung tissue, and enter the bloodstream. Here they can have many effects. Making blood platelets sticky is one—and is probably the mechanism by which fine and ultrafine particles contribute to heart disease.

In spite of their vanishingly small size, fine and ultrafine particles come in many varieties. They may sport highly complex architectures or be as simple as dots of elemental carbon (soot). And particles need not be solids. Many take the form of superfine droplets. Some are combinations of different liquids within a single droplet. Some are solids wrapped with liquid jackets. Some are liquids suspended inside solids—like chocolates with a liqueur center. Fine and ultrafine particles can be metals, oils, acids, hydrocarbons, or all of the above. They are mixtures of mixtures. The EPA regulates particles above 2.5 micrometers but does not attempt to regulate the fine or ultrafine ones. They are simply too hard to measure.

The third reason for uncertainty is that air is a transmutational medium. As in an alchemist's flask, the atmosphere concocts new materials from the ingredients placed into it. Recent evidence suggests that some of the major carcinogens in air are synthesized when organic chemicals released from various sources react with each other and are transformed into entirely new substances. Many may not even be identified yet. Currently, a team of U.S. and Chinese researchers are looking at air pollutants that may be morphing into carcinogens on their way across the Pacific Ocean from Asia to the United States. Thus, a simple laundry list of air emissions—such as the Toxics Release Inventory—cannot account for the presence of all the cancer-causing agents to which we are exposed.

Of these various pollutants that appear, literally, out of thin air, the most notorious is ozone. This molecule is created naturally up in the stratosphere from the interaction of ultraviolet radiation with oxygen. The resulting layer of ozone protects us from excessive UV exposure, and, thus, with good reason, we concern ourselves with its disappearance. At the earth's surface, however, ozone is a noxious, unnaturally occurring irritant to eyes and lungs. With equally good reason, people who live in cities during the summer concern themselves with rising levels in its daily parts-per-million concentration. (Molecules of ground-level ozone are too heavy to rise into the upper reaches of the atmosphere and take the place of their faltering compatriots, upon whose presence all life depends.)

Although it's a major ingredient of urban smog, ground-level ozone is not emitted into the air by a known polluting source. Instead, it's created when sunlight catalyzes a reaction between two kinds of vapors: nitrogen oxides, which are emitted from tailpipes and smokestacks, and volatile organic compounds, which rise into the air when houses are painted, cars refueled, roads paved, and clothes dry-cleaned.

In the classic sense of the word, ozone is not a carcinogen. Nevertheless, in the complex unfolding of cellular events that typifies carcinogenesis, ozone seems to play a supporting part. A powerful poison, ozone causes inflammation of the airways and thereby interferes with the body's ability to sweep foreign particles—some of which may be carcinogenic—out of the lungs. Ozone also hampers the activities of the lungs' macrophages. Part of the immune system, these amoebalike scavengers offer a first line of defense against a variety of pathogens and foreign substances. In studies of laboratory animals, ozone appears to magnify the effect of other lung carcinogens and influence the carcinogenic process itself. Lung tumors in ozone exposed mice have distinctly different genetic mutations than do those from mice that have breathed clean air.

In the attempt to understand how substances in air may contribute to cancer, the issue of ozone raises some vexing questions: How do we assess exposure to airborne pollutants that emerge from chemical recombinations of other airborne pollutants? How do we quantify the cancer-causing potential of a substance that enhances the cancer-causing potential of other substances? How many cancer deaths do we chalk up to ozone? What is the body count?

The most recent estimate we have of cancer diagnoses attributable to air pollution derives from a 2009 EPA investigation called the National Air Toxics Assessment. Although it does not take into account the questions raised above, this report looked closely at air contaminants within all counties within the United States and concluded that all of us—all 285 million U.S. residents—have elevated cancer risks from exposure to air pollution. This risk is not evenly distributed, however. Counties within southern California, Ohio, West Virginia, and parts of Indiana and Illinois had significantly higher risks for air pollution–related cancers than, say, North Dakota. The average of the estimated risks from all the 500 counties within the United States is 36 in a million, meaning that for every one million of us, 36 people will contract cancer from breathing. The EPA researchers didn't carry out the math any further, but I will. Thirty-six times 285 is 10,260 cancer patients.

~~~

With a five-year survival rate of only 15 percent, lung cancer is so swiftly fatal that we rarely hear stories of its victims. While those diagnosed with breast cancer form support groups, write books, lobby Congress, and organize rallies, gala benefits, and races-for-the-cure, those with lung cancer tend to vanish quietly from our midst. The small public presence afforded them is usually a posthumous one. Guilt and blame also silence lung cancer patients, who are seen as having brought about their own misfortune.
~~~

Whether we hold individual consumers or corporate producers responsible, the primacy of tobacco in the epidemiological portrait of lung cancer is indisputable. Smoking is the dominant cause of lung cancer. Nevertheless, there is more to the story of lung cancer than cigarettes. If other factors seem minor in comparison, it is only because tobacco is such a major killer. But lung cancer among nonsmokers is the sixth most common cause of cancer death in the United States. Not all of these deaths are cigarette independent. About 20 percent (three thousand deaths each year) are thought to be attributable to secondhand tobacco smoke. So shocking is this statistic that it has rightfully prompted substantive changes in laws governing smoking in workplaces, airplanes, restaurants, and other public domains. However, the majority of nonsmoking lung cancers remain unexplained. While air pollution is not the only possible cause, it is one that unavoidably affects us all, and it is one that may interact with and potentiate the effects of other factors. On these bases alone, this topic deserves thoughtful inquiry. Several lines of evidence suggest its role may be significant.

The first comes from doctors' offices. Oncologists who specialize in lung cancer are reporting increasing numbers of nonsmokers among their patients, as well as increasing cases of a specific kind of lung malignancy not strongly associated with tobacco in men. Called adenocarcinoma, it is distinguishable from oat cell and squamous cell carcinomas, both strongly linked to smoking.

Meanwhile, epidemiologists have been focusing on understanding the urban factor in lung cancer. Studies from the United States and Europe consistently find higher rates of lung cancer among nonsmokers living in urban areas than among those living in the countryside. Areas with chemical plants, pulp and paper mills, and petroleum industries also show elevated rates of lung cancer. Truck drivers and other workers who inhale diesel exhaust on the job have higher rates of lung cancer. And a cohort study of five thousand chimney sweeps in Sweden found increased mortality from lung cancer and other tumors that was not explainable by smoking habits but that was related to exposure to carcinogenic soot. In Utah, where smoking rates are very low, researchers compared lung cancer rates in two counties that were similar in all respects, except that one them then became home to a steel mill. Initially, lung cancer rates were indistinguishable, but within fifteen years, cancer mortality in the more polluted county was higher. On a larger scale, a prospective study of nearly a half-million U.S. residents found that the risk of death from lung cancer was higher in parts of the country where density of particles in the air was higher.

As ports of entry, the long tunnels and spongy rooms of the lung are only the first place where airborne carcinogens meet human tissues. Those carcinogens absorbed across the lung's membranes are carried in the bloodstream and deposited throughout the body. Much less is known about the relationship of these contaminants to other forms of cancer, but the question is provoking increasing interest.

By-products from the burning of fossil fuels are under particular suspicion. Breast cancer, as we have seen, was first linked to potential sources of air pollution in Long Island. Subsequently, associations have been found between exposure to traffic exhaust during puberty and risk of early-onset breast cancer. Perhaps not coincidentally, a growing body of evidence suggests that tailpipe emissions have estrogenic activity. Air pollutants may also alter breast density in ways that raise the risk for breast cancer. A 2007 review of the literature concluded that the risk of breast cancer associated with exposure to engine exhaust and other aromatic hydrocarbons is roughly equivalent in magnitude to some of the well-known and well-established risks for breast cancer, such as late age at first childbirth and sedentary lifestyle. Corroborating evidence comes from the laboratory: members of a chemical family of combustion by-products called aromatic hydrocarbons—of which benzo[a]pyrene is one—cause breast cancer in animals. According to researchers at Albert Einstein College in New York, aromatic hydrocarbons inhaled by the lungs can become stored, concentrated, and metabolized in the breast, where the ductal cells become targets for carcinogens.

Bladder cancer, too, has been linked in several studies to air pollution. The strongest evidence comes from Taiwan, where researchers found positive associations between air pollution, especially from petrochemical plants, and the risk of dying from bladder cancer. An investigation of bladder cancer deaths among children and adolescents in Taiwan found that almost all those afflicted lived within a few miles of three large petroleum and petrochemical plants.

That caught my attention.

Closing a lethal circle, air pollutants have also been implicated in promoting the spread of cancer from other organs *to* the lung. For example melanoma-afflicted mice who breathed air polluted with nitrogen dioxide developed more tumors in their lungs than those who breathed clean air. They also died sooner. Not lung cancer per se, these tumors are metastases, secondary growths from cancer cells shed from the original tumor. They are carried by the blood to the lungs, where, like planted seeds, they take root. As cancer patients know, metastases to the lung are an ominous sign. The first capillary bed encountered by blood leaving most other organs, the lung is the most common place for cancer metastases to take hold. And somehow—at least in mice—breathing nitrogen dioxide seems to facilitate this process.

The pathologist Arnis Richters, of the University of Southern California, believes that at least two sinister mechanisms are at work here. First, nitrogen dioxide impedes so-called killer T cells, whose function, among others, is to rid the body of wandering tumor cells. Second, nitrogen dioxide causes blisters to form deep in the lung's airy chambers, where such errant cells can then become trapped. "Since many cancer patients have circulating cancer cells," says Richters, "it is possible that noxious air pollutants may play a more important role in dissemination of cancer than is realized at the present time."

Thus, like its chemical offspring, ozone, nitrogen dioxide raises thorny questions about causality. Nitrogen dioxide is not, as far as we know, a carcinogen. And yet, for those of us who have had cancer, its presence in air may affect our chances of surviving our disease.

I am driving through my favorite section of town, the old neighborhood north of the post office. It was Pekin's original town site, but I like it for a different reason. All the east-west streets have women's names, and not just ordinary ones: [illegible] Amanda, Charlotte, Susannah, Minerva, and the one I, as a child, most revered—Ann Eliza. All of these streets dead-end at the river.

I have my sister's two sons in tow, and we stop to pick up treats at Patsy's Bakery. It is a Sunday morning at the end of February, the first warm day in months. The church parking lots are all full. We head for the river, zigzagging as we go so we can hit all the girl streets. As we do, the smell that is almost always present here gets stronger. I'm half thinking about the research papers I've been studying.

> These results are consistent with findings of my previous analyses and provide further evidence that air pollution is a moderate risk factor for lung cancer.

I have thought a lot about how to describe this smell, but I cannot. I can smell it more acutely now than I could when I lived here, although it is probably less potent than in earlier decades. It is still too familiar for words. A complicated smell, it seems to contain more than one odor. It is . . . pungent. After watching the barges and the fishing boats for a while, we swing onto Route 9 and head over the Pekin bridge.

> Overall, the studies suggest that emissions from some types of industries may increase the lung cancer risk for the surrounding population.

On the west bank of the Illinois, the floodplain spreads out like a dance floor in a pool hall. It's mostly corn and bean fields now—all the way up to the power plant and the coal piles and the access roads and the railroad yards and the bait shops. I love this river valley and the bluffs that rise above it, the backdrop of my childhood. I want the boys to love these landscapes, too—but with full knowledge rather than denial, in the terribly difficult way that one is asked to love alcoholic parents: not abandoning them to wretchedness, not enabling their self-destruction, nor pretending there is no problem. I don't know how to explain this to my young nephews, but maybe I don't have to. Maybe we adults need only demonstrate an attitude of passionate attention about where we live.

"Let's check out the west bluff." I turn right onto Route 24 and then left onto a narrow side road, steep as any mountain grade, and downshift. Even in second gear the car stutters, and what's left of our doughnuts tumbles backward into our laps. Suddenly, we are in thick woods that draw a curtain on the river's floodplain behind us.

> In conclusion, it is difficult to interpret the epidemiological evidence on ambient air pollution and lung cancer.

At the top of the ramparts, we emerge back into the sunlight and find ourselves in Tuscarora, an unincorporated scattering of homes that follow the topographical contours of the bluff. These eventually give way to more corn and bean fields—a mirror image of the east bluff. Not so, claim my two companions, who believe the west bluff is hillier and therefore more fun to drive on. I'm not persuaded, but it's a good argument.

> Many toxic chemicals are not routinely monitored in ambient air in Illinois, and little is known about ambient concentrations or the relative importance of various sources of these chemicals.

In the nineteenth century, many good people—medical doctors and officers of the government among them—believed that infectious diseases were brought on by bad air. These were followers of the miasma theory of disease causation. Air was thought to be "corrupted" when it passed over decaying organic matter—swamps, sewage, dead bodies. Breathing the poisonous odors emanating from such places was, according to the miasma school of thought, injurious to the body and had the power to trigger terrible physical maladies.

The miasma theory was eventually supplanted by its rival, the germ theory. But before its drift into obscurity, miasma's disciples managed to usher in significant reforms in public health policy, namely closed sewage systems, clean drinking water, and deep burial of the dead. These did much to deter epidemics even before the real, microscopic causes of illness were discovered. The miasma theory, although mistaken, saved many thousands of lives.

Old descriptions of the Illinois wilderness, including the valley I am now driving through, provide detailed observations of its air—how sweet or rank were its odors, how salubrious or potentially pestilent its breezes.

> Whether the excess risk of lung cancer can be attributed to urban air pollution cannot be determined conclusively, but it is suggested that it at least contributes to the risk.

I follow Route 24 east over the McClugage Bridge. Here the river is nearly three-quarters of a mile wide, and at the exact center we cross from Peoria back into Tazewell County. This border fascinated me as a child. A line in the middle of a river! How did they know exactly where it was? Stories about my childhood are fascinating to my current company, including my old fear of drawbridges and my enchantment with tugboats. And the fact of their fascination fascinates me. *This is where we come from.*

Still, I wonder if I should be bringing the boys down here at all. The younger one has asthma—as does his mother, who developed it as an adult. A secretary at the local grade school, Julie is astonished, she says, at the number of children with inhalers in their backpacks. On the days that she must send unusually high numbers of wheezing children home from school, my sister has begun to take note of what the weather is like, which way the wind is blowing, how the air smells, and how labored her own breathing is. Perhaps it is once again time, we both agree, to look at the environment to understand what ails us. Perhaps it is time to risk being right for the wrong reason—as did our predecessors who successfully prevented the spread of infectious disease by cleaning up pollutants in the absence of complete knowledge about the microbes they contained.

Increases in childhood asthma and the clustering of lung cancers around cities with dirty air are telling us something. Suppose we do nothing until the exact mechanisms are elucidated, until exposures are definitively ascertained, until the precise combination of all pollutants and their specific interactions with each other and with the tissues of our respiratory airways are exhaustively understood. Then are we not mimicking those who, at one time, could just as well have claimed that there was not sufficient reason—on the grounds that science had not yet identified any specific biological agent responsible for cholera—to keep human excrement out of the drinking water?

We begin the long climb up the bluff. At the top is the brokenhearted town of Creve Coeur, Pekin's smaller, meaner, drunker brother. On the other side of Creve Coeur is the road home. I celebrate by opening the sunroof.

"Aunt Sandy, when did you get this car?"

"Honey, I got it when a friend of mine in Boston was sick and needed to go to the doctor a lot. Do you remember when I told you about that?"

"She died, right?"

"Yes, she did."

"You had cancer, too, didn't you?"

In the absence of other data, it would be advisable to avoid excessive and prolonged exposure to such agents.

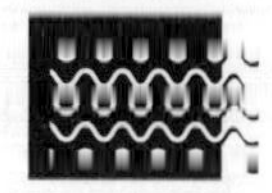

water

"And the fish suspending themselves so curiously below there and the beautiful curious liquid
And the water plants with their graceful flat heads, all became part of him."

—Walt Whitman,
"There Was a Child Went Forth"

My mother grew up by the Vermilion River and my father alongside Lake Michigan. I have, therefore, no familial connection to the Illinois River, no handed-down tales to pass on. In getting to know this river I was raised beside, I've relied as much on library research as on my own observations. These sometimes tell two different stories.

In one archival photograph from the early 1900s, four men and two boys stand on the river's edge beside what looks to be an immense pile of stone butterflies. The solemn fellow in the foreground holds one of them, wings spread, in the palm of his upraised hand. The others in his outfit stand in the background, stiff and expressionless as fence posts. These men are, in fact, mussel gatherers showing off their catch. They will sell their heap of shells to one of fifteen button factories that line the shores of the lower Illinois River.

By 1948, the last one had closed. Pollution and overharvesting killed off the mussels, and plastic replaced mother-of-pearl in the production of shirt buttons. The species depicted in the photograph are as alien to me as the process of turning them into objects of human attire.

In 1948, diving ducks also began to disappear from the Illinois. Ring-necked duck, canvasback, ruddy duck, and lesser scaup: these are species I learned to identify from stuffed specimens and in distant field sites. I have been trained to recognize their patterns of coloration and differently pitched calls, but I do not recognize them as fellow inhabitants of the river system I grew up in—although it served for centuries as the flyway for their migrations.

My newlywed parents began building their house on the bluff in 1955. In this same year, the valley's population of scaups ("highly social . . . note the purplish gloss on head . . . shows bold white stripe on secondaries . . . calls are short low croaks") plummeted to zero. Researchers attribute their disappearance to the synchronous demise of the river's fingernail clams. Likely poisoned by organochlorine contamination of the river's sediments, they had served as the ducks' major food source. The clams have never come back, either.

Dabbling ducks, such as wigeons ("pale grey head and bluish bill") and gadwalls ("rarely congregates . . . call, very low and reedy"), feed on the seeds of aquatic plants. Their departure from the Illinois corresponds to the arrival of herbicides. As agriculture became increasingly mechanized and chemically dependent, the flow of silt and weed killers from surrounding fields created waters barren of all such vegetation. Wild celery, coontail, and sago: these species, according to old accounts, once flourished in the quiet, shallow waters of Peoria Lake. They vanished completely in the 1950s, along with the birds that ate them. I don't know how to recognize these plants.

The story of the fish begins fifty years earlier. At the turn of the twentieth century, over two thousand commercial fishers worked the Illinois and supplied their harvests to markets as far away as Boston. Special fishing trains also carried sport fishers to and from Havana, the river town just downstream of Pekin. As measured by pounds of fish caught per mile of stream, the Illinois was considered the most productive inland river in North America.

This remarkable fecundity was a gift of geology. Much of the Illinois flows through a floodplain left behind by the ancient Mississippi. This flat pan of ground allowed the river to spread out a far-flung web of interconnected backwaters—the perfect nursery, spawning grounds, and winter refuge for fish. In the summer, periodic droughts firmed up the bottom, improving conditions for vegetation. The plants, in turn, deterred the wind from stirring up sediment during periods of spring and autumn flooding, when the river poured itself into the twisting sloughs, swales, potholes, marshes, and subsidiary lakes that surrounded it.

Then came the Chicago Sanitary & Ship Canal. This part of the story is a kernel of central Illinois lore. The S&S Canal opened January 17, 1900, and effectively connected Lake Michigan to the Illinois River, creating a continuous navigational route down to New Orleans. This is the meaning of the second *S* ("ship"). The first *S* ("sanitary") refers to the flushing of Chicago's wastewater into this canal and, from there, through the Des Plaines River and into the Illinois. Consequently, the level of the Illinois rose considerably. Backwaters flooded and stayed flooded. Bottomland groves of pin oak and pecan trees died. A wave of industrial pollution moved slowly and inexorably south (reaching Pekin around 1915), and downstream residents protested vociferously. Finally in 1939, the U.S. Supreme Court was moved to reduce by one-half the diversion of water into the Illinois. In the meantime, locks and dams began shaping the river into a series of stepped navigational channels. By World War II, the river resembled its present configuration. Straightened, leveed, drained, and dammed, the Illinois River became a sewage canal for industry and a barge canal for shipping—S&S. A report published the year I turned seven features a photograph of Illinois River fish with open sores and fins eroded down to stumps.

The federal Clean Water Act of 1972 brought a modicum of improvement to the Illinois River. As annual amounts of industrial waste released into the river declined, water quality improved. The long-term ecological effects, however, are less clear. Like a cloth already frayed, the river shows signs of continued damage even at lower levels of stress. Recovery has been uneven, at best. Mussels have returned to some parts of the river, and fin erosion is a less common problem in fish. On the other hand, the level of pesticide contamination remains high, and aquatic plants have [illegible].

In the Upper Illinois, fish advisories continue to caution sport anglers to severely limit—or eliminate—their consumption of fish known to contain high levels of cancer-causing chemicals. These warnings, most strict for children and women of reproductive age, are especially emphatic about the danger of eating large fish, in which the amplifying effects of biomagnification have had the longest time to operate. The bigger the fish, the more concentrated the poison.

This, then, is the river I know. Isolated from its floodplain by levees, factories, and farms, the Illinois River flows alone. Barge convoys, big as football fields, suck the river into their wakes and then send it crashing against the banks. The accompanying tugs churn the river like egg beaters, constantly resuspending toxic materials. These include vintage chemicals—PCBs, DDT, dieldrin, chlordane, heptachlor—as well as more contemporary pollutants contributed by industrial discharges, chemical spills, and farm runoff. The resulting waves slosh silt and poison into whatever fish-spawning backwaters still exist.

More than 350 different spills of hazardous substances into the waterway were reported between 1974 and 1989 alone. The continued absence of bottom-dwelling animals makes the aquatic biologist Doug Blodgett of the Illinois Natural History Survey in Havana suspect that such spills remain frequent. Killing as it goes, each spill creates a toxic pulse that moves through a given section of stream within hours. Once-a-month monitoring does little to detect most of these transient accidents. Such spills are, of course, in addition to routine industrial discharges.

The fastest way to get to the Illinois River from my parents' house is to follow Derby Street into Normandale. This was the route I used in high school—unbeknownst to my parents, who considered the riverfront dangerous.

Derby Street itself is an avenue of nostalgia and munitions. Store fronts with names like Karen's Kountry Kottage and Grandma's Feather Bed alternate down the block with various gun and ammo shops. One is notable both for the missiles on display in the parking lot and for the half of a Jeep (with GI mannequin positioned in the driver's seat) mounted trophy-style on the side of the building. Actually getting to the river from here is tricky. It requires a stroll through the subdivision, a climb over chain-link fencing, and a firm decision to ignore No Trespassing signs. Then, abruptly, there is the water—brown, familiar, blank.

Silence is comfortable here. The river embraces silence. The Illinois River seemed to me, as a teenager, not so much dangerous, or even endangered, as reassuring.

Standing here now, aware of all that is not here, I know myself as a natural historian of ghosts. Since 1908, twenty species of fish have disappeared from this river. One in every three native amphibian species have been completely, or almost completely, extirpated from the state. In their extinction, they join one in every five crayfish and more than half the species of mussels. The river reminds me of a poem by Robert Frost that asks, "What to make of a diminished thing?" and provides no answer.

Public drinking water is regulated nationally by the Safe Drinking Water Act, which became law in 1974. The act directs the EPA to set maximum contaminant levels that represent the highest limits allowable by law of particular toxic substances in public water supplies. In this way, all public drinking water is monitored on a regular, ongoing basis. Different contaminants have different legal limits. For example, there is one maximum contaminant level for the herbicide atrazine (3 parts per billion) and another for the dry-cleaning fluid perchloroethylene (5 parts per billion). The maximum contaminant level for PCBs is 0.5 parts per billion, while those for the banned pesticide chlordane and the PVC feedstock vinyl chloride each stand at 2 parts per billion. The legal limit for the phthalate plasticizer DEHP is 6 parts per billion. Since 1999, amendments to the law require water utilities to inform consumers, in their water bill and at least once a year, what pollutants have been detected in their drinking water and whether water quality standards have been violated. According to these reports, 10 percent of the nation's drinking water systems are out of compliance with EPA standards.

For at least two reasons, this figure may underrepresent the magnitude of the problem.

First, as with pesticide regulations, the legal limits for each chemical have been arrived at through a compromise between public safety and economics. Maximum contaminant levels are not solely a health-based standard. Instead, they take into consideration cost and the ability of available technology to reduce contaminants to particular levels. These then become the legal benchmark. For many chemicals, two numbers exist: the enforceable maximum and the health-based, maximum-contaminant-level *goal.* The enforceable values for the carcinogens benzene, vinyl chloride, and trichloroethylene, for instance, have been set at 5, 2, and 5 parts per billion, respectively. Their maximum-contaminant-level goals, however, are all zero.

Like an accountant who proficiently measures and records individual values but fails to sum the results, this system of regulating contaminants in water suffers from a constricted one-chemical-at-a-time vision. It ignores exposures to combinations of chemicals that may act in concert. The carcinogen arsenic, for example, occurs naturally in some groundwater and has a maximum contaminant level. However, if water that contains arsenic is also laced with traces of herbicides, dry-cleaning fluids, nitrates, or industrial solvents—even at concentrations well below their respective legal limits—the resulting mixture may well pose hazards not recognized by a laundry list of individual exposure limits. Exposure to one compound may decrease the body's ability to detoxify another, for example.

This system of regulation is also blind to the many common chemicals in drinking water that have no maximum contaminant levels. As of

2009, enforceable limits had been established for only 90 contaminants. All the rest are unregulated and therefore not routinely monitored. For example, of the 216 chemical pollutants identified as breast carcinogens in animals, at least 52 are found in drinking water, but only 12 of them are regulated under the Safe Drinking Water Act. At least 20 breast carcinogens known to be in drinking water are not regulated. Also federally unregulated are pharmaceuticals and ingredients found in personal care products, such as shampoo, makeup, insect repellants, and deodorants. Furthermore, when maximum contaminants do exist, they are not always set with cancer in mind. For example, the maximum contaminant level for nitrates—anhydrous ammonia's calling card—is 10,000 parts per billion (10 parts per *million*) for finished drinking water. But this standard was set to protect formula-fed babies from developing anemia in response to acute nitrate poisoning. It does not reflect new knowledge about the ways in which nitrates are converted in our digestive tract to carcinogenic N-nitrosamines (a transformation we discussed in Chapter Seven). Recent evidence suggests that nitrates, at levels below their legal limit in drinking water, may raise cancer risk.

To the question, then, of whether drinking water is regulated on the basis of contemporary scientific knowledge, the answer is no. Perhaps most revealing of all is the fact that regulation for some contaminants is based on the annual average of four quarterly measurements. In other words, drinking-water standards are violated only when the yearly average concentration of said contaminant exceeds its maximum contaminant limit. A one-time transgression does not automatically create a violation. This distinction is important in the Midwest, where herbicide concentrations in drinking water drawn from rivers and streams often reach hair-raising levels during the spring quarter, the months of planting and rain. In 1995, in the first study of its kind, researchers sampled water from faucets in kitchens, offices, and bathrooms every three days from mid-May through the end of June in communities throughout the Corn Belt. Herbicides turned up in the tap water in all but one of twenty-nine towns and cities. Atrazine exceeded its maximum contaminant level in five cities, including Danville, Illinois, where its concentration in water reached six times the legal limit. (Danville is southeast of Pekin, near the Indiana border. My Uncle Jack grew up there.) A 2009 study showed that not much has changed since then: atrazine in fifty-four public drinking water [illegible] concentrations above the maximum contaminant level.

Biologically speaking, we live only in the present. Our bodies do not respond to contaminants on the basis of averages; they must cope the best they can with the load of contaminants already received as well as with those streaming in at any given moment. If, during the period of April through June, a girl living in rural Illinois drinks enough weed killer to overwhelm her body's ability to detoxify it, and if, as animal evidence suggests, these chemicals are capable of altering development of her breast tissue, then the damage has been done, regardless of what happens during the months of August, October, or January.

Many researchers now believe that exposure to even minute amounts of carcinogens at certain points in early development can magnify later cancer risks greatly. What are the implications for the unborn child who happens to reach one of these key points at the same time a bevy of farm chemicals in the local water supply is reaching its peak? What are the implications for the adolescent girl whose breast buds start to form during this particular quarter of the calendar year?

Exposure to waterborne carcinogens is more commonplace than many people realize. In the same way that intake of airborne pollution involves the food we eat as well as the air we breathe, intake of contaminants carried by tap water involves breathing and skin absorption as well as drinking. These alternative routes are especially important for the class of synthetic contaminants called volatile organics—carbon-based compounds that vaporize more readily than water. The dry-cleaning solvent perchloroethylene is a common one. Most are suspected carcinogens.

We have already seen in Chapter Eight how volatile organic compounds combine with nitrogen oxides to create poisonous ground-level ozone, a major air contaminant. As a contaminant of tap water, they present additional dangers. Volatile organics are easily absorbed across human skin and enter our breathing space when they evaporate. The higher the water temperature, the greater the rate of evaporation. Humidifiers, dishwashers, and washing machines all transform volatile waterborne contaminants into airborne ones, as does cooking. These sources of exposure are thought to be particularly worrisome for infants, and for women home all day engaged in housework.

The simple, relaxing act of taking a bath turns out to be a significant route of exposure to volatile organics. In at least two studies, the exhaled breath of people who had recently showered or bathed contained elevated levels of volatile organic compounds, including chloroform. In fact, a ten-minute shower or a thirty-minute bath contributed a greater internal dose of these volatile compounds than drinking half a gallon of tap water. Showering in an enclosed stall appears to contribute the greatest dose, probably because of the inhalation of steam.

The particular route of exposure profoundly affects the biological course of a contaminant within the body. The water that we drink and use in cooking passes through the liver first and is metabolized before entering the bloodstream. A dose received from bathing is dispersed to many different organs before it reaches the liver. The relative hazards of each pathway depend on the biological activity of the contaminant and its metabolic breakdown product, as well as on the relative sensitivity of the various tissues exposed along the way.

The importance of showers as an exposure route to drinking water contaminants was the result of an accidental discovery. In 1984, investigators who were looking into the dumping practices of an electroplating company in Rockford, Illinois, uncovered a plume of chlorinated solvents in the groundwater. A municipal well and 150 private wells were contaminated. Levels varied but in some cases exceeded five hundred parts per billion. Southeast Rockford was thus catapulted onto the Superfund National Priorities List. (The electroplating company specialized in brass fittings for caskets.)

Five years later, researchers found elevated levels of these same chemicals in the air space of homes receiving water from the affected wells—and in the blood of their human occupants. Curiously, blood levels correlated more closely with household air levels than with actual water levels. Air levels, in turn, were roughly correlated with length of "shower run times." These results were based on a small study population and therefore have low statistical power. However, they support the notion that inhalation contributes more significantly to overall body burden of volatile organic compounds than drinking—even when water contamination is dramatic. Bottled water, by itself, is not the answer.

In 1990, epidemiologists looking at cancer maps of Illinois noticed high death rates from bladder cancer in the northwest portion of the state. Further investigation confirmed the excess incidence but found it was largely confined to one town within one ZIP code. That town was Rockford.

~~~

Some time ago, I came across a survey form used in 1918 by Illinois inspectors of wells and cisterns. Among its many questions, one required measuring the distance between the water source under inspection and all possible sources of pollution. Singled out for specific mention were feedlots, privies, stables, cesspools, and "dumping grounds for slops." The survey also inquired whether small animals could fall in at the top and, most important, whether any cases of typhoid fever had ever been attributed to use of this water.

The survey's approach to protecting drinking water seemed to me to show remarkable foresight. The thrust of its questions reflect an understanding about the relationship between the safety of drinking water and the kinds of activities that go on near the source of that water. "What care is taken in collecting and storing water?" "State general condition of health of those using water." "Is the drainage from all these places toward or away from the Well, Spring, or Cistern?" "If there is any other possible
~~~

source of pollution, state it." Apparently, somewhere in the transition from the age of bacterial contagion to the age of chemical carcinogens this type of consciousness was lost.

A lengthy report on groundwater quality in my hometown was released in [illegible]. It contains a detailed description of contamination in two of Pekin's seven drinking-water wells. Located near the river, both wellheads are close to various industrial sites, many underground storage tanks, and the local sewage treatment plant. The chemicals detected in the water—tetrachloroethylene and 1,1,1-trichloroethane—could have migrated in from any number of possible sources. Both are suspected carcinogens. The report's assessment team expresses specific concern about a site on Second Street once occupied by Valley Chemical and Solvents Corporation. Upon closing its doors in 1989, Valley Chemical left behind a shameful trail of soil and water contamination. Its property was, you might say, a dumping ground for slops.

The city of Pekin responded swiftly to this report. Committees were established and city ordinances proposed. A course on groundwater protection was even added to the public grade school curriculum. Before all this flurry of activity, however, the initial reaction was one of astonishment. "Nothing has been done through the years to protect that aquifer," the mayor of Pekin admitted in the newspaper. "Nobody really ever thought about it. We always had good water and nobody ever thought that would change."

~~~~~~

"Have any cases of cancer ever been attributed to use of this water?" is a more difficult query than one asking about typhoid fever cases. No one is more knowledgeable about the nature of this difficulty than Kenneth Cantor, an environmental epidemiologist and senior scientist at the National Cancer Institute. Cantor has studied the relationship between water pollution and human cancer for much of his career.
~~~~~~

In the introduction to a review of the subject, Cantor and his colleagues note that the various studies of cancer and chemical contamination of drinking water are "impossible to summarize in a straightforward way" because they typically involve exposures to poorly defined mixtures of chemicals over unknown periods of time. Widespread chemical contamination of drinking water may be an unintentional, ongoing human experiment, but it is one that runs without the benefit of controls or experimental design.

The most out-of-control, run-amok experiment in drinking water contamination ever set in motion may be Camp Lejeune. In the early 1980s at this U.S. Marine Corps Base Camp in North Carolina, two drinking water systems were discovered contaminated with two different industrial solvents: TCE (trichloroethylene) and PCE (perchloroethylene). Other A-list contaminants also started turning up—benzene, toluene, vinyl chloride—and at levels not trivial. In some cases they were hundreds of times over their maximum contaminant levels and approached the highest concentrations ever measured in a public drinking water well. The people so exposed were not just Marines, as the base was a temporary home to many young families, including mothers who drank the water while pregnant. Contamination appears to date back to the 1950s, meaning that a million people may have been exposed over a decades-long period before Camp Lejeune became a Superfund site in 1989. Those million people, including the children conceived and raised at Camp Lejeune, are now scattered across the country, making meaningless the epidemiologist's usual task of looking for a geographic cluster of cancers.

Federal studies were commissioned. And the results—which largely revealed nothing—have been fiercely contested by both scientists and former Marines alike. In April 2009, the Agency for Toxics Substances and Disease Registry, admitting omissions and scientific inaccuracies, took the unusual step of retracting its 1997 study that had concluded the contaminated water was not linked to health problems on the base. Meanwhile, the National Research Council's long-awaited 2009 report concluded that it could not conclude: too many years had gone by. There was no historical information about who used the water, for how often, and for how long. There were difficulties in modeling drinking water distribution throughout the system, as the contaminated wells had cycled on and off over the years. The evidence that the solvents in question were inherently carcinogenic was conflicting. And most eyebrow-raising of all, the NRC concluded that no other future study, no matter how it was designed, could ever get to the bottom of the matter. In effect, it declared all further investigation to be a waste of time. "It cannot be determined reliably whether diseases and disorders experienced by former residents and workers at Camp Lejeune are associated with their exposure to contaminants in the water supply because of data shortcomings and methodological limitations, and these limitations cannot be overcome with additional study."

Is this report an honest appraisal of the limits of science? Or is it an excuse to turn a blind eye? These questions are being asked both by other scientists—some of whom believe that further research is warranted and that not-yet-deployed methodologies could help unravel the mystery—and by the Marines themselves. One of them is a male survivor of breast cancer, which is an exceedingly rare disease among men. (In the United States, a man's lifetime risk of breast cancer is one-tenth of 1 percent.) As of July 2009, he had located seventeen other former Marines, all men, with breast cancer, all of whom had been stationed at Camp Lejeune. "It was almost a relief to find out my cancer actually came from somewhere," he said. "I'm not just some idiot who got breast cancer for no reason." Soon after, twenty-five more male Marine veterans with breast cancer and Camp Lejeune histories came forward. Further investigations are planned.

Other studies of cancer and contaminated drinking water have dispelled the notion that cancer is a disease of no reason. None of them are perfect. Most of them are ecological in design, that is, they simply describe patterns of association between health problems and environmental problems. All together, however, they tell a consistent story. And all together, these studies may have the power to make cancer patients living in communities with contaminated water see themselves as "not just some idiot." They certainly have had that effect on me.

As we saw in Chapter Four, bladder cancer mortality was elevated among men (but not women) living near Pennsylvania's infamous Drake Superfund site, an old chemical dumping ground full of known bladder carcinogens. On Cape Cod, high rates of bladder cancer and leukemia were associated with living in homes serviced by vinyl-lined water pipes that were leaching perchloroethylene. Recall also from Chapter Four the nationwide study in which cancer mortality was found to be elevated in U.S. counties where drinking water was contaminated by leaking hazardous waste sites.

Similar studies have been conducted in other settings, both urban and rural. In New Jersey, researchers found associations between volatile organic compounds in municipal water and leukemia among women (but not men). In Iowa, lymphoma rates were elevated in counties where drinking water was drawn from dieldrin-contaminated rivers. In Massachusetts, childhood leukemias in the industrial town of Woburn were linked to a pair of water wells contaminated with chlorinated solvents. In North Carolina, a cancer cluster in the rural community of Bynum was linked to consumption of river water contaminated upstream with both agricultural and industrial chemicals. This study is particularly compelling because, once the normal latency period for cancer is factored in, the sudden increase in cancer deaths that emerged in the 1980s corresponds closely with the time of peak exposure to known carcinogens in the river (1947 to 1976). Likewise in Woburn, the surge in childhood leukemias coincides with a period of known water contamination and abates a few years after the imputed wells closed down. (Public outcry about the plight of Woburn's children played a direct role in the creation of the Massachusetts Cancer Registry. It also inspired the book and film *A Civil Action*.)

Corroborating evidence also comes from abroad. In a study from China, liver cancer was strongly associated with drinking water from ditches containing agricultural chemicals. In Germany, excess cases of childhood leukemia in villages near uranium mines have been tentatively linked to radium-contaminated drinking water. And in Finland, high rates of non-Hodgkin lymphoma were discovered in a rural community where water was contaminated by chlorophenols, probably from local sawmills. Used for treating lumber, chlorophenols are related chemically to the phenoxy herbicides, which are also linked to non-Hodgkin lymphoma.

In a practice that began early in the twentieth century, the city of Chicago began in 1908 to pour chlorine into wastewater before sending it downstream. In the same year, the waterworks of Boonton, New Jersey, became the first to add chlorine to water intended for drinking. Chlorination proved a cheap, effective means of halting waterborne epidemics during World War I. By 1940, about 30 percent of community drinking water in the United States was chlorinated, and at present, about seven of every ten Americans drink chlorinated water.

Over the past three decades, dozens of studies have emerged that link chlorination of drinking water to cancer. "Mounting evidence suggests an association of bladder cancer with long-term exposure to disinfection byproducts in drinking water," asserts Kenneth Cantor. While the data are strongest for bladder cancer, suggestive evidence also exists for cancers of the rectum, colon, and esophagus.

Upon hearing this news, many otherwise even-tempered individuals may feel tempted to throw up their arms in frustrated despair as though they had just been asked to choose between death by cancer and death by cholera. Happily, this is not our predicament. Far less gloomy options are available—as we will see below. They will not be realized, however, unless we recognize the hazards created by the approach presently used to combat disease pathogens in our drinking water and, with this knowledge, insist on safer practices.

Chlorine gas is a noxious poison. However, the problem with chlorinated drinking water does not lie with chlorine itself. Rather, the problem begins when elemental chlorine spontaneously reacts with organic (carbon-based) contaminants already present in water. Their organo-chlorine offspring are known as disinfection by-products. More than six hundred different water disinfection by-products have been discovered. But only a sliver of these have been tested for carcinogenicity. One disinfection by-product for which there is troubling evidence is MX. MX is a complicated organochlorine whose full chemical name—3-chloro-4-(dichloromethyl)-5-hydroxy-2(5H)-furanone—is so unwieldy that it is referred to only by the science fiction-y designation Mutagen X, or, more typically, by the hip-sounding abbreviation MX. Unique to drinking water, MX was flagged as a mammary gland carcinogen in rodent bioassays. As its name suggests, it is characteristically capable of damaging chromosomes in ways that trigger genetic mutations. MX is not regulated in drinking water and not monitored routinely. Thus, asking if MX might be playing a role in the story of breast cancer among U.S. women is an unanswerable question. We have no exposure data. A 2002 nationwide survey of chlorination by-products in drinking water samples did reveal that MX was found in levels higher than previously reported, with the highest levels collected from utilities using chlorine dioxide or chlorine for primary disinfection.

In contrast to MX, trihalomethanes and haloacetic acids, two subgroups of disinfection by-products, are regulated and monitored. Chloroform is the most common and well-known member of the first group. As with any waterborne volatile compound, our route of exposure to trihalomethanes and haloacetic acids is threefold: ingestion, inhalation, and skin absorption. (Indeed, trihalomethanes appear as one of the major chemical culprits in the bathing studies already discussed.) All trihalomethanes and haloacetic acids are regulated as two groups, regardless of the individual components of the mixture. The maximum contaminant level of total trihalomethanes is 80 parts per billion. The maximum contaminant level of haloacetic acids is 60 parts per million. In the EPA's table of drinking-water standards, along the rows labeled "Total Trihalomethanes" and "Total Haloacetic Acids" and under the column titled "Potential Health Effects" are the words *risk of cancer.*

Many studies all telescope into these three words. One of the most ambitious of these investigations was led by Kenneth Cantor himself more than twenty years ago. His research team personally interviewed nine thousand people living in ten different areas of the United States. Individual histories were then combined with water utility data to create a lifetime profile of drinking-water use for each respondent. In the final analysis,

> bladder cancer risk increased with the amount of tap water consumed, and this increase was strongly influenced by the duration of living at residences served by chlorinated surface water. . . . There was no increase of risk with tap water consumption among persons who had lived at places served by nonchlorinated ground water for most of their lives.

These results have been subsequently corroborated.

Giving people cancer in order to ensure them a water supply safe from disease-causing microbes is not necessary. Part of the solution lies in making wider use of alternative disinfection strategies. These include granular activated charcoal (which binds with contaminants and removes them) and ozonation (which bubbles ozone gas through raw water to kill microorganisms). Both techniques have been used successfully in many U.S. and European communities. And compared to chlorine, ozone is a more effective assassin of the diarrhea-inducing, water-borne parasite *Cryptosporidium*.

Part of the solution is to direct research dollars toward identifying and characterizing the full suite of water disinfection by-products. Choosing the most enlightened method of purifying water depends on knowing what hazardous chemicals are being created during the process. Half of the more than six hundred disinfection by-products so far ascertained were discovered by a single team of researchers overseen by chemist Susan Richardson at the EPA's National Exposure Research Laboratory. She and her coworkers are now hard at work looking for others; she believes there [illegible] in efforts not yet identified. More crucially, Richardson is attempting to reveal which water treatment scenarios are linked with which toxic chemicals and, in turn, which toxic chemicals are linked to cancer. Along the way, she has discovered that ozonation creates disinfection by-products of its own. Will they prove as hazardous as chlorination by-products? There are many pieces to this particular public health puzzle, and Richardson's work is on the forefront of the effort to find and assemble them.

Part of the solution lies in directing a spirit of urgency, inventiveness, and ingenuity toward the development of altogether new approaches to the sanitizing of water. No doubt many technologies await discovery, requiring only the devotion of resources and a collaboration of creative minds to bring them into existence.

Finally, part of the answer lies in keeping carbon-based contaminants out of drinking water in the first place. This last dictum is doubly important. Less organic content means fewer trihalomethanes. Less organic content also knocks down the number of microorganisms, thereby reducing the amount of chlorine needed for disinfection. Tellingly, water from lakes, rivers, and reservoirs generates more trihalomethanes upon chlorination than does water drawn from aquifers. This is because, in general, surface water carries more organic matter than groundwater. Some of the progenitors of trihalomethanes are natural and unavoidable: decaying leaves, fallen feathers, and grains of pollen, for example. These all contribute to the total carbon load in a body of water. But many others are neither natural nor unavoidable: sewage, chemical spills, industrial discharge, soot and other fallout from air pollution, agricultural runoff, and motor oil, for instance. Drastically reducing these inputs would go a long way toward solving the problem of disinfection by-products—as well as other grosser forms of water contamination.

This part of the solution requires that water utilities and the water-consuming public become vigilant about the protection of watersheds and aquifers. Kenneth Cantor puts it this way: "Disinfection is the final barrier against transmission of waterborne pathogens, but it must not be the sole barrier. Source protection is required."

Source protection means more than keeping swimmers out of reservoirs and erecting fences around wellheads. In some regions, this kind of protection will require new thinking about agriculture, which needs to substitute the techniques of organic farming for practices that pour soil and pesticides into river systems. In regions where cattle feedlots and hog farms periodically send lava flows of manure into watersheds, it will require new thinking about animal husbandry. In other areas, it will require new thinking about industry. Manufacturers must find safer alternatives to organic solvents and other synthetic, carbon-based chemicals that are released into water directly, fall off barges in transit, waft into the air only to rain down elsewhere, or eventually worm their way into water via landfills and dump sites. Finally, in all regions, protection of water supplies will require new thinking on the part of individual citizens, who are asked to assume the frightening cancer risks that others have decided, on their behalf, are acceptable.

Back at the waterworks, additional improvements are possible. For example, making chlorination the last step of water treatment, rather than the initial one, lowers the amount of trihalomethanes generated, especially if the water is carefully filtered through granulated charcoal first. Artificial membranes can remove a slew of contaminants, including pesticides and solvents. Water can also be aerated to allow volatile organic compounds, including trihalomethanes, to vaporize. Because they transfer contaminants from drinking water to other environmental media, I consider these kinds of solutions less beneficial than a comprehensive strategy of primary prevention. Aeration sends waterborne organic compounds into the atmosphere where we can inhale them, and filters and membranes fill with toxic chemicals, which must go somewhere. Even while providing some immediate respite from exposure through tap water, these technological shell games keep carcinogens in circulation.

In 1910, a New Jersey court examiner declared that chlorination left "no deleterious substances in the water." He was wrong. Nevertheless, it is clear that the disinfection of drinking water with chlorine has prevented widespread contagion and death, even as it has also contributed to the burden of human cancer. I do not advocate a ban on the chlorination of drinking water. But neither do I believe we should blindly continue old disinfection practices as though our bodies and our water supplies still existed in the world of ninety years ago. In 1910, chloroform was not considered a deleterious substance. When its toxicity was later recognized, its use as a surgical anesthetic was phased out. We need not be forced to drink it now as the price for contagion-free water.

~~~

At dead center, in the channel used by barges, the Illinois River is about as deep as the deep end of a swimming pool. If you dove to the bottom here, you would first pass through a flocculent layer of silt many feet deep. Underneath this fluffy mass is the clay trough of the riverbed. If you could somehow continue the descent, drilling down through this foundation, you would eventually find yourself once again in the water—the water under the water—which is held between the glittering sand grains of the Sankoty Aquifer.

This underground basin not only lies beneath the river but stretches out for miles along its east flank and extends south toward Havana. It occupies what was once the valleys and snaky tributaries of the ancient Mississippi River—before they were bulldozed by glaciers. The Sankoty Aquifer is the source of Pekin's drinking water.

Technically speaking, an aquifer refers not to the groundwater it holds but to the collection of grit, gravel, clay, and rock the water flows through. The Sankoty ranges from 50 to 150 feet thick and consists mostly of quartz sand grains ranging in size from dust to marbles. They are said to be distinctly pink. Sankoty sand grains are also, I'm told, neatly sorted and stratified by size, indicating they were once carried along and then deposited by the flow of a melting glacier. The resulting outwash is referred to as a valley train. Aquifers can include any porous and permeable material, such as unsorted scrap gouged up and laid back down by the glacial ice itself—this would be called till—or layers of wind-blown silt, which geologists have named loess.
~~~

Regardless of materials, aquifers come in a few basic varieties. Bedrock aquifers are covered by a lid of impermeable materials. Artesian aquifers, often lying on a slant, are under hydrostatic pressure. Water-table aquifers are like uncovered pots with rain and melting snow periodically dribbling in through the overlying soil. Their surface, the water table proper, rises and falls with seasonal fluctuations in precipitation. The Sankoty is a very large water-table aquifer.

The chemical contamination of the Sankoty Aquifer is an ongoing story with no identifiable beginning, no defining catastrophic event, and nothing that could reasonably be called a resolution. Even before the 1993 assessment roused Pekin's residents into action, there were signs of trouble. As part of a 1989 survey, the Illinois Environmental Protection Agency discovered "a substantial level" of 1,1,1-trichloroethane in one of Pekin's drinking-water wells and low levels of benzene and tetrachloroethylene in another. A year later, the Illinois EPA issued two groundwater contamination advisories after accidents at loading docks sent into the public water wells of Creve Coeur a variety of gasoline additives and crude oil derivatives. Then, in the spring of 1991, high river levels contributed to the flow of chemical contaminants into a community drinking well in north Pekin.

This last discovery was particularly unsettling. The interchange between groundwater and surface water is normally a one-way affair with aquifers, recharged by rain, emptying themselves into the rivers and streams that lie across them. Barring a flood, flow of river water into groundwater is not supposed to happen. Abrupt increases in river volume or heavy pumping from wells can, however alter the direction of these unseen currents and possibly divert surface water into the underground world of aquifers. Water levels inside certain Sankoty wells, for example, appear to fluctuate in tandem with lock-and-dam operations on the Illinois River, implying a more reciprocal communion between these two bodies of water than was once presumed.

In light of this and other ominous realizations, the results of the Illinois EPA groundwater assessment actually appear quite mild. Industrial chemicals turned up in a few discrete locations, but the field team found no signs of aquiferwide contamination. Indeed, the authors of the study expressed surprise at not stumbling on an even bigger problem, especially after they reviewed the history of industrial practices within the local area: "Any contaminant that could have been produced, almost certainly was produced. And yet, we do not find widespread contamination of the groundwater environment."

It is an eerie paradox, and it is difficult to know how alarmed or reassured to feel. Certainly there is danger in breathing sighs of relief too soon. Just as the presence of a single cockroach in the kitchen sink speaks of the hundreds more behind the wall, periodic detections of contaminants in groundwater aquifers are often harbingers of widespread contamination yet to come. Groundwater flows both leisurely—sometimes only inches per year—and smoothly as it moves along the pores and cracks of the underground landscape. Without speed or turbulence, dispersion of chemical contaminants is also slow. Over time, an intermittent detection in one well can eventually become a constant detection in several. Moreover, even the merest trace of contamination can portend a serious problem if what is being detected is the bottom edge of a falling curtain of chemicals slowly moving its way down the aquifer's overlying substrate.

As a general rule, contamination in lowland areas of discharge—where aquifers give up their water to rivers and streams—is considered a lesser problem than contamination in the upland areas of recharge where aquifers receive rain and snow from the atmosphere. Areas of recharge are the headwaters of aquifers, and contamination here can fan out and fill the whole. In either case, once groundwater becomes contaminated, little can be done to remedy the problem. In contrast to the alfresco run of surface water, groundwater has no oxygen to hasten the breakdown of chemical contaminants nor open air to facilitate the evaporation of solvents and other volatile organics. Contamination lingers in the still, watery vaults of aquifers.

~~~~~~
~~~~~~

With great pomp and flair, the city of Pekin passed its proposed groundwater protection ordinance in 1995. It has since been hailed as a model for the state. Essentially, the statute regulates land use in three recharge areas, each a narrow stretch of ground a mile or so long where the groundwater underneath flows into the city's wells. Additionally, the ordinance draws a 2,000 foot ring of protection around each of the wellheads. Inside these seven circles, the city restricts, and in some cases prohibits, the siting of new businesses that handle large quantities of hazardous materials. Existing businesses are largely grandfathered, although some have promised to make improvements.

Ever since the ordinance was drafted, personal knowledge about recharge, discharge, depth to water table, glacial deposits, and other details of hydrogeology has become a matter of civic pride in Pekin. Groundwater maps overlaid with a grid of the city's streets have appeared on the front page of the newspaper so that residents can locate themselves vis-à-vis the aquifer. The west end of Derby Street lies within a recharge area, for example, as do sections of Sabella, Charlotte, and Henrietta Streets. The east bluff does not. The owner of a gas station on Fourth Street, upon discovering himself inside one of the protected zones, pledged to install double-walled tanks to prevent leaks. He even praised the ordinance after attending a public hearing. "It's great, it should have been done years ago, and nobody paid attention to it."

Pekin's ordinance is a candle in the dark. It has sparked open discussion about the relationship between health and the environment, and it has lit in people's hearts new respect for the body of water they walk over and drink from. Its ability to safeguard the Sankoty Aquifer against the toxic activities that continue to go on ninety feet above it, however, is not at all clear. Toxic runoff from storm sewers empties into several creeks and at least one lake that overlay the recharge zone. And as the district superintendent of the water company points out, the rain itself contains pollutants. No local zoning laws can legislate against pesticide-laced raindrops or solvent-contaminated snowflakes falling in a recharge zone. Solutions to these problems need to be hammered out in chambers larger than small-town city councils.

In the meantime, industrial chemicals and pesticides persist in mak- ing cameo appearances in [illegible] sometimes [illegible] They include benzene, perchloroethylene, 1,1,1-trichloroethane, the phthalate [illegible] and a couple of [illegible] This brings us to the present moment. We know the story of the Sankoty Aquifer begins with a glacier. We know that someone, sooner or later, ends up drinking whatever poisons are spread on the earth above it. What happens next is the part of the story that is still unwritten.

About 40 percent of Americans draw their water from aquifers. The rest drink from rivers, lakes, and streams. Of course, ecologically speaking, everyone drinks from aquifers: all running surface water was at one time groundwater, aquifers being the mothers of rivers. As Rachel Carson pointed out, contamination of groundwater is, therefore, contamination of water everywhere. And because the human body is 65 percent water by weight, contamination of water everywhere is, therefore, contamination of people everywhere.

In January 2008, the Illinois Groundwater Protection Program released its comprehensive report on the status of the state's aquifers, based on analyses of groundwater data collected over eighteen years. The authors found "a statistically significant increasing trend of community water supply wells with volatile organic compound detections per year. More importantly, this data show an increasing trend of groundwater degradation."

In April 2009, the U.S. Geological Survey released its comprehensive report on the status of the nation's domestic drinking-water wells. It finds synthetic organic chemicals—nitrates, herbicides, insecticides, solvents, disinfection by-products, ingredients of gasoline, refrigerants, and fumigants—in more than half of the 2,100 wells it sampled. Concentrations of most contaminants were low but usually occurred in mixtures. The authors note that the toxicity of mixtures can be greater than that of any single contaminant, but "available human-health benchmarks do not allow full assessment of the potential health effects of contaminant mixtures because benchmarks have been established for only a few specific mixtures and are not yet available for all contaminants."

Groundwater provides no archival photographs to consult. It offers no shores to walk along, no reflective surfaces to peer into, no fish, bivalves, otters, or game birds to inquire about. Our relationship to aquifers is deeply biological, but it is not visual.

I once descended to the bottom of a well in order to look at groundwater in its natural habitat. This happened in Hawaii, where drinking water is drawn from a flattened lens of rain that is trapped between volcanic rock and ocean water. (Freshwater floats on salt.) At the Halawa pumping station, I rode a cable car down a 300-foot shaft to arrive in a blasted-out cavern filled with water. It was very dark and very quiet.

Illinois, I believe, would provide a more exotic underground landscape. In the descent, one could examine the mashed remains of preglacial forests, bluffs, dunes, islands, and cliffs. The bedrock floor would be inlaid with the trenches of ancestral riverbeds. The water table's undulating ceiling would offer a subdued reflection of the overlying topography.

Cultivating an ability to imagine these vast basins beneath us is an imperative need. What is required is a kind of mental divining rod that would connect this subterranean world to the images we see every day: a kettle boiling on the stove, a sprinkler bowing over the garden, a bathtub filling up. Our drinking water should not contain the fear of cancer. The presence of carcinogens in groundwater, no matter how faint, means we have paid too high a price for accepting the unimaginative way things are.

ten

fire

"this world in peace
this laced temple of darkening colors
it could not have been made for shambles"

—John Knoepfle, "Confluence"

"It's different country down here," my mother says, and I agree.

We're following the river valley south into Mason County on a winding unmarked road everyone around here calls the Manito Blacktop. To catch it, you take Route 29 past Normandale, continue on by the distillery, and turn right at the federal prison. From here the blacktop appears to be an access road for the power plant, but once you drive through the silver forest of transformers and towers, with the Illinois River on your right, you know you're on a real road heading out of town.

The houses along the blacktop have a scattered, haphazard feel to them—as though the river gave permission to strew one's possessions about. Propane tanks, extra cars, and satellite dishes are parked in sloping yards alongside signs advertising the sale of garden produce. The soil here is very porous, a fact made evident by the irrigating center pivots that lie over the fields like the skeletons of enormous bats. These draw water up from the Sankoty Aquifer and spray it in all directions. Where the soil is too sandy for standard-issue field corn, the fields sport green beans, peas, sweet corn, popcorn, cucumbers, pumpkins, or melons. Indeed, this is one of the few pockets of Illinois that still grows food for people to eat (now referred to as "specialty crops"). The little village of Manito announces itself as the Imperial Valley of the Midwest, and then the road banks and curves away to the right. [illegible] on the right-angled Illinois prairie east of here where farmhouses are spaced across the landscape like battleships assigned to their own black-dirt square of sea and farmers depend only on rain to get them through.

I'm assuming, anyway, these difference are the ones my mother is referring to. We continue south and west toward Havana. It is September 1994. A convergence of various forces—political, historical, and personal—form the purpose for this trip. They all have to do with garbage incinerators.

The political reasons started with an obscure law that was passed at the end of the 1980s. Called the Illinois Retail Rate Law, this legislation required electrical power companies to purchase—at the retail rate charged to customers—any and all electricity generated by trash-burning incinerators. The utilities then recouped their loss of profit through tax credits. Overnight, garbage incinerators became the most heavily subsidized development project in Illinois history.

In their aim to attract incinerator builders and investors to the state, the lawmakers had succeeded splendidly. Previously, Illinois was home to one operating garbage incinerator (in Chicago), but as of 1994, twelve others were in various stages of study, siting, or construction. Six were under consideration in central Illinois, one of them in Havana, where siting had been approved by the city council the previous December. Specifically, the plan was to build a 15-acre incinerator—handling eighteen hundred tons of garbage per day—in a popcorn field adjacent to the railroad's coal docks. Trash would be hauled in by boxcar from Chicago. In order to turn the steam-driven, electricity-generating turbines, three million gallons of water a day would be drawn up from the aquifer, requiring a doubling of Havana's pumping capacity.

The historical circumstances surrounding the summer and fall of 1994 ran counter to the immediate political ones. On a national level, successful recycling efforts had taken the pressure off of landfills, and recyclers were now competing with incinerators for trash. Incinerators themselves were proving costly. Columbus, Ohio, was preparing to shut down its energy-generating incinerator in the wake of intractable economic and environmental problems. Albany, New York, had become plenty vexed at its own garbage burner the previous January. Already facing expensive upgrades in order to meet air pollution standards, the incinerator had covered a fresh snowfall with a black layer of soot after a series of pollution control mishaps. It was quickly mothballed.

Meanwhile, emerging environmental research indicated that trash incinerators routinely released troubling amounts of toxic and carcinogenic pollutants, including dioxin. In addition, several studies published in 1994 had demonstrated that dioxin is harmful at far lower exposures than anyone ever suspected. Even at a few parts per trillion, dioxin was capable, it seemed, of profoundly altering biological processes. Also in the fall of 1994, the EPA released a three-thousand-page draft reassessment of dioxin and was now soliciting public commentary and reaction. Three years in the making, the study reaffirmed dioxin's classification as a probable human carcinogen.

The draft report also announced three other findings. First, dioxin's effects on the immune system, reproduction, and infant development are much more significant than previously thought. Second, there is no safe dose below which dioxin causes no biological effect. Third, quantities of dioxin and dioxinlike chemicals present in most people's bodies are already at or near levels shown to cause problems in animals. Finally, the report identified incineration—of both medical waste and common household garbage—as the leading source of dioxin emissions in the United States and food (meat, dairy, and fish) as the immediate source of 95 percent of the dioxin found in the bodies of the general population.

In the fifteen years since the release of this draft report, the case against dioxin has strengthened. A known endocrine disruptor, dioxin now appears on the Stockholm Convention's list of chemicals slated for worldwide abolition. The International Agency for Research on Cancer upgraded dioxin from a probable to a known human carcinogen in 1997.

The U.S. National Toxicology Program followed suit in 2001. The most potent carcinogen ever known, dioxin is the only substance in the Toxics Release Inventory whose annual emissions are tallied in grams rather than in pounds. Nevertheless, as of July 2009, the EPA's draft report is still undergoing revisions and has not yet been published in final form. Its current incarnation is a 2003 draft that is awaiting rewrites suggested by the National Research Council in 2006.

In September 1994, I had not yet seen the first draft, whose release had been rumored for months. Dorothy Anderson, a pediatrician and the president of the Mason County Board of Public Health, had a copy. It is to her house my mother and I are headed.

Our personal interest in the incinerator issue is simple. A plan for a trash incinerator nearly identical to this one—introduced by the same developer and backed by the same set of corporate investors—is under consideration by the village board of Forrest, eighty miles northeast of Pekin, in Livingston County. This facility is to be built not in the village itself but out in Pleasant Ridge township, three miles north of town. My mother knows precisely the section of cornfield they have in mind. It's exactly one mile south and three-quarters of a mile east of her brother's farm. She also knows that Roy has thrown his hat in with a group of farmers organizing to oppose the siting.

~~~~~~

No matter how you look at it, scooping garbage into an oven and setting it afire is an equally primitive alternative to digging a hole in the ground and burying it. The former contaminates air; the latter, groundwater.

The relative popularity of these two options has waxed and waned over the decades. In 1960, about one-third of the nation's trash was burned in incinerators. Because of serious air pollution, these were later phased out in favor of landfills. In the 1980s, incinerators, now sporting high-tech pollution-control devices and designed to generate electricity, staged a comeback, their promoters referring to them as "waste-to-
~~~~~~

energy" or "resource recovery" plants. Considered state-of-the-art in its day, the incinerator that was being planned for Pleasant Ridge in 1994 was of this type and was portrayed to the citizenry as an environmentally enlightened solution to the trash crisis. Its air pollution control devices were said to be superior in design to those of the primitive incinerators that came before, its projected dioxin emissions well within the "acceptable" range. Nevertheless, it would be considered a dinosaur today; others like it, cursed by all-too-frequent accidents and pollution-control problems, have already accepted early retirements. Indeed, the entire fleet of 1980s and 1990s incinerators are now widely considered passé, and their reputation is badly damaged. A $55-million waste-to-energy facility in Rutland, Vermont, closed after only nine months when its state air quality permit was revoked. An incinerator of a similar vintage in Detroit has run for twenty years—but cost taxpayers over $1 billion and, at the same time, prevented investments in recycling. Its future is deeply uncertain in light of mounting financial losses and community opposition.

In spite of this high failure rate, trash incineration is now being promoted once again, this time repackaged as a source of "renewable" energy. These new incinerators have state-of-the-art engineering, involving gasification, pyrolysis, and plasma incineration technologies. Undoubtedly, their dioxin emissions to air would be considerably less than those of the antiquated incinerators of yore (including the 1994 one that could, right now, be seeding dioxin into the air over the fields my great-grandfather first plowed). But technology that relies on burning plastics and other synthetic materials cannot eliminate the creation of toxic pollutants—although it may improve on ways of capturing and concentrating them in the ashes, char, and slag that are left behind and require disposal. Increased dioxin capture just means more dioxin going into the ground and less into air. Indeed, the recent trends in reported dioxin emissions from the Toxics Release Inventory show just that: a waning of dioxin releases into air offset by an increase of dioxin releases onto land.

In short, no matter how improved or what they are called, incinerators present two problems that landfills do not. First, incinerators only transform garbage; they don't provide a final resting place for it. There remains the question of where to put the ashes. Second, these cavernous furnaces create, out of the ordinary garbage they are stoked with, new species of toxic chemicals. In addition to producing electricity, they generate hazardous waste.

The first problem with incinerators flows from a primary law of physics. Most of us at one point or another in our education probably had it memorized: matter can neither be created nor destroyed. Every single atom fed into an incinerator survives. If 1,800 tons of garbage per day go in, 1,800 tons per day also come out, albeit in a chemically altered form. Some of this matter rises as gas or tiny particles and is released into the air as stack emissions. (Much of the gas is carbon dioxide.) The rest of it is captured as ash, which requires disposal.

In 1993, John Kirby, the developer responsible for the downstate Illinois incinerator proposals, showed newspaper reporters a jar of ashes weighing 3.7 pounds. This is all that remains, he boasted, of an average person's weekly 40 pounds of garbage after it has been run through a waste-to-energy facility. Landfilling 3.7 pounds, Kirby pointed out, creates less of a volume problem than burying all 40. In this he is certainly correct, but by extension, the containment of 3.7 pounds means 36.3 pounds of garbage are sent up into the sky. Incinerator advocates cannot have it both ways: fewer ashes means less required landfill space but more air emissions; more ashes means less air pollution but creates a bigger disposal problem. The indestructibly of matter reigns supreme.

Moreover, the process of burning concentrates into the ash whatever hazardous materials are present in the original refuse. Heavy metals, such as mercury, lead, and cadmium, for example, are not destroyed by fire. Occurring as ingredients in household batteries, light bulbs, paints, dyes, and thermometers, they are absolutely persistent. Air pollution control depends on the ability of an incinerator's cooling chambers to condense these metals onto fine particles, which are then caught in special filters.

Once again, the irony of trade-offs becomes readily apparent: the less air pollution, the more toxic the ash. An incinerator burning eighteen boxcars of trash per day, for example, produces about ten truckloads of ashes per day. The trucks must then rumble out onto the highways, hauling their poisonous cargo through all kinds of weather. Once ensconced in special burying grounds, incinerator ash, of course, presents a hazard to groundwater.

The second problem is more an issue of chemistry than of physics. Somewhere between the furnace and the top of the stack, on the papery surfaces of fly ash particles, in the crucible of heating and cooling, carbon and chlorine atoms rearrange themselves to create molecules of dioxins and their closely related organochlorine allies, the furans.

There are many dozens of dioxins and furans, but, as with snowflakes, their individual chemical configurations are all variations on a theme. Recall that benzene consists of a hexagonal ring of carbon atoms. This ring can then be studded with chlorine atoms. Two chlorinated benzene rings bonded directly together form a polychlorinated biphenyl, a PCB. By contrast, two chlorinated benzene rings held together by a single atom of oxygen and a double carbon bond are called a furan. A pair of chlorinated benzene rings linked by two oxygen atoms form a dioxin. There are 135 furans and 75 dioxins, each with a different number and arrangement of attached chlorines.

Dioxins and furans behave similarly in the human body, and they all to some degree elicit the set of biological effects described earlier. The most poisonous by far, however, is the dioxin known as TCDD. This particular molecule bears four chlorine atoms, each bonded to an outer corner. Because these points of attachment are located on the carbon atoms numbered 2, 3, 7, and 8, its full name is a mouthful: 2,3,7,8-tetrachlorodibenzo-p-diozin. Imagine looking down from an airplane window at a pair of skydivers in free fall, both hands joined together. Their geometry provides a reasonable impersonation of a TCDD molecule: the divers' linked arms represent the double oxygen bridge, their bodies the benzene rings, and their splayed, outstretched legs the four chlorine atoms.

TCDD is scary because it is so stable. The symmetrical arrangement of its chlorine legs prevents enzymes—ours or any other living creature's—from breaking TCDD apart. In human tissues, TCDD has a half-life of at least seven years. As we shall see, this particular geometry also allows TCDD admission into a cell's nucleus and access to its DNA.

Incineration is not the only source of dioxins and furans. They can also form spontaneously during the manufacture of certain pesticides—especially phenoxy herbicides and chlorophenols—and during the bleaching of paper products, for example. What all three of these processes have in common is chlorine. Dioxin is synthesized when certain types of organic matter are placed together with chlorine in a reactive environment. Such conditions are created by combinations as banal as newspapers plus plastic wrap plus fire.

In the inferno of an incinerator, many common synthetic products may serve as chlorine donors for the spontaneous generation of dioxins and furans: paint thinners, pesticides, household cleaners. Because it is half chlorine by weight, PVC plastic (polyvinyl chloride) is a major source of chlorine and, during incineration, serves as a feedstock for the unintentional manufacture of dioxin. PVC in the waste stream can take many forms: construction debris, the interiors of junked cars, old shower curtains, worn-out shoes, broken toys, cracked garden hoses, leaky air mattresses. Or the cut-up, discarded credit cards that had once been used to purchase all of the above.

~~~~~~

Dorothy feeds us slabs of freshly baked bread, garden tomatoes, and great hunks of cheese. She and my mother have already figured out they belong to the same church, a discovery that creates a spirit of common purpose between them.

Now the three of us are sitting at her kitchen table discussing the ambitions of John Kirby—the incinerator developer, state lobbyist, and entrepreneur—whose plans affect us all so deeply. Dorothy's perspective is straightforward. She is a practicing Methodist, a practicing physician, and the head of the board of public health. Opposing the Havana incinerator is both an affirmation of her spiritual beliefs and an act of preventive medicine.
~~~~~~

"If nothing else, I have a charge to keep."

Places like Havana, she asserts, are especially vulnerable to the designs of incinerator developers, who entice rural communities with promises of jobs and lucrative "host fees." These often exceed a small town's entire annual budget.

Dorothy ticks off the statistics. Mason County has 15 percent unemployment, lots of teenage pregnancy, and high rates of infant mortality. It is one of the state's poorest counties. One in four children live in poverty.

"So Kirby offered Havana a million dollars," she shrugs, as if to say, end of story.

We are quiet for a while.

Dorothy brings out apples and paring knives. My mother can core and quarter an apple faster and more perfectly than anyone else I know. Doctor Dorothy, blonde mother of four, is pretty swift at apple dissection herself, I notice. Somehow I am botching mine. I silently convince myself that I suffer from such problems only when in the presence of my mother. A chunk with a stem attached shoots across the table. Mom takes what's left in my hand and makes quick work of it. Then she begins to speak.

Forrest, she says, is in similar straits. Included in the package of temptations presented to residents out there is the promise of a new school library. A school referendum to do the same had just failed. As in Havana, Kirby's proposal had torn the tight-knit community asunder. Teachers found themselves on the opposite side of the issue from parents, farmers from grain elevator operators, ministers from their parishioners, village board members from village board members. The farmer who had given the incinerator developers the option on his land was now sunk in remorse. His brother had joined the opposition. I hear in her voice her worry for her brother Roy, who is a tax assessor for the township and a farmer.

"Neighbors aren't talking to neighbors. Even some family members aren't on speaking terms anymore. It's the money talking."

Dorothy's theory is that a small, but possibly fatal, slipup by Kirby in Havana may explain his activities in Forrest. In December 1993, after a rancorous siting hearing, the city council of Havana voted five to two in favor of building the incinerator. The group opposing the project, of which Dorothy is a part, then appealed the council's decision to the Illinois Pollution Control Board.

[illegible] abounded, their lawyer argued. From the start, Kirby and the mayor had been working hand in hand on this project. Furthermore, one of Kirby's several companies had flown city council members to Boston, where they toured an incinerator located near Cape Cod that was to serve as a model for the one in Havana. Incinerator opponents were not invited. Hence, they were denied crucial knowledge obtained by trip participants.

The newspaper had represented the tenor of the Pollution Control Board hearing with this piece of testimony on the details of the infamous Boston trip:

LAWYER: Was it a big plane, little plane?

COUNCIL MEMBER: Oh, any plane would have been big to me. I never been on one.

LAWYER: And did they take you to supper, Mr. Thomas?

COUNCIL MEMBER: Oh, yes, they had everything you ever wanted to eat there. I had crab something one night and lobster the next. It was real, real, real nice. You know, I've never had lobster before.

LAWYER: So you were treated first class everywhere you went?

COUNCIL MEMBER: Well, it wasn't just me. All of us were treated that way.

In June, the Pollution Control Board ruled in favor of incinerator opponents, overturning the city council's decision—at least temporarily. Kirby was in the process of appealing. In the meantime, he may be pursuing Forrest as a way of hedging his bets, Dorothy suggests. We know that he recently ferried a delegation from Forrest out to Massachusetts, but this time he wisely extended his invitation to include incinerator opponents.

I close my eyes. The board's decision brings to mind an image of an unarmed man standing before a line of tanks. How long can the ruling hold? The Havana incinerator proposal had been officially declared dead twice before, and both times it had come roaring back. He had shifted his proposal from county board to city council. He had found new investors.

~~~

The image of a giant incinerator sitting out in the silence of Pleasant Ridge cornfields, fleets of ash trucks and refuse-filled railcars forever coming and going, was disorienting, virtually impossible to accept.

Even the newest, fanciest incinerators send traces of dioxins and furans into the air. These molecules cling to bits of dust and sediment. As they move downwind, they sink back to earth or are washed out with rain. Here they coat soil and vegetation—corn, beans, hay, watermelons, whatever. These chemical contaminants are then consumed by us directly or are first concentrated in flesh, milk, and eggs of farm animals. A number of European studies have documented elevated levels of dioxin in the milk of cows grazing in pastures near municipal incinerators, for example. These studies have been subsequently confirmed. High levels of dioxin are found in food and dairy products produced near incinerators. People who live near waste incinerators have higher levels of dioxin in their blood than do members of the general population, with incinerator workers at particularly high risk of exposure.

Every day people all over the nation sit down to a meal that originated from the central Illinois countryside. The question of whether or not to construct an incinerator in the midst of all this agriculture seemed to me in 1994 like a national issue—one the whole country should vote on. It still is. Because our main route of exposure to dioxin is in food, all of us have a stake in the question of whether or not generating electricity by lighting garbage on fire is a genuine form of renewable energy—as is now being claimed—or a wolf in sheep's clothing. Instead, decisions on incinerator sitings are too often made by a handful of small-town city councils desperate to shore up their communities' economies. This was true in 1994, and is still true today.
~~~

Ascertaining dioxin's contribution to human cancers is one of the more frustrating challenges for public health researchers. Because dioxin is so potent at such vanishingly small levels, exposure is expensive to measure. Because it is so widely distributed, there remain no populations to serve as unexposed controls. Because dioxin so often rides the coattails of other carcinogens, confounding factors abound.

Animal studies provide a complex set of clues. In the laboratory, dioxin is an unequivocal carcinogen. As the dioxin researcher James Huff has noted, "In every species so far exposed to TCDD . . . and by every route of exposure, clear carcinogenic responses have been found." These include cancers of the lung, mouth, nose, thyroid gland, adrenal gland, lymphatic system, and skin. Dioxin also causes liver cancer in rats and mice, but it does so more often in females. Female rats whose ovaries have been removed, however, tend not to develop liver cancer when exposed to dioxin. On the other hand, they are far more likely to succumb to lung cancer. Clearly, an organism's own internal hormones modulate dioxin's carcinogenic powers, but through some unclear means.

We now know that dioxin perturbs many hormone systems and can do so through multiple mechanisms. It can alter the metabolism of hormones. It can alter their transport through the bloodstream. It can increase the number of hormone receptors, thereby leaving an organism more sensitive to its own hormones. Dioxin also alters a plethora of growth factors, including vitamin A, interferon, and interleukon. Dioxin can even affect the mobilization of calcium and impact serum triglyceride levels.

These particular effects do not just vary by species, they also vary by timing of exposure. Dioxin is now understood to be a developmental tox-

tant as well as an all-purpose carcinogen. Experiments with lab animals show that exposures in early life can permanently alter, for example, the structure of the mammary gland. Rats exposed to dioxin before birth develop breasts with more terminal end buds remaining in adulthood. (Recall from Chapter Seven that the pesticide atrazine has a similar effect.) Rather than promptly complete their pubertal metamorphosis into lobules, the terminal end buds within the breasts of dioxin-exposed rats remain suspended in an immature state. You could say that they do not know when to put away childish things. This dawdling pace of sexual maturation widens the window of sensitivity to carcinogenic damage. And so, through this mechanism, exposure to dioxin before birth predisposes rats to breast cancer.

And, as with atrazine, the impact of altered development has implications beyond cancer. Rats exposed prenatally to dioxin grew breasts with rudimentary architecture: they had fewer ducts with fewer branches that were less deeply anchored within the mammary fat pad. And when exposed rats became pregnant themselves, their breasts did not respond by growing larger quickly in preparation for making milk. Moreover, dioxin muffled the activity of their milk-making genes so profoundly that, as new mothers, the exposed rats could not provide sufficient milk for their offspring. Their pups all died.

Epidemiologists studying dioxin have focused on human populations exposed in the workplace or through chemical accidents. Several report an association between dioxin exposure and overall cancer incidence, but, with the exception of soft-tissue sarcomas (tumors that arise in muscles, fat, blood vessels, or fibrous tissues), no one particular cancer stands out. A 1991 study of five thousand TCDD-exposed workers employed at twelve U.S. plants, for example, showed significant elevations in overall cancer mortality.

Several intriguing studies have come from Germany. In a 1990 study, researchers found excess cancer mortality among workers known to be heavily exposed to TCDD during a 1953 explosion at a German chemical plant. In another cohort study, researchers found elevated cancer deaths among workers at a dioxin-contaminated chemical plant in Hamburg. When compared to other workers, chemical plant employees with twenty or more years of employment suffered twice the cancer mortality, and women workers showed elevated breast cancer mortality. Likewise, a 1996 [illegible] study [illegible] significant increase in cancer mortality among more than twenty-four hundred German workers involved in manufacturing herbicides known to be contaminated with TCDD. In both studies, cancer risk rose with level of exposure.

One of the largest studies to date is still in progress. In July 1976, an explosion at a pesticide-manufacturing plant near Seveso, Italy, released a dioxin-suffused cloud of chemicals into the air. Within a few days, leaves fell from trees, birds and other animals died, and children developed skin lesions. Since then, the epidemiologist Pier Alberto Bertazzi and his colleagues have been monitoring the blood and health of some forty-five thousand men, women, and children in and around Seveso, who are thought to have received the highest dioxin exposure in a human population.

As of 1993, Bertazzi had found excesses of certain cancers among inhabitants in Zone B, the second-most dioxin-contaminated area. Compared to the general population, Zone B residents had three times the rate of liver cancer. Rates of leukemia, multiple myeloma, and certain soft-tissue sarcomas were also elevated. In contrast to the findings in Germany, breast cancer incidence among the women of Zone B was initially lower than normal. So were their rates of uterine cancer.

By 2008, residents in Zone A had significantly elevated mortality for leukemia, multiple myeloma, and non-Hodgkin lymphoma. They also had lower circulating levels of antibodies of a type called immunoglobulin G. The higher their level of dioxins in blood, the lower their blood levels of immunoglobulin. This trend echoes results from the laboratory: among rats, dioxin inhibits immunoglobulin secretion and decreases resistance to parasites and infection. All together these recent findings indicate that the immune system is one of dioxin's most sensitive targets.

Follow-up studies also showed that the earlier findings on breast cancer were falsely reassuring: when age of exposure was factored in, high dioxin levels in blood did raise the risk for a breast cancer diagnosis. While overall mortality from breast cancer was neither higher nor lower among Zone A or B women, among those who were young (less than forty years old) at the time of the explosion, high dioxin exposure doubled the risk for a breast cancer twenty-five years later.

These results are consistent with recent findings from Michigan, where dioxin is found in floodplain soils in and around the town of Midland. This contamination is likely a historic legacy from chlorine-based industrial practices at the Dow Chemical Company's facilities there. Studies show that living on properties with dioxin-contaminated soils and eating locally caught fish have led to elevated blood dioxin levels. Moreover, using GIS mapping and space-time statistics, researchers were able to observe breast cancer clusters associated with soil dioxin. In other words, there was a higher burden of breast cancer in neighborhoods near dioxin-contaminated areas.

∿∿∿∿∿

I spent a lot of time in the fall of 1994 driving around the back roads of Illinois to various incinerator meetings. Some of these took place in school gymnasiums, and others in farmhouse kitchens. The assorted communities considering Kirby's proposal—from Forrest to Beardstown—were in various stages of deliberation. From what I could see, the same conversations were happening everywhere. It was as though someone had handed each of these towns identical scripts. Some were just beginning Act I (where civic leaders circle cautiously around the proposal, and a few members of the constituency issue admonishments along the lines of "beware of Greeks bearing gifts"). Some were already at the end of Act II (where the town becomes a house divided, predictions of ruin fly from both camps, and lifelong friendships dissolve into enmity). The whole drama had a foreordained feel to it, but I couldn't begin to guess how it would end.

Havana had been at it the longest, and the plot there had thickened more than once. The feasibility study, for example, was supposed to clarify once and for all what the incinerator's health risks would be. Instead, it further deepened the lines of [illegible] conducted by an independent team of scientists but paid for by Kirby's company—the city of Havana couldn't afford it—the study was released in the spring of [illegible]. The team concluded that dioxin would be emitted at acceptable levels, predicted no significant effect on human health or wildlife, and recommended project approval.

Two rebuttals swiftly followed—one commissioned by private citizens, and the other by the county farm bureau. The first, also authored by an independent, university scientist, criticized the feasibility study for downplaying food consumption as a route of exposure, for miscalculating wind direction, and for ignoring the fact that people already have background levels of dioxin in their tissues. The second one claimed the feasibility study exaggerated the incinerator's economic benefits.

The county farm bureau and the Central Illinois Irrigated Growers Association officially declared themselves opposed. Their position was given new credence when a national popcorn buyer announced that it would reconsider its long-standing commitment to Mason County if the incinerator threatened the popcorn crop.

By midsummer of 1992, Ban the Burn signs had sprouted up in lawns, and a convoy of tractors pulling anti-incinerator floats joined the annual Fourth of July parade. The sign proclaiming God Recycles and the Devil Burns caught the most attention and disapproval—both because of its religiosity and because the chamber of commerce was presumed pro-incinerator and they were the sponsors of the parade.

The specific arguments and counterarguments were peculiar to each community, but even these shared common elements. There was, for example, the Hypocrisy Argument. Incinerator opponents, according to incinerator advocates, had no right to object to the possible environmental risks when in fact (a) their own unrecycled garbage was being carted off to some other community for disposal, and (b) they were heavy users of pesticides. From a letter to the editor in Havana:

> I want to write and say please give it a chance. You say garbage from other places will be brought in there to burn. May I ask you where is our garbage being hauled? . . . I challenge any of you that is against the incinerator to prove to me that all the toxics you say will go in the air and pollute our air is any more toxic than all the sprays, plow downs, herbicides and all the hundreds of chemicals farmers are using now days

Incinerator opponents also cried hypocrite. Incinerator advocates, they countered, had no right to claim they were only looking out for the well-being of the community when (a) they were actually seeking to line their own pockets and (b) would allow ruination to befall a neighbor's farm in the process. From a letter to the editor in Forrest: "What has become of our Christian virtues? What has happened to 'Love thy neighbor as thyself'? Why has the wonderful, Christian community of Forrest sold out to the god of money?"

There was also the Trust Argument. Pro-incinerator folks emphasized that village board and city council members had been freely elected for the purpose of pursuing opportunities for community growth. The community, therefore, needed to trust them to do their job. If the naysayers were unhappy, let them run their own slate of candidates next time. Opponents located trust elsewhere. They pointed out that future generations were dependent on the current generation to safeguard the environment, with failure to do so being a betrayal of trust. Moreover, public servants, even freely elected ones, were susceptible to the seductive influences of a developer who stood to make a personal fortune if the project went through. The goings-on of the whole crew required rigorous surveillance. One busy citizen even wrote in to report that Kirby could hardly be counted a trustworthy steward of air quality, as he, along with the mayor, was a smoker.

Finally, there was the Question of Risk. Incinerator proponents tended to couch risk as opportunity and risk-taking as bravery. Without risk, the community would die: "Will Forrest reject a move forward for the more complacent choice of slow death? Nothing new is without problems. The dead have no worries."

Those opposed saw risk as recklessness. One pollution control mishap, one seven-axle explosion, one overturned ash truck and a brisk wind—and the community would regret its decision forever. Even if accidents never happened, they argued, the emissions that were now said to [illegible]. Nobody really knew what was in garbage anyway. How could developers be so sure what would come out of the stack?

Dioxin is an agent provocateur. It works its evil in part by inciting cells to certain actions that increase their susceptibility to damage by other carcinogens. One of dioxin's known tricks is to induce cells to step up production of a group of enzymes called cytochromes P450, which serve the important function of metabolizing toxic substances. Sometimes, however, the first step of this conversion transforms a harmless chemical intruder into something truly dangerous. As we have seen with beluga whales in Chapter Six, it is very often the metabolic breakdown product, rather than the parent material, that goes on to wreak carcinogenic havoc. By virtue of its shape, dioxin is protected from deconstruction by these same enzymes. Hence, it remains powerful at faint concentrations.

Its effect on cytochromes P450 may also explain why dioxin is linked with so many different kinds of cancer. If, as it now seems, dioxin aids and abets a whole assortment of carcinogens—some associated with one set of cancers and some with another—then different people should develop different ailments, depending on their specific prior exposures, hormonal status, and stage of life. Dioxin may bring on liver cancer in some dioxin-exposed individuals, for example, and hasten the progression of lymphoma in others.

We also know with some confidence how dioxin stimulates P450 production in the first place. Once a dioxin molecule leaves the bloodstream and slips into the interior of a cell, it binds to a naturally occurring protein called the Ah (aryl hydrocarbon) receptor. This complex is subsequently shuttled into the nucleus, the walled-off chamber within every cell that contains the DNA. Once here, the duo attaches to and turns on a particular set of genes. These activated genes then send out instructions for the manufacture of particular enzymes, namely, cytochromes P450.

Genes coding for P450 enzymes are not dioxin's only target, but less is known about the others. They seem to include genes that regulate growth, as well as genes responsible for the regulation of and sensitivity to certain hormones. Each is relevant to the expression of cancer.

Seven of the 75 dioxins, 10 of the 135 furans, and 11 of the 209 PCBs have the ability to bind with the efficient little Ah receptor. (TCDD does so most tightly.) Apart from allowing industrial chemicals access to our genes, what is its function? Why does it exist? With what naturally occurring agent is an Ah receptor supposed to make contact? No one actually knows, but the picture is beginning to clarify. When unbothered by dioxin, the Ah receptor appears to control the expression of a whole network of genes that help guide embryonic development. In mammals, it plays a role in genesis of the face, head, kidney, and heart. Conversely, when dioxin is present, the Ah receptor signals to an entirely different set of genes—those that govern detoxification—to increase their activity. In these and other ways, early-life exposure to dioxin may permanently alter the course of development.

~~~

In all the driving around I did that fall, I never ran into Mr. Kirby. He would be in one town while I was in another. We probably passed each other on the highway. Everyone I spoke with said he was a nice guy. The pleasantness of Kirby's personality was probably the only issue on which there was complete agreement. Published photographs showed a sizable man with snow-white hair, a farmer's leathery complexion, and a tendency toward colorful dress.

He was in fact a farmer—or at least he had grown up on a farm and had once owned one. His poultry operation was so successful that the
~~~

Springfield paper once featured it in a full-page profile. Someone else actually ran the farm, but it was Kirby's children who sorted and packed the [illegible]. These kinds of details impressed people. Kirby was a Korean War veteran and had been handpicked by his father to go to college. At one time he had been a school principal.

He also was what my mother called a wheeler-dealer. He entered politics by becoming assistant to the state school superintendent. He ran (unsuccessfully) for state auditor and then (briefly and unsuccessfully) in the Republican primary for U.S. senator. Kirby became influential with the governor and counted Ronald Reagan and Everett Dirksen among his friends. In the 1970s, Kirby was a special consultant to the developers of a racetrack. The newspapers called him brilliant. The racetrack was never built.

Apparently, Kirby got into the trash business the old-fashioned way—by buying trucks, landfills, and transfer stations and then selling them for a profit. Later, he focused on putting together agreements for other trash haulers and then incinerator operators. Here is where I lost the thread of exactly what Kirby did for a living. It involved some combination of lobbying government, negotiating with insurance brokers, acquiring permits, attracting venture capital, and otherwise dealing with revenue bonds and investment companies. I wasn't the only one who felt mystified by his profession. The *Pekin Daily Times* asked him who he thought John Kirby was, and John Kirby replied, "I think I'm a guy that pretty well knows what he wants to do."

He also said, "There's no question there'll be, by the turn of the century, five or six 1,800-ton-a-day incinerators" in Illinois.

~~~~~~

Hanging above the stairway in my parents' house is an aerial photograph of the Maurer family farm, circa 1950. You must walk by it without even a sidelong glance of interest unless you truly want to hear my mother's accompanying disquisition. This involves explaining—in the
~~~~~~

reverent authoritative tone she reserves for talking about the farm—what each and every outbuilding was used for, how all three generations who lived here had six children apiece, and why Roy and Pop slept out in the milk house during the scarlet fever outbreak. (The house was quarantined; they needed to sell the milk.)

But of course you express curiosity about the photograph—which appears to depict an entire village rather than a single farm—and so my mother is duty-bound to point out the following. First, the house. Built in 1908, it features two stairways and eight bedrooms. All the kids were born in the downstairs one. To the north, out to the road, is the orchard.

To the south is the garden, the chicken house, the tool shed, the smokehouse, the cob house, and the threshing-machine shed. The garden was always organic and still is today. The toolshed housed the discs, harrows, plows, manure spreaders, and planters. The cob house held corncobs and coal for the stove. The threshing-machine shed is the one with the tall doors to accommodate the massive contraption that separated oats from straw. The chicken house had a special room for setting hens.

The rest of the animals were quartered on the east side, where lie the barnyard, the barn, the milk house, the shop, the pig house, and the corncrib. To the west are another orchard, a special pasture for calves, and a windbreak of catalpa trees. Beyond all that are the 160 acres of the Maurer spread in Pleasant Ridge township—four miles north of Forrest, just west of Route 47, along County Road 1200 N, in southeast Livingston County.

Even I have a hard time perceiving the ridge in Pleasant Ridge township, which is about as flat as they come. The open earth lies like a black sail from horizon to horizon.

A month after our meeting with Dorothy, I am driving home to Pekin from an evening meeting with a group of Pleasant Ridge farmers and am full of pie. One does not leave a convocation of Illinois farmers without eating pie.

Out here I am known as "Kath-urn's girl" (my mother's name being Kathryn). This fact makes even more peculiar the realization that I just

[illegible]

will balance the rosy presentation that Kirby's local collaborator, the Forrest Development Corporation, sponsored the month before. All this education anticipates the referendum in November.

It's late and I'm just at the crossroads, the junction of Routes 47 and 24, when the radio begins playing the opening strains of a symphony. It's modern and orchestral, but the feeling is that of an enormous choir singing. What is it? I realize immediately where I need to be to listen to this music and so execute a U-turn and head back north.

Three miles later, I pull the car over, turn on the flashers, turn up the volume, unroll the windows, and walk straight out onto the section of land containing the eighty-acre rectangle—optioned, annexed, disputed, despised—where the incinerator is to be sited. The music follows me.

Earlier tonight, I spoke with the farmer who works this field. The drainage is troublesome, he said, and he mentioned another problem, too—the specifics of which I have already forgotten. Now, he rues the day he ever complained about it. "I'd give anything just to keep plowing that field forever."

The music plays on. It is sad but somehow glorifying. (Months later I will learn that it is Vaughan Williams's *Fantasia on a Theme by Thomas Tallis.*) I walk as far as I can without straying out of earshot and then lie down, mindful of broken cornstalks. I realize for the first time how much the events of the last month—all the research, the strategizing, the attempts to predict the future—have exhausted me.

The music's loveliness makes me realize other things, too. Whatever the facts about the incinerator were, the truth was that it was obscene. And so was arguing about exactly how many picograms of dioxin could acceptably contaminate these fields, the bodies of the people who plow them, and the flesh of their hogs, turkeys, and garden vegetables. So was discussing exactly how many thousands of gallons of water a day could be pumped out of this land in order to burn garbage. So was manufacturing substances that are poisonous when incinerated and undegradable when buried. And so the rapacity of subsidizing incineration over recycling in the first place.

In the 1990s, between [illegible] and [illegible] incinerators operated at any given time in the United States. They handled about 17 percent of the nation's trash. Now it is about 12.5 percent. Any respectable recycling program would easily put them all out of business.

The concept of Zero Waste is now gaining momentum. It seeks to end our dependencies on both incinerators and landfills through investments in waste reduction, reuse, recycling, and composting programs. By necessity, Zero Waste looks to law and technology to redesign products in ways that make them recyclable and nontoxic. In spite of its unending indecision on how regulate dioxin, the EPA has in fact released another report—in final form—on the generation of municipal waste that's quite pointed: even absent a plan to detoxify materials in the waste stream, the vast majority of what's called garbage, says the EPA, can be reused, recycled, or composted.

Perhaps incinerator emissions are no worse than the injection of agricultural chemicals into these same fields—incinerator proponents certainly had a point there—but two obscenities do not cancel each other out. This land is bordered on one side by the north fork of the Vermilion River, and on the other by the south fork. The incinerator will pollute the watershed no matter which way the wind blows. The fish of the Vermilion are already contaminated with persistent organic pollutants (chlordane, DDT, dieldrin, heptachlor, and aldrin). Still, one of its tributaries manages to support a small population of river redhorse, an endangered fish species.

The music lifts and falls like a human voice.

I try to imagine it out there somewhere, the river redhorse—whatever it looks like—resting quietly in a current of water, even as I am resting here on the earth. A scud of clouds covers over the stars and then blows by. A melody carried from one instrument to another is transformed and then submerged.

Some of the farmers I ate pie with tonight are no doubt responsible for the contamination of Vermilion River fish, but in a larger sense, so are we all. Now we had the chance to work together to prevent a problem rather than abate the damage later. It was a start.

I send my thoughts out to our farm, which lies a mile to the north. It is a greatly abridged version of its earlier self. Except for the odd sheep or two, the animals are all gone, and so is the metropolis of barns, shops, sheds, and corncribs. But the fields are still there, and the house is still there, flanked by a few silver grain bins and my Aunt Ann's organic garden. When the music ends, I'll be able to see its lights as I walk out of this field.

~~~

On November 9, 1994, the results of the incinerator referendum in Forrest showed 466 against and 406 for. Some members of the Forrest Development Corporation vowed to proceed anyway, but Kirby demurred. "We're apprehensive about committing to the project if that support for it is soft. We don't want to have to fight a battle every time we want a sewer extension."

The following September, an appellate court in Springfield, Illinois, upheld unanimously the decision of the Illinois Pollution Control Board regarding the unfair siting approval of the incinerator in Havana. The judges cited both a Massachusetts trip paid for by Kirby's corporation and the improper influence of that corporation on the hearing officer.

On January 11, 1996, the Illinois General Assembly repealed the Retail Rate Law. According to the governor, "Most communities do not want the incinerators. And it is time we stopped asking our taxpayers to subsidize them."

On January 25, 1996, John Kirby died of malignant mesothelioma—a form of lung cancer linked to asbestos exposure—in a Springfield, Illinois, hospice.

On March 4, 2006, my one hundred-year-old grandmother died of old age, my eighty-year-old Uncle Roy stopped plowing for good, and the
~~~

Maurer farm in Pleasant Ridge went up for sale. Assisted by The Land Connection (see Chapter Seven), a group of organic farmland investors purchased the house and the land and are shepherding it through its three-year transition to a certified organic farm. It is currently being tenant-farmed by a neighbor, Scott Friedman, who was already interested in learning more about organic techniques. The new owners share profits with him through a crop share lease. They all have studied closely my mother's aerial photo of the farm during its glory days, which now hangs, poster-sized, in The Land Connection's offices. There are plans to rebuild the orchard. There is talk of a wind turbine. The last time I drove by the farm, in early spring, its fields were blowing with brilliant green waves of organic winter wheat undercropped by alfalfa. Whoever the new permanent farmer of this land turns out to be, I hope to stand someday with him—or her—on the wrap-around porch. I'll tell the story of Mr. Kirby, and then we'll turn toward the southeast horizon where, look, no incinerator rises from the earth.

eleven

our bodies, inscribed

Among forest trees, size and age can be remarkably dissociated. Seedlings germinating in deep shade are often swiftly overtaken by those sprouting up in light-filled spaces nearby. Saplings browsed by a passing deer lose height relative to neighbors less palatable. By these and other means, senior members of a forest community can grow old beneath a canopy of younger trees.

Field ecologists, therefore, rely on tree-ring analysis to reconstruct the history of forests. I once spent a summer in Minnesota engaged in this kind of work, which begins with pressing the bit end of a hand borer against the bark of a tree at chest height, leaning against it with all one's weight, and slowly turning the handle until the steel threads have chewed into the flesh beneath and have wound themselves straight into the tree's exact center. A slender wand of cool, damp wood is then extracted with the narrowest of spatulas, sealed in an envelope, and, along with an assortment of other tree cores, taken back to the laboratory to be read.

These cores are banded with colored rings, each representing a season of growth. An experienced dendrochronologist (which I am not) can identify in the subtler patterns of these circles not only age but also periods of changing light levels, insect plagues, drought, flood, or fire. An individual tree carries within its own body an ecological chronicle of the entire community.

In this, people are not so different. Our bodies, too, are living scrolls of sorts. What is written there—inside the fibers of our cells and

Body burden refers to the sum total of these exposures and encompasses all routes of entry (inhalation, ingestion, and skin absorption) and all sources (food, air, water, workplace, home, and so forth). In the case of fat-soluble, persistent chemicals, body burdens provide a measure of cumulative exposure. Some of these exposures occurred in infancy, others in adolescence, and still others in adulthood. In the case of chemicals quickly metabolized and excreted, the body burden is an index more akin to a press release than a biography. It reports on the status of immediate and ongoing exposures to particular contaminants at single points in time.

Body burden analysis typically requires extensive sampling. This task is most easily accomplished during an autopsy or by examining archived human tissues. In Japan, researchers examined a variety of industrial contaminants in preserved fat collected from men who had died between 1928 and 1985. The highest concentrations of DDT, PCBs, and chlordane were found in samples collected during these chemicals' respective periods of maximum production, import, and use. For living people, total exposures are more often derived from measurements taken in a less invasive manner. Blood, urine, breast milk, exhaled air, semen, hair, tears, sweat, and fingernails have all been used for this purpose.

Measuring the levels of pollutants in people—usually in blood or urine—is called biomonitoring. The data so generated provide undeniable proof of exposure. Biomonitoring is the drug test of the ecological world, and, as such, it answers one simple question: "Is it in me?" Biomonitoring does not rely on computer modeling, interviews, questionnaires, diaries, or childhood memories of riding bicycles through the DDT fogging trucks and eating fish caught from lakeside docks. It does not require reconstructing the names of all the chemicals you handled on that [illegible] through the Toxics Release Inventory. It is impervious to wishful thinking. As one of its leading proponents describes it, biomonitoring is "a reality check on health in a hazardous environment."

As a tool of public health, however, the usefulness of biomonitoring data depends on two additional factors: sampling a representative cross section of the population and doing so for a long enough period of time that trends emerge. For example, biomonitoring over several decades demonstrated that gasoline was once a leading source of lead exposure for U.S. children. When laws were passed in the 1970s that gradually phased lead out of gasoline, the average lead levels in children's blood began falling in tandem. Between 1976 and 1980, gasoline lead declined by 55 percent. So too did the lead in children's bodies. By contrast, the models that had been generated prior to the new law predicted that banning lead in gasoline would drop children's lead levels by only a trivial amount. So convincing was the biomonitoring data, the EPA was moved to speed up the timetable for abolishing lead in gasoline, a goal that was finally reached in 1991.

More recently, biomonitoring data has shown us that banning smoke in public places truly results in less smoke inside of us: during the 1990s, median levels of cotinine in the blood of nonsmokers fell by 70 percent. (Cotinine, the metabolic byproduct of nicotine, is a sign of exposure to second-hand smoke.) Biomonitoring also demonstrates that half of all U.S. children are, nevertheless, still exposed to cigarette smoke. Smoking cessation efforts have miles to go.

In 1999, the Centers for Disease Control began assessing a variety of environmental chemicals in the bodies of a representative sample of the U.S. population—about 5,000 people from 15 geographic locations. The suite of chemicals so assessed has expanded over the years from 27 to 148 and now includes, among other things, heavy metals, pesticides, dioxins, PCBs, polycyclic aromatic hydrocarbons, flame retardants, and two endocrine-disrupting ingredients in plastics: phthalates and bisphenol A.

The most recent report from the CDC's survey held several surprises. First, it turns out that virtually all U.S. residents have brominated flame retardants coursing through their veins, and at levels 2–35 times higher [illegible] endocrine-disrupting persistent organic pollutants similar in structure and behavior to PCBs (with bromine taking the place of chlorine in the doubled-ringed molecule.) Unlike PCBs, though, levels of brominated flame retardants fall, rather than rise, with age: children have higher concentrations than teenagers, and teenagers have higher levels than adults. Second, bisphenol A, an estrogenic mimic, is also ubiquitous in the bodies of U.S. residents—92.7 percent of us pee bisphenol A in our urine—and again, children have higher concentrations than grown-ups. Third, women of reproductive age have remarkably high levels of endocrine-disrupting chemicals—precisely the group in which one would wish to find the lowest levels. There is also happy news: first-time pregnant women have, as a group, decisively lower levels of persistent organic pollutants in their blood than pregnant women tested fifty years earlier. PCBs, in particular, fell dramatically. As with the lead and cotenine time trends, the lesson to be drawn here is the same: bans work to lower exposure.

Biomonitoring is all the rage. Advances in analytical chemistry have made detecting trace amounts of chemicals easier and cheaper than it was only a decade earlier. With the falling price tag has come a shift toward measuring pollution directly in people rather than attempting to calculate human exposure indirectly by measuring pollution in air, water, and soil and then constructing elaborate computer models. The ongoing Agricultural Health Study is biomonitoring farmers and their families. The Breast Cancer and Environmental Research Centers are biomonitoring girls for chemicals possibly linked to breast cancer or to early puberty, which itself is a risk factor for breast cancer. Researchers at the Silent Spring Institute are looking for endocrine-disrupting chemicals in Cape Cod women's blood as well as in the dust in their houses. Similarly, the National Children's Study plans to combine biomonitoring data with environmental monitoring data as it follows one hundred thousand U.S.

children from prebirth through adulthood. California has passed legislation to establish the first state-based biomonitoring program, and several other states have followed suit.

However indisputable the results of biomonitoring, the interpretation of those results is full of argument. All by itself, biomonitoring provides no threads to the past—it cannot reveal sources or routes of exposure. It cannot tell us where all the bisphenol A is coming from. Nor does biomonitoring allow us a glimpse of the future—it cannot say what health problems will arise because of particular exposures. The data offer no predictions as to whether the universal exposure of U.S. children to bisphenol A will raise their cancer rates as adults. Thus, biomonitoring data is most meaningful when examined together with the results of other investigations.

Sometimes biomonitoring studies can be cleverly designed to provide their own interpretation. A 2009 Korean study, for example, showed that children living near a steel mill had hydroypyrene in their urine. Hydropyrene is a chemical metabolite of polycyclic aromatic hydrocarbons, which steel mills are known to emit. But hydrocarbons have many potential sources of exposure, including the consumption of charred meat. So investigators also biomonitored children living farther away from the mill. The children near the mill had significantly higher levels of urinary hydropyrene than children living remotely. Moreoever, their higher levels fluctuated with the seasons—rising in the summer and falling in the winter months—which paralleled the seasonal changes in ambient pollution levels near the steel mill. The temporal-spatial patterns in the biomonitoring data strongly implicate the mill as the polluter of the children.

~~~

The human body is an endless construction site where demolition and renovation occur simultaneously and continuously. Different tissues carry on this work at different rates; the lining of the stomach is entirely
~~~

overhauled every few days, while a complete restoration of the bones' internal scaffolding requires years. All tissues replace themselves through the orderly process of cell division—mitosis—in which one cell splits in half and becomes two. Damaged and aged cells slated for removal undergo a programmed form of death known as apoptosis. All this activity is coordinated through an elaborate system of communication.

A certain amount of supervision is provided by surrounding tissues. Chemical signals from neighboring cells—called growth factors—regulate the pace of cell division. And we know that marching orders sometimes arrive from distant headquarters. These often take the form of hormones, as when estrogen from a woman's ovaries causes the cells in her breast to begin dividing.

Wherever the signal's origin, mitosis begins inside a circle within a circle: the nucleus of the cell where the DNA is quartered.

The first step is the doubling of each of the strands of DNA, the chromosomes. Their duplication will enable both daughter cells to receive a complete set. For this task, a crew of enzymes creates an exact replica of each original chromosome (which is split in half lengthwise and used as a template for its own duplication). Lying side by side, the two identical strands are then cuffed together and come to resemble a gangly letter H or sometimes a stout V.

Humans possess forty-six individual chromosomes, each consisting of a curly DNA ladder and each bearing many thousand genes. Once all forty-six gene-studded chromosomes have been so copied, a dance begins. The nuclear membrane disintegrates. The chromosomal couples move to the center of the cell and form a vertical line. Fine threads called spindle fibers extend horizontally from opposing ends of the cell and attach to each member of a pair. The fibers contract. Simultaneously, the twinned chromosomes pull apart, their midpoint connections giving way as the left and right halves of the Hs and the Vs are towed through the watery protoplasm to opposite poles. Just as the cell begins to pinch in half, a membranous curtain closes around each new grouping of single-stranded chromosomes, and they are once again cloistered within a nucleus. They will remain there, directing the synthesis of proteins, until the mitotic cycle begins anew and once again releases them.

Cancer is mitosis run amuck. Instead of reproducing in careful, methodical fashion, cancer cells carry on replication and division despite a myriad of directives designed to restrain such activity. Cancer cells are dancers deaf to the choreographer. They are builders in flagrant disregard of zoning ordinances and architectural blueprints. They are drivers who pass on the right and bomb through four-way stops. They are Cells Gone Wild. They are defiant, disobedient, unstable, chaotic, and in the view of many cancer biologists, almost purposeful in the ways they disrupt cellular biochemistry. For example, normal cells growing in culture will stop dividing once they touch each other. Not so cancer cells. Uninhibited by contact, they will keep on proliferating over the top of their neighbors. Their ability to do so is fueled by their knack for making their own growth factors. Normal cells rely on neighboring cells to send growth factors, which function as invitations to initiate cell division. Cancer cells invite themselves. Recent research also suggests that cancer cells have the ability to recruit normal cells to work for them. In this view, tumors are not just homogeneous balls of bad cells. Rather, they are composite tissues, with cancerous and normal cells coexisting in a complex society. But the malignant cells are the ones running the casino.

Besides a propensity for limitless growth, cancer cells are known for two other traits: invasiveness and primitivism. The ability to invade other tissues distinguishes cancer from other freakish outgrowths, such as warts. This facility operates at both a local level—cancer also ignores property lines—and a distant one, as when cancer cells are shed from the primary tumor and seeded throughout the body as metastases. Destroying healthy tissue and clogging vital passageways, metastases are what make cancer deadly.

By primitive, biologists mean that the tissues created by cancer appear to have reverted back to some earlier, cruder, unformed stage of development. They no longer bear much resemblance to the differentiated structures of which they were originally a part. Typically, the hard lump in the breast that turns out to be a malignancy is a direct descendant of one of the smooth, flat cells that wallpaper the interior surface of the slender mammary ducts. But, microscopically, the tumor's mass of cells no longer looks anything like [illegible] less a tissue resembles its previous, respectable, specialized self, the more virulent the cancer. Along with runaway growth and the propensity to spread, this tendency to devolve into an immature, unrecognizable state is the result of a long accumulation of injuries.

A cancer cell, then, is made, not born. And there are several pathways to its creation. The best-understood route involves a series of incremental changes to chromosomal DNA. Some of these DNA alterations can be inherited, but the vast majority are acquired during the lifetime of an individual when genes perfectly healthy at the time of conception become damaged. This process itself can happen through numerous pathways. Routine errors made during DNA replication are one. Sabotage by carcinogens is another. About twenty-five thousand different genes are strung along our chromosomes. To contribute to cancer, at least some of these encounters with carcinogens involve genes that help govern cell division.

These growth-regulating genes come in two basic varieties. The first group is the proto-oncogenes. In their normal state, these bits of DNA convey messages that encourage cell division. When mutated, however, proto-oncogenes lose their prefix and go over to the dark side. As oncogenes, they ratchet up the rate of growth to hyperactive levels. Working on exactly the opposite principle are the tumor suppressor genes. Normally, they dampen the rate of cell division. In some circumstances—as when signs of DNA damage are about—they actually halt mitosis altogether and thereby nip in the bud the possible genesis of cancerous growth. Loss or inactivation of tumor suppressor genes may contribute to the birth of a tumor.

Nearly three hundred cancer-related genes have so far been described, with hundreds more on the list of suspects. Different kinds of cancers are associated with different kinds of mutated genes. The cells of most colon tumors, for example, turn out to contain both hyperactive oncogenes and nonfunctional tumor suppressor genes. One specific tumor suppressor gene located on chromosome 17 has been fingered in several big-ticket malignancies, including cancers of the lung, breast, colon, esophagus, bladder, brain, and bone. Indeed, alterations of this gene, named p53, may be involved in as many as half of all human cancers. Much as a gunshot wound indicates what kind of firearm was used in an assault, the particular nature of the p53 mutation often suggests the type of carcinogen responsible for the damage. Cigarette smoke leaves one kind of lesion, ultraviolet radiation another, and exposure to vinyl chloride a third. The mutational spectrum of this gene is so broad that the lung tumors from uranium miners can sometimes be distinguished from the lung tumors of smokers simply by looking at the specific location of the mutation. Breast tumors frequently display p53 mutations in a spectrum resembling that seen in lung tumors and varying across geographic regions.

Harm can befall growth regulator genes through a whole variety of pathways. Benzo[a]pyrene can adhere to a section of chromosome and, in so doing, create a DNA adduct. Like bits of chewing gum stuck to a strand of hair, adducts can cause mistakes to be made during the next cycle of DNA replication. Other carcinogens interfere with cell division directly, by, for example, disabling the spindle fiber apparatus, which causes chromosomes to pull apart improperly. By these and other means, daughter cells can end up receiving mutated oncogenes and/or missing or impaired tumor suppressor genes. Alterations in DNA repair genes can abet the process. These normally function to fix chromosomes vandalized by mutating agents or damaged accidentally during the normal course of mitosis. An injury to a repair gene is, therefore, a treacherous event, as it can lead to the accumulation of genetic lesions of all kinds and, eventually, loss of genomic stability. Fortunately, tumor formation is lengthy and complicated, often requiring decades to unfold. It is also capable of being arrested at many points along the way.

In the language of cancer biology, the making of a cancer cell involves three overlapping stages: initiation, promotion, and progression. To become a full-blown malignancy, a cancer cell must pass through them all.

The first rite of passage, initiation, is characterized by small structural alterations in the cell's DNA strands. Arising spontaneously or resulting from an encounter with a carcinogen, these modifications—like tiny [illegible]—are swift, permanent, and subtle. A small hole here. An incon[illegible] under the microscope, indistinguishable in shape and appearance from their undamaged counterparts. Nevertheless, many initiated cells meet an early demise through the winnowing action of apoptosis. Any agent that interferes with the genes that oversee programmed cell death is, then, a coconspirator in the story of cancer because it permits damaged cells to continue along the pathway to tumor formation.

The immune system also plays a role in the selective destruction of incipient cancer cells, which presumably reveal their hand by exhibiting biochemical traits recognizable as abnormal. At what specific stage immune cells begin to mount a reaction is not entirely clear. It is known that certain environmental contaminants, including dioxin, suppress human immunity and that immune suppression is associated with several kinds of cancers, most notably leukemias and lymphomas.

Molecular biologists refer to initiated cancer cells as *transformed*. On a cellular level, carcinogenesis is complete at this point. To cause the malignant symptoms of cancer, however, a transformed cell must escape detection by the immune system and advance to the next stage, *promotion*, which requires additional exposures to cancer-stimulating substances. Unlike initiation, promotion unfolds over a long period and may involve no actual mutations. Promoters are cancer accelerators. In general, cancer promoters encourage cells to divide by altering their genetic activity. Genes that are normally quiescent, for example, may become activated. Estrogen is a cancer promoter for transformed cells bearing estrogen receptors. As demonstrated in lab animals, many organochlorine compounds also play this role. The good news is that these effects wane when such agents are removed from the body.

Quite often, cancer promoters perturb an intricate communications pathway known as signal transduction. This system consists of a team of protein molecules relaying messages back and forth between the perimeter of the cell and the heartwood of the nucleus. Signal transduction proteins play a key role in the timing and coordination of cell division. Promoting agents can affect the production and behavior of these courier molecules without permanently damaging the genes that code for their manufacture. The result is an expanded cluster of transformed cells.

Like initiation but unlike promotion, the progression stage very often involves exposures that inflict physical injury to the DNA molecule. Mutations pile up. Chromosomes fall into disrepair and become increasingly unstable. Substances that act at this stage bestow onto the cells they cripple some of cancer's most fearsome abilities: the capacity to spread and invade, enhanced sensitivity to hormones, and a knack for attracting blood vessels to the growing mass of tumor cells. Some researchers believe that arsenic, asbestos, and benzene can function as cancer progressors.

New evidence suggests that progression may also occur by routes other than physical injury to genes. In this regard, agents that disrupt communications networks that help maintain tissue architecture are now considered Persons of Interest. Fundamentally, a tumor represents badly organized tissue. Its continued growth depends on its ability to renegotiate the terms of its relationship with the framework of connective tissue that surrounds it and provides support. The biological name for this tissue is stroma (from the Greek word for "a mat to sit down upon"). A new line of thought in carcinogenesis focuses attention on how chemical exposures affect tissue architecture within a nascent cancerous growth and within the stromal microenvironment that surrounds it.

Agents that contribute to cancer do not all fall neatly into the categories of initiator, promoter, and progressor. Some, like radiation, are complete carcinogens that can play all three roles by themselves. Others appear to behave as promoters at low doses and complete carcinogens at higher levels, and they may also interfere with apoptosis. Still others initiate at low doses and promote and progress when their concentration in the body rises.

More profoundly, there is a growing appreciation among cancer researchers for the possibility that some cancers may arise through mechanisms that don't fall into any of these three time-honored categories at all. Chemicals with the power to disrupt development—dioxin and atrazine are two examples—can leave an organism with a larger pool of initiated cells that are innately predisposed to cancer. Thus, many leading researchers are now calling for expanding the classic battery of tests used to identify chemical carcinogens to include assays that would pinpoint developmental toxicants as well. Substances that disrupt development, along with those that disrupt tissue architecture, are part of a growing group of agents now referred to as cancer enablers.

These shifting biological possibilities bring with them many social implications. First, they explain why no safe dose of a carcinogen can be found. They also explain why similar exposures can pose very different degrees of danger to different people. The trace presence of a cancer-promoting pesticide in drinking water, for example, may represent absolute hazard to those whose breast, prostate, colon, or bladder tissue has already been damaged by some prior event. Or to those rare few born with a mutated gene that predisposes them to cancer. Or to fetuses, infants, and children, whose tissues are still under construction. Adults whose genetic material has suffered less previous damage may more successfully ward off the effects of promoting agents—as would those lucky persons who happen to possess a set of genes that allows for especially efficient detoxification and excretion of promoting substances.

The implications become even broader when we consider the dozens of known and suspected carcinogens to which we are routinely exposed and which may work in concert, or cumulatively anywhere along the cancer continuum. In rats, for example, DDT acts to accelerate tumors induced by a chemical called 2-acetamidophenanthrene. Exposure to either one alone, by contrast, is not capable of causing tumors to progress to a detectable level.

In the words of the veteran cancer biologist Ross Hume Hall, "Too often cancer research has focused on finding the last straw. It's time we looked at all the straws."

Recently, capturing the attention of cancer biologists are two other possible pathways to cancer: chronic inflammation and abnormal epigenetic regulation. The first affects the ecology of all the tissues surrounding the cell. The second affects the flickering patterns of genetic signaling within the cell. The growing awareness that both processes could be playing critical roles in transforming cells—or shepherding already transformed cells toward something truly villainous—has not yet upended the traditional dogma that locates the genesis of origin in genetic mutations. But it someday might. This science is shifting under my feet as I write, and is very much worth watching.

Inflammation is not part of the elegant, tightly regulated branch office of immunity that practices sophisticated acts of espionage and dispatches assassins (in the form of antibodies) to take out targeted foreign enemies. Rather, inflammation belongs to the more primitive and generic branch of the immune system known as innate immunity. It's involved whenever wounds create openings for pathogens to enter the body, and its weapons include histamines and prostaglandins, along with plenty of heat, swelling, and blood flow. It deploys them all swiftly and without prejudice. The result is lots of collateral damage—the tools it uses to destroy microbes are also toxic to surrounding tissues—but the upside is you don't die of a staph infection.

Chronic inflammation, which has no beneficial role, occurs when this reactive system fails to recognize victory and march back to the barracks or when it turns on civilians and begins attacking the body's own tissues. The result is an escalating cycle of tissue damage that requires repair. And with the need for extra cell division, chances for a malignant transformation presumably increase. This is the original hypothesis anyway.

Within the past decade, cancer biologists have learned something even more troubling about inflammation: the immune cells so involved sometimes assist, rather than attack, developing tumors. In an ultimate act of treason, this assistance can include the procurement of blood vessels to nurture the cancer cells within the tumor. There are several circumstantial lines of evidence implicating inflammation as a cancer-enabling condition: Inflammatory cells are sometimes found loitering near tumors. Some cancer cells can be made to grow faster when tissues are inflamed. Some cancers arise without accumulating genetic mutations. Many cancer cells display overactive prostaglandin genes (COX-2 genes). And infections that induce inflammation are known to increase the risk for certain cancers. (The risk of stomach cancer rises with infection by H. pylori bacteria, which is also a cause of ulcers. The risk of liver cancer rises with infection by the hepatitis C virus.)

The metabolic stress created by obesity can also induce chronic inflammation and can do so in ways that interfere with insulin signaling. One outcome may be diabetes. Another may also be an increased risk for cancer. These results do not mean that the immune system plays only a turncoat's role in the story of cancer. Indeed, there is still plenty of evidence to affirm the opposite: that immune cells labor loyally to identify newborn cancer cells and destroy them while they are still destroyable. In other words, not all the cops are working for the criminals. What the new evidence does suggest is that immunity is complicated and potentially corruptible. It means that the search to understand the origins of cancer must be expanded to include a close look at substances that alter immune functioning. What is the upstream cause of chronic inflammation? No one knows. The latest thinking posits a "failure in the establishment of immunoregulatory networks." To me, this means we have more reason than ever to scrutinize chemicals that are known to target the immune system or disrupt its development.

Let's move again inside the cell, away from the landscape of whole tissues, and focus once more on the beaded strands of chromosomes within the nucleus. Attached to these chains of DNA are various bobbles that play a key role in gene expression. Some are simple four-atom methyl groups and others are fancier proteins called histones. They are supposed to be there. Their job is to turn genes on and off, silencing the genes whose messages are not needed at the moment and coaxing the rest into loud song. They conduct the genetic choir. Their study is called epigenetics, which has been aptly described as "a code written over the top of our DNA sequence." Disrupted epigenetic regulation is now understood as another pathway to cancer, especially if the disruption affects the components that direct the genes for cell growth. For example, in some brain tumors, the gene that should have actively damped down tumor formation is no longer functional. It's not broken; it's just been rendered mute by an abnormal epigenetic change.

It may seem a question of only intellectual interest to ask whether a cancer arises from a direct mutation of a gene or from an alteration to the director of that gene. However, the difference matters greatly for at least two reasons. First, because epigenetic changes are more reversible than mutations. And second, because we may miss whole categories of carcinogens if we test only for the ones that can trigger mutations.

Environmental epigenetics examines how environmental exposures influence the epigenome—the code on top of the code. What the results of this nascent field of study emphasize is, once again, the vulnerability of early life. Prenatal life and infancy—and perhaps again in puberty—are times when the epigenome is moving rapidly over the genome, programming and imprinting it with instructions. Indeed, normal epigenetic regulation is what makes development possible: immature cells become differentiated and assume their grown-up functions when methyl groups silence long strings of genes that are no longer needed for the specific tasks of, say, a bone marrow cell or a breast duct—and activate those that are. When epigenetic regulation is disrupted at the beginning of life, the process of differentiation can be thrown off course in ways that may raise the risk for many diseases, including cancer. Epigenetics is an object lesson in how the timing makes the poison. It also shows us why it is time to retire the crotchety nature-versus-nurture controversy: the epigenome guides the genome and in turn responds to messages from the outside environment. Nature and nurture—genes and environment—feed each other and dance together.

~~~~~~

They have been compared to footprints, fingerprints, graffiti, and stigmata. They have been hailed as decoding tools by which to read the
~~~~~~

body. They are biological markers, and, defined most plainly, they are indicators of physical damage caused by the interplay between human cells and environmental carcinogens. As such, biological markers do what biomonitoring alone cannot: they serve as both signals of past exposure and predictors of future cancers.

Adducts, the sticky chemical tags of mutation-inducing chemicals that adhere to DNA, are a well-described biomarker. Unlike epigenetic methyl groups, which are also attached to the chromosome, DNA adducts are not supposed to be there. They are interlopers. They break things. As discussed in Chapter Six, the tissues of beluga whales living in contaminated stretches of the St. Lawrence River display high concentrations of benzo[a]pyrene adducts. Similarly, in rats, researchers consistently find tight correlations between exposure to chemicals known to cause cancer and the concentration of adducts in the DNA of certain tissues. DNA adducts, then, are a validated biomarker: we know that their presence indicates exposure, and, in turn, we know that the exposure is a risk factor for cancer.

Until very recently, the search for biomarkers has largely focused on those made by genotoxic carcinogens, that is, chemicals that cause cancer through genetic injury. Thus, the biomarkers so far identified are mostly signs of bona fide genetic damage. DNA adducts are such a biomarker. A strand of DNA that has been nailed by an adduct will invariably make an error when replicating during mitosis. The adduct is not the mutation per se, but it serves as a billboard for it.

With a growing awareness of the importance of epigenetic forces at work in the creation of a cancer cell, new interest has arisen in the search for epigenetic biomarkers. These would represent signs of disrupted gene activity. As a hypothetical example, a methyl group that silences a tumor suppressor gene might allow the pace of cell division to speed out of control in ways that predispose for cancer. At this writing, the technical ability to test for epigenetic markers far outstrips the ability to validate them. The brave new world of high-throughput assays now makes possible a search for thousands of molecular alterations in hundreds of different samples all at one time. Such tests will soon allow researchers to ascertain rapidly how a particular chemical exposure alters gene activity, metabolic processes, or total protein production within a cell. What's less clear is how these alterations predict, or don't predict, the genesis of cancer.

There are, nevertheless, a couple of intriguing new studies worth mentioning.

The first is a study from Spain that found that low levels of pesticide mixtures could trigger changes in gene expression in normal human breast cells so exposed. The second concerns the Greenlandic Inuit. Inuit people living in Greenland carry some of the highest body burdens of persistent organic pollutants of any people on earth. Thanks to global distillation and long Arctic food chains (see Chapter Eight), their bodies have become repositories for chemicals produced and deployed throughout the northern hemisphere. A 2008 study of the Inuit found that increased levels of DDT and PCBs were associated with decreased levels of DNA methylation. That is, individuals with higher levels of contaminants had fewer methyl groups. Low methylation in the laboratory is associated with chromosomal instability because genes that would normally be silenced are active instead. The long-term health consequences are not known, nor is the biological mechanism by which persistent organic pollutants interfere with methylation. Many of these chemicals are proven endocrine disruptors, which may alter estrogen receptor activity in ways that can derange the methylation patterns. But no one really knows.

By contrast, our understanding of genotoxic biomarkers (those that cause mutation) is far deeper. And although many molecular biologists are now racing toward the new frontier of epigenetics, we mustn't lose sight of the old knowledge. In many cases, we have not yet fully acted on it. An example would be a classic study published in the mid-1990s by molecular biologist Federica Perera at Columbia University. It reveals in striking detail an unbroken molecular chain of causation from ambient air to malignant cellular transformation as it takes place in one of the most polluted regions on earth: Silesia, Poland.

Hard up against Poland's southern border, Silesia is blanketed with chemical plants, foundries, steel mills, coal mines, and cookeries (the great ovens that distill coal into coke for steelmaking). The cancer death rate is also impressively high here, persuading Dr. Perera to examine Silesian DNA closely. Perera and her coworkers focused on polycyclic aromatic hydrocarbons, such as benzo[a]pyrene, which are released into Silesia's air in great abundance, mostly as by-products of coal and coke burning. Simply measuring their airborne concentration turns out not to be a reliable indicator of individual human exposure because polycyclic aromatic hydrocarbons are not only available for inhalation but also stick to skin (and are absorbed) and insinuate themselves into food (and are ingested). Moreover, these carcinogenic contaminants are handled differently by different people, depending on genetic and other factors that affect metabolism and detoxification.

The proof is in the cells' pudding. Perera found that the DNA of Silesian coke workers and Silesian city dwellers bore similar loads of polycyclic aromatic hydrocarbon adducts. These levels were two to three times higher than among rural folk. Perera also discovered a pronounced seasonal effect: the number of adducts rose during the winter months, when coal burned for domestic heating adds to the burden of aromatic hydrocarbons contributed by industry. Moreover, the level of adducts was correlated with the presence of chromosomal mutations known to be affiliated with lung cancer. Together with studies showing that people with lung cancer carry higher burdens of polycyclic aromatic hydrocarbon adducts on their DNA than people without the disease, Perera's findings "strongly suggested that severe air pollution could indeed help induce lung cancer." A response to these findings is long overdue.

~~~~~~

When I tell people that I had bladder cancer at age twenty, they usually shake their heads. If I go on to mention that cancer runs in my family, they usually start to nod. She is from one of those cancer families, I can almost hear them thinking. Sometimes, I just leave it at that. But, if I am up for blank stares, I'll let them know that I am adopted and go on to
~~~~~~

describe one study of cancer among adoptees that found correlations within their adoptive families but not within their biological ones. ("Deaths of adoptive parents from cancer before the age of 50 increased the rate of mortality from cancer fivefold among the adoptees. . . . Deaths of biological parents from cancer had no detectable effect on the rate of mortality from cancer among the adoptees.") At this point, most people become very quiet.

These silences remind me how unfamiliar many of us are with the notion that families share environments as well as chromosomes or with the concept that our genes work in communion with substances streaming in from the larger, ecological world. What runs in families does not necessarily run in blood. And our genes are less an inherited set of teacups in a glass-enclosed china cabinet than they are plates used in a busy diner. Cracks, chips, and scrapes accumulate. Accidents happen.

My Aunt Jean died of bladder cancer. Raymond and Violet both died of colon cancer. LeRoy is currently under treatment. These are my father's relatives. About Uncle Ray I remember very little, except that he, along with my dad, was one of the less loud of the concrete-pouring, brick-laying Steingraber brothers. Aunt Jean laughed a lot and once asked me to draw a pig so she could tape it to her refrigerator door. Red-haired Aunt Vi cooked magnificent dinners, was partial to wearing pink, and was married to a man truly untempted by silence. Together, she once remarked, the two of them sure knew how to enjoy themselves. Her widowed husband, my Uncle Ed, is now being treated aggressively for prostate cancer. Nonetheless, at last report, he was busy building a shrine to his wife out in the backyard. When it comes to expressions of grief, my father's side of the family tends toward large-scale construction projects.

The man who was to be my brother-in-law was stricken with intestinal cancer at the age of twenty-one. He cleaned out chemical drums for a living. Three years before Jeff's diagnosis, I was diagnosed with bladder cancer, and three years before my diagnosis, my mother learned she had metastatic breast cancer. That she is still alive today is a topic of considerable wonder among her doctors. Mom is matter-of-fact about this, although she will, if prompted, drily point out that she has outlived her oncologist and three of her other doctors, two of whom died of cancer.

My mother was first diagnosed in 1974, a year that is considered an anomaly in the annals of breast cancer. Graphs displaying U.S. breast cancer incidence rates across the decades show a gently rising line that suddenly zooms skyward, falls back, then continues its slow ascent. The story behind the blip of '74 has been deemed a textbook lesson in statistical artifacts.

In this year, First Lady Betty Ford and Second Lady Happy Rockefeller both underwent mastectomies. The words *breast cancer* entered public conversation. Women who might otherwise have delayed routine checkups or who were hesitant to seek medical opinion about a lump were propelled into doctors' offices. The result was that a lot of women were diagnosed with breast cancer within a short period of time, my mother among them.

When I, at age fifteen, inquired why my mother was in the hospital, the answer was "Because she has what Mrs. Ford has." When my mother, at age forty-four, questioned whether a radical mastectomy was necessary, she was told, "If it's good enough for Happy, it's good enough for you."

Back at home, a new fixture appeared on the dresser in my parents' bedroom: a bald Styrofoam head. It had come with the wig—which it dutifully wore when my mother didn't—and it remains in my mind as the most vivid image of her illness. Its features were peculiar. It lacked ears. Its closed eyes and too-small nose were half formed, as though worn smooth by water. It wore the serene, expressionless face of someone drowned or unborn.

Not that the rest of us were any more demonstrative. My father vanished into his workshop. I became the heroine of homework and long walks. My twelve-year-old sister wrote protracted, angry manifestos—and then tore them up into small fragments. These were secretly reassembled and read by our mother, who steadfastly believed that an atmosphere of normalcy was health promoting.

Some twenty years later, Mom and I sit out on my Boston balcony, drinking iced tea. I describe some medical decisions that I am facing. She provides calm, thoughtful advice—as I knew she would. Finally, I ask her about all those years of chemotherapy, surgeries, and bad news. Did she feel supported during that time?

She looks away. "Too much sympathy would have weakened me." It isn't exactly an answer to my question, and I want to ask what she means. But I don't.

My sister and I sit out in her backyard, drinking beer and watching her boys chase fireflies. I realize—as though for the first time—that she had seen her mother, sister, and fiancé all in treatment for cancer by the time she was old enough for college. I ask her about this.

"It just kept happening." Julie says, ticking off the chronology of diagnoses we both have memorized. "You and I quit talking for a while. Dad and Mom quit talking. We all got very quiet."

"That's how I remember it, too. Everybody lost their vocabulary." I want to ask her about Jeff's death and about the Styrofoam head. But I don't.

twelve

ecological roots

Among a pair of identical twins I know, both have identical cowlicks. However, one is deathly allergic to walnuts and the other not at all. A study published in 2005 offers a possible explanation for why. Twins, who are natural genetic clones of each other, turn out to have different epigenomes. That is to say, the set of instructions that overlie the instructions of the genome—the patterns of methyl groups and histones attached to their identical chromosomes—is not identical.

Moreover, in a phenomenon called *epigenetic drift*, twins become more different with time. Younger twins are more alike than older twins. As twins age, their genetic activity becomes increasingly distinguishable from one another. However much the two siblings continue to resemble each other physically, their "gene-expression portraits" look less and less alike as the years go by. What's more, epigenetic markers are more dissimilar among twins who have spent less of their lives together than among those who spend more years in close proximity. These findings, say the authors, underline the significant role that environmental factors play in translating shared genes into biological differences.

Identical twins are the mirror image of adoptees. Adopted children and the people they share the dinner table with share none of their genomic material. And so I wondered, as I read this study on twins, if the reverse process might not also take place: Growing up together in the same home, do adopted siblings, like my sister and me, experience something like

The single paper on cancer among adoptees was published in 1988 in Denmark. This study, as mentioned in the previous chapter, reports tight associations between death by cancer among adoptees and death by cancer among their adoptive parents. Since then, to my knowledge, no further cancer research on adoptees has been conducted. This is not an oversight. Epidemiologists seeking to understand the relative contributions of genes and environment in creating risks for cancer consider adoptee studies "the most powerful design" because adoption, by definition, segregates genetic origins from environmental influences. The problem is the secrecy that surrounds adoption in the United States—codified as sealed records and irretrievable birth certificates—which makes impossible the task of obtaining medical information about the biological relatives of adoptees. Even for adoptees. (Only a few states have opened adoption records for adopted adults, and the recent phenomenon of open adoption, is, needless to say, not retroactive.)

That leaves twins. Looking at differences in the cancer incidence among pairs of identical and fraternal twins can isolate—as much as any study can—the role of DNA in cancer causation. What recent studies show is that inherited genetic factors make only a minor contribution to susceptibility of most cancers. "The environment has the principal role in causing sporadic cancer." This was the conclusion of a large study of twins in Scandinavia, which has excellent cancer registries as well as long-standing twin registries. The risk of an identical twin developing the same cancer as their sibling was between 11 and 18 percent. For fraternal twins, who are no more genetically alike than other siblings, the risk was 3 to 9 percent. To be sure, this difference of twofold or more represents the power of genes in determining cancer risk. But if the genetic inheritance were the main driver of risk, you would expect to see cancer concordance between identical twins approach 100 percent rather than not-quite-even 20 percent.

The twin studies are part of a larger body of evidence that emphasizes the rarity of so-called cancer genes—inherited genetic defects that, all by themselves, strongly predispose their hosts to cancer. This is a recent insight. During the 1990s, the story looked different. As the Human Genome Project was racing to decode the all the DNA in a human cell and identify the location of the genes thereon—a project mostly completed in 2003—many cancer researchers hoped and believed that a new day was at hand in our understanding of the genetic underpinnings of cancer. What has emerged instead is a picture that is both simpler and more complicated than predicted. Humans have far fewer genes than the one hundred thousand that researchers had forecast. In fact, we possess only about twenty-five thousand, roughly the same number as a mouse. It also turns out that we share many genetic elements (such as families of proteins) with creatures scattered far across the web of life, including plants, flies, and worms. The human genome is a humbling, but unlonely, portrait of ourselves.

The complexity arises from two independent lines of inquiry. First, there is the emerging evidence that accumulations of genetic errors, not just one, are behind the birth of any single tumor. Second, it looks increasingly likely that regulation of those genes by epigenetic factors is where much of the story lies. If so, then potentially hundreds of genes collectively contribute to the risk of cancer, and they do so through intercourse with the larger ecological world. And they do so differently at different points of development. And we don't yet know where many of these genes are located. And we don't know when their critical windows of vulnerability are open.

In sum, we have far fewer total genes than we thought we had, but many more of those genes are involved in cancer than we had once presumed, and their involvement or noninvolvement is modulated by environment factors.

However surprising these results, they make unsurprising the results of several other studies. The Swedish Family-Cancer Database—the largest data set of its kind in the world—shows that family history of cancer plays a modest role in the risk of disease. According to a 2008 analysis, prostate cancer shows the strongest link to family history: 20 percent of men so diagnosed had a father or brother previously diagnosed with the same disease. For breast cancer, the familial proportion was 13.6 percent. For colon cancer, 13 percent. For bladder cancer, 5 percent. (Another way of expressing these figures is to say, for example, that 95 percent of all bladder cancer patients in Sweden have no parent or sibling who share this identity.)

Here in the United States, much attention is paid to two rare breast cancer genes, BRCA 1 and BRCA 2, which do in fact confer very high risks for the disease. Genetic tests are available to women who want to learn if they are carriers, and, for many women, there is real value in the knowledge. But even here, genes are not destiny. Among women who are carriers of BRCA 1 and BRCA 2 mutations, 30 percent never go on to develop the disease. Moreover, there are generational differences: of BRCA mutation carriers born before 1940—whose early childhoods predate the dawn of the synthetic chemical industry—only 24 percent developed breast cancer by age fifty. Of those born after 1940, 67 percent had breast cancer by age fifty. These results indicate that even in the exceptional situation in which the inheritance of a single mutated gene strongly predisposes for cancer, environmental factors can modulate risk.

Similarly, for pancreatic cancer, the most lethal cancer of them all, family history and environmental exposures interact with each other. Pancreatic cancer does have a familial component, as revealed by the Johns Hopkins National Familial Pancreas Tumor Registry. But, as explained by a 2009 study, family history can interact multiplicatively with exposures to asbestos, radon, and second-hand cigarette smoke (especially during childhood) to result in an earlier age at diagnosis. A multiplicative interaction is a sign of synergism: each risk factor makes more potent the other and, all together, their cumulative effect is more than the sum of their parts. Family history without environmental exposures delays onset of disease and improves survival. You get more birthdays.

Let me speak plainly here. I enjoy dazzling science. Genomes and epigenomes intrigue me. And I would like, as an adopted person, to have access to my genetic history. It would surely offer me some perspective as I navigate medical decisions and attempt to convince my insurance company to pay for certain procedures. I felt the absence of knowledge about my medical history most acutely when I became suddenly anemic in my 30s and was put through a battery of screening tests that finally uncovered the presence of a large, precancerous adenoma in my colon. There are indeed familial types of colon cancer, and knowledge of a positive family history changes the protocols of when, how frequently, and by what method a person should be screened. For this and other reasons, I support adoption reform and open records. (Aside: with or without a family history of colon cancer, if you are over fifty and have not yet received a colonoscopy, close this book and go make your appointment.)

But whatever my own personal and professional curiosities about genetics, the evidence seems clear to me: training a spotlight on ancestry focuses us on the one piece of the cancer puzzle we can do nothing about. The age-adjusted cancer incidence rate is not higher now than it was fifty years ago because we have sprouted new cancer genes. Rare, heritable genes that predispose their hosts to cancer by creating special susceptibilities to the effects of carcinogens have undoubtedly been with us for a long time. But the ill effects of some of these genes might well be diminished by lowering the burden of environmental carcinogens to which we are all exposed. Even in the case of hereditary colon cancer, for example, what is passed down the generations is a faulty DNA repair gene. Its human heirs are thereby rendered less capable of coping with environmental assaults on their genes or repairing the spontaneous mistakes that occur during normal cell division.

~~~~~~
~~~~~~

In 1983, I took a train home from Michigan to Illinois to see my family for the holidays—and to keep an appointment at the hospital.

The scheduling of cancer checkups is always an elaborate decision. The calendar date must avoid [illegible]. Monday or Tuesday appointments are best; otherwise, one risks waiting through the weekend for the results of a laggard lab test or delayed radiology report. It's also best if these appointments fall within a hectic deadline-filled month so that frenetic activity can preclude fretfulness. During the years I was a graduate student, this meant the ends of semesters, which explains why some half-dozen Christmas carols now remind me of outpatient waiting rooms. This particular appointment was destined to yield test results described as "unremarkable" (my favorite medical adjective). More remarkable was my journey there.

Something about the landscape changes abruptly between northern and central Illinois. I am not sure what it is exactly, but it happens right around the little towns of Wilmington and Dwight. The horizon recedes, and the sky becomes larger. Distances increase, as though all objects are moving slowly away from each other. Lines become more sharply drawn. These changes always make me restless and, when driving, drive faster. But since I am in a train, I close the book I am reading and begin straightening the pages of a newspaper strewn over the adjacent seat.

That is when my eye catches the headline of a back-page article: SCIENTISTS IDENTIFY GENE RESPONSIBLE FOR HUMAN BLADDER CANCER. Pulling the newspaper onto my lap, I stare out the window for awhile. It is only early evening, but the fields are already dark, a patchwork of lights quilted across them. I look for signs of snow. There are none. Finally, I read the article.

Using mice, researchers at the Massachusetts Institute of Technology had transformed normal cells into cancer cells using the DNA from a human bladder tumor. Through this process, they located the segment of DNA responsible for the transformation, and they were able to pinpoint the exact alteration that had caused a respectable gene to go bad.

In this case, the mutation turned out to be a substitution of one unit of genetic material for another in a single rung of the DNA ladder.

Namely, at some point during DNA replication, a double-ringed base called guanine was swapped for the single-ringed thymine. Like a typographical error in which one letter replaces another—*snow* instead of *show*, *block* instead of *black*—the message sent out by this gene was utterly changed. Instead of instructing the cell to manufacture the amino acid glycine, the altered gene now spelled for valine.

Guanine instead of thymine. Valine instead of glycine. I look away again—this time at my own face superimposed over the landscape by the window's black mirror. If, in fact, this mutation was involved in my cancer, when did it happen? Where was I? Why had it escaped repair? I had been betrayed. But by what?

Nine years later, other researchers would determine that this point mutation alters the structure of proteins involved in signal transduction—the crucial line of communication between the cell membrane and the nucleus that helps coordinate cell division. And many other discoveries have followed from there. Besides the oncogene just described, two tumor suppressor genes, p15 and p16, have also been discovered to play a role. Their deletion is a common event in transitional cell carcinoma, the kind of cancer I had. Mutations of the famous p53 tumor suppressor gene, with guest-star appearances in so many different cancers, have been detected in more than half of invasive bladder tumors. Also associated with transitional cell carcinomas are surplus numbers of growth factor receptors. Their overexpression has been linked to the kinds of gross genetic injuries that appear near the end of the malignant process. At the beginning of the malignant process, changes in epigenetic methylation patterns appear.

The nature of the transaction between these various genes and certain bladder carcinogens has likewise been worked out in the years since a newspaper article introduced me to the then-new concept of oncogenes. Consider that redoubtable class of bladder carcinogens called aromatic amines—present in cigarette smoke; added to rubber; formulated as dyes; used in printing; and featured in the manufacture of pharmaceuticals and pesticides. *O*-toluidine is a member of this group. The first reports of invasive bladder cancers among workers in the synthetic dye industry were published in 1895. (Recall also Wilhelm Hueper's dogs, described in

We also know quite a lot about what happens after that. We know that men exposed to aromatic amines are more susceptible to contracting bladder cancer if they also have variations in genes named GSTM1, GSTT1, and NAT2. We know that gene variations can alter the function of enzymes whose job it is to repair DNA damage in the lining of the bladder. We've identified the three main types of repair needed after exposure to bladder carcinogens. We know that failure to conduct these three repairs can lead to genomic instability and thereby to cancer.

We also now know that aromatic amines are gradually detoxified by the body through a process called acetylation. Like all such processes, it is carried out by a special group of detoxifying enzymes whose actions are controlled and modified by a number of genes. People who are slow acetylators have low levels of these enzymes and are at greater risk of bladder cancer from exposure to aromatic amines. Members of this population can be readily identified because they bear significantly higher burdens of adducts than fast acetylators at the same exposure levels. These genetically susceptible individuals hardly constitute a tiny minority: more than half of Americans and Europeans are estimated to be slow acetylators.

Very likely, I am one. You may be one, too.

So we know a lot about bladder cancer. Bladder carcinogens were among the earliest human carcinogens ever identified, and one of the first human oncogenes ever decoded was isolated from some unlucky fellow's bladder tumor. More than most malignancies, bladder cancer has provided researchers with a picture of the sequential epigenetic and genetic changes that unfold from initiation through promotion to progression, from precursor lesions to increasingly more aggressive tumors.

And yet all this knowledge about mechanisms has not translated into an effective campaign to prevent the disease. The overall incidence rate of bladder cancer increased 10 percent between 1973 and 1991 and has stayed up. Nor has all this knowledge about mechanisms produced dramatic breakthroughs in treatment: the five-year mortality rate from bladder cancer has inched downward by only 5 percent over the past two decades. The hope, nevertheless, is that the deluge of recent findings about the epigenetic and genetic alterations associated with bladder cancer will someday produce reliable urinary biomarkers that could be used as a tool for early detection of disease recurrence. A few of these are already in use, and their specificity and accuracy will surely improve with time. About this, I am willing to be patient. I make use of these tests.

Significantly less than half of all bladder cancers are thought to be attributable to cigarette smoking, which is the largest known risk factor for this disease. The rate of cigarette smoking has been falling, and with it, the lung cancer rate (among men). Yet there is no parallel decline in bladder cancer incidence. Perhaps bladder cancer has a longer lag time than lung cancer. Again, I'm willing to be patient. But in the meantime, the question still remains: What is causing bladder cancer in the rest of us, the majority of bladder cancer patients, for whom tobacco is not a factor?

When I wrote the first edition of this book, this paragraph contained the following sentence: *The Toxics Release Inventory disclosed environmental releases of the aromatic amine* o*-toluidine that totaled 14,625 pounds in 1992 alone.* Here is the update: In 2007, the Toxics Release Inventory disclosed environmental releases of the aromatic amine *o*-toluidine that totaled 16,536 pounds.

Two other aromatic amines also appear in the 2007 inventory of releases.

What all the new knowledge about the molecular pathways of bladder cancer has *not* inspired, apparently, is a considered evaluation of known and suspected bladder carcinogens. In addition to aromatic amines, they include, at the very least, pesticides, solvents, arsenic, polycyclic aromatic hydrocarbons, and water disinfection by-products. Do their mixtures create interactive effects? What are our various routes of exposure? Does the total daily burden of our exposure to bladder carcinogens exceed the metabolic capacities of the half of us who are slow acetylators? These questions remain, to my knowledge, largely unaddressed by anyone. The possession of a defective carcinogen-detoxifying gene would matter less in a culture that did not tolerate carcinogens in air, food, and water.

And more immediately, why—a century after some of them were so identified—do powerful bladder carcinogens like aromatic amines continue to be manufactured, imported, used, and released into the environment? Why have safer substitutes not replaced them all? What is the position of the oncology community, the urology community, the cancer research community, and the public health community on this issue?

Bladder cancer is the fourth most common cancer among men. It ranks eighth among women, but our survival rates are worse. It has the highest recurrence rate of any cancer. It requires lifelong surveillance and is therefore one of the most expensive cancers to treat. Among Medicare patients, bladder cancer racks up the highest per-patient costs of any cancer. In 2008, 68,810 people were diagnosed with bladder cancer and 14,000 people died. About this, I am not willing to be patient.

~~~

One of the obstacles that prevent us from addressing cancer's environmental roots is the word *lifestyle*. Risks of lifestyle are not independent of environmental risks. Our lives are played out within the ecological world, after all. We eat food—we choose our diets—but that food also has particular environmental origins. And yet public education campaigns about cancer consistently emphasize lifestyle and downplay the environment, or subsume the latter into the former. I collect the colorful pamphlets on cancer that are made available in hospitals and clinics. In the 1990s, when I was teaching college biology and also spending many hours in doctors' offices, I began to compare the descriptions of cancer
~~~

in the tracts displayed in the skinny, silver racks above the magazines in my doctor's waiting room with the chapter on cancer provided in my students' genetics textbook. Here is what I found.

On the topic of how many people get cancer, a pink and blue brochure published by the U.S. Department of Health and Human Services offered the following:

> Good News: Everyone does not get cancer. 2 out of 3 Americans never will get it.

Whereas, according to *Human Genetics: A Modern Synthesis*:

> One of three Americans will develop some form of cancer in his or her lifetime, and one in five will die from it.

(Since these materials were published, the proportion of Americans contracting cancer has risen from 30 to 40 percent.)

On the topic of what causes cancer, the brochure states:

> In the past few years, scientists have identified many causes of cancer. Today it is known that about 80% of cancer cases are tied to the way people live their lives.

Whereas the textbook contends:

> As much as 90 percent of all forms of cancer is attributable to specific environmental factors.

In regard to prevention, the brochure emphasizes individual choice and responsibility:

> You can control many of the factors that cause cancer. This means you can help protect yourself from the possibility of getting cancer.

Because exposure to these environmental factors can, in principle, be controlled, most cancer could be prevented. . . . Reducing or eliminating exposures to environmental carcinogens would dramatically reduce the prevalence of cancer in the United States.

The textbook goes on to identify some of these carcinogens, the routes of exposure, and the types of cancer that result. In contrast, the brochure emphasizes the importance of personal habits, such as sunbathing, that raise one's risk of contracting cancer. Thus, in my students' textbook, vinyl chloride is identified as a carcinogen to which PVC manufacturers are exposed, whereas in the brochure, occupations that involve working with certain chemicals are called a risk factor. The textbook declares that "radiation is a carcinogen." The brochure advises us to "avoid unnecessary X-rays." Both emphasize the role of diet and tobacco.

In its ardent focus on lifestyle, the Good News brochure is typical of the educational pamphlets in my collection. By emphasizing personal habits rather than carcinogens, they frame the cause of the disease as a problem of *behavior* rather than as a problem of *exposure* to disease-causing agents. At its best, this perspective can offer us practical guidance and the reassurance that there are actions we as individuals can take to protect ourselves. (Not smoking, rightfully so, tops this list.) At its worst, the lifestyle approach to cancer is dismissive of hazards that lie beyond personal choice. A narrow focus on lifestyle—like a narrow focus on genetic mechanisms—obscures cancer's environmental roots. It presumes that the ongoing contamination of our air, food, and water is an immutable fact of the human condition to which we must accommodate ourselves. When we are urged to "avoid carcinogens in the environment and workplace," this advice begs the question. Why must there be carcinogens in our environment and at our job sites? Why are we still manufacturing carcinogens at all?

Cancer is not the first disease to inspire this kind of message. In 1832, at the height of an epidemic, the New York City medical council announced that cholera's usual victims were those who were imprudent, intemperate, or prone to injury by the consumption of improper medicines. Lists of cholera prevention tips were posted publicly. Their advice ranged from avoiding drafts and "crude vegetables" to abstaining from alcohol. Maintaining "regular habits" was also said to be protective. Decades later, improvements in public sanitation would bring cholera under control—without any understanding of the biological mechanisms or the molecular pathways to cholera, or even knowing the identity of the causative agent. (The pathogen responsible would finally be isolated by the bacteriologist Robert Koch in 1883.)

Of course, the behavioral changes urged by the 1832 handbills were not all without merit: uncooked produce, as it turned out, *was* an important route of exposure, but it was a fecal-borne bacteria that was the cause. Not a salad-eating lifestyle.

The orthodoxy of lifestyle has loosened somewhat since I started my pamphlet collection in 1990. For years, it was rare to encounter the word *carcinogen* or *environment* in American Cancer Society materials. Now the words are sometimes there. A friend with leukemia who lives in Iowa reports that an educational pamphlet he recently picked up in his hematologist's office made indirect reference to pesticides in the section about the possible causes of his disease. But the word did not recur in the section on how to prevent it. (As in, "avoid pesticides.")

What the American Cancer Society offers the public about leukemia on its Web site confirms this disparity. Pesticides are alluded to in the description of causes, but the section on disease prevention does not mention the word again and certainly does not advocate for organic farming. Nor does it mention that formaldehyde exposure is linked to leukemia or offer green chemistry as a preventive approach. What it does offer is this:

> Although many types of cancer can be prevented by lifestyle changes to avoid certain risk factors, there are very few known risk factors for chronic lymphocytic leukemia (CLL). Most CLL patients have no known risk factors, so there is no way to prevent these cancers.

This refrain is repeated in its primer on childhood leukemia:

> Although many adult cancers can be prevented by lifestyle changes . . . there is no known way to prevent most childhood cancers at this time. Most adults and children with leukemia have no known risk factors, so there is no way to prevent their leukemias from developing.

These are not entirely incoherent statements. Risk factors apply to populations, not to individuals. But within these defeatist statements is the presumption that cancer prevention is the responsibility only of individuals. As though our larger systems of industry and agriculture had nothing to do with it. As though the laws that permit or disallow the use of carcinogens and untested chemicals were irrelevant to our situation.

By contrast, cancer organizations in other nations seem far less bewildered about how to prevent cancer. Recall that the Canadian Cancer Society openly supported legislation to ban the cosmetic use of pesticides precisely because of troubling links between pesticide exposure and childhood cancers. As of April 22, 2009—Earth Day—hardware and garden stores across Ontario were required to remove 245 chemical bug and weed killers from their shelves. Beginning on that day, as part of a program to decrease toxic exposures to chemicals linked to cancer, residents of Ontario could no longer use pesticides for lawns and gardens, and stores could no longer sell them. (Pesticides for use on termites, mosquitoes, and poison ivy were exempt from the ban.)

In Europe, the French cancer research organization ARTAC (Association for Research and Treatments Against Cancer) has called for a ban on all products that are "certainly or probably carcinogenic" in an international declaration on chemical pollution called the Paris Appeal. Signed in 2006 by thousands of scientists and physician groups representing two million European doctors, the document makes explicit the connections between the chemicals responsible for climate change and the chemicals responsible for polluting people. Among the declaration's many whereas-es, are the following:

> . . . Whereas the global incidence of cancers is on the rise worldwide; whereas since 1950 the incidence of cancers among the populations of highly industrialized nations has increased steadily; whereas anyone, young or old, can be affected by cancer; whereas chemical pollution, the magnitude of which remains to be assessed, could largely contribute to the onset of cancer. . . .

The document's primary architect is Dominique Belpomme, a medical oncologist. In urging the United Nations to take up the recommendations put forth in the appeal, Belpomme emphasized a geographic pattern associated with a startling new trend within the European Union: the rising cancer rate among newborns. In response to the Paris Appeal and other pressures, the European Commission announced, in June 2009, that it is launching the European Partnership for Action Against Cancer with an aim toward coordinating cancer prevention and early detection efforts across the various nations of the European Union. Along with lifestyle factors, the partnership will look at occupational and environmental factors.

One conceptual obstacle that has blinded public health agencies here in the United States from appreciating the role of the environment in the burden of cancer is a quarter-century-old pie chart. This is not just anybody's chart. It was part of a paper entitled "The Causes of Cancer" that was originally commissioned by Congress and subsequently published in the *Journal of the National Cancer Institute* in 1981. In it, authors Richard Doll and Richard Peto attempt to ascribe fractions to each cause of cancer.

are apportioned only a tiny sliver, totaling 6 percent for both slices. The report's conclusion, that the majority of cancer deaths could be avoided by improving diet and eliminating smoking, was repeated in 1996 by the Harvard Center for Cancer Prevention. The Harvard report adjusted the numbers a bit, but duplicated the earlier estimates on pollution and workplace carcinogens: 6 percent.

Both the pie and the conclusions drawn from it have been extraordinarily influential, as agencies have used them to formulate cancer control policies and educational programs. Lifestyle became the bull's-eye of cancer prevention efforts, while targeting of environmental factors, perceived as a trivial contribution to the cancer problem, was seen as inefficient. The state health department in Illinois, for example, reprinted the original 1981 chart years later in a county-by-county analysis of cancer incidence and concluded, "many persons could reduce their chances of developing or dying from cancer by adopting healthier lifestyles and by visiting their physicians regularly for cancer-related checkups." The fact that Illinois is a leading producer of hazardous waste, a heavy user of pesticides, and home to a number of Superfund sites was neither mentioned nor considered. Nor was the fact that Doll and Peto had restricted their original analysis to white men.

It's easy to see that the chart is outdated. Since 1981, smoking rates and alcohol consumption are down; obesity rates are up. And, most of us have developed by now, I hope, an awareness that the experiences of white men do not universally represent us all—and certainly not children, for whom cancer rates are rising more rapidly than for adults.

But more profoundly, the simplistic accounting system itself—assigning percentages of cancer to isolated, independent risk factors, each presumed to function alone—is fatally flawed. How do we account for malignancies, such as certain liver cancers, to which both drinking and job hazards contribute? Or lung and bladder cancers where both job hazards and smoking conspire? Should the effects of pesticide residues be tallied under "pollution" or under "diet"? What about pollution's indirect effects—such as hormone disruption or immune suppression—which act to augment the danger of risk factors across the board? What about toxicants that reprogram developmental pathways in the very young in ways that leave them more susceptible to cancer later on? What if a mixture of pollutants disrupts metabolic pathways and leads to obesity? (There is evidence for this.) What if obesity is associated with a greater retention of pollutants? (And there is evidence for this, too.) Interactions between risk factors aside, how can the environment's cancer fraction be estimated at all when the vast majority of industrial chemicals in commerce have never been tested for their ability to cause cancer?

Many public health figures, including epidemiologist Richard Clapp at Boston University and oncologist Dominique Belpomme himself, have pointed out the folly of attributing specific fractions of cancer risk to single causes. Now that we recognize that most cancers arise from the accretion of many small genetic and epigenetic changes over long periods of time—likely aided and abetted by inflammatory processes, endocrine disruption, changes in tissue architecture, and alterations in cellular signaling—it is very likely that any given tumor has multiple origins. Attributing percentages of cancer to a single cause is an obsolete practice that no longer reflects our contemporary understanding of the complex web of causation underlying and surrounding the disease.

The American Cancer Society nevertheless reprises this approach—and recycles the original numbers—in its *Cancer Facts and Figures 2008*. It dutifully attributes 30 percent of all cancer deaths to smoking; 35 percent to food, physical activity, and obesity; and 6 percent to occupational and environmental pollutants, as though we still lived in the one mutation/one cancer world of 1981 where endocrine disruption, epigenetics, GIS mapping, biomarkers, and biomonitoring did not yet exist. Stuck inside this time capsule, we cannot see what the new science shows us: that cancer is an ecological disease with upstream interactions of a complexity that preclude the assignment of a single cause. Cancer is a river whose headwaters is not a single source but a network of streams that both fan out and reconnect.

During the last year of her life, Rachel Carson discussed before a U.S. Senate subcommittee her emerging ideas about the relationship between environmental contamination and human rights. The problems addressed in *Silent Spring*, she asserted, were merely one piece of a larger story—namely, the threat to human health created by reckless pollution of the living world. Abetting this hidden menace was a failure to inform common citizens about the senseless and frightening dangers they were being asked, without their consent, to endure. In *Silent Spring*, Carson had predicted that full knowledge of this situation would lead us to reject the counsel of those who claim there is simply no choice but to go on filling the world with poisons. Now she urged recognition of the right to know about poisons introduced into one's environment by others and the right to protection against them. These ideas are Carson's final legacy.

The process of exploration that results from asserting our right to know about carcinogens in our environment is a different journey for every person who undertakes it. For all of us, however, I believe it necessarily entails a three-part inquiry. Like the Dickens character Ebenezer Scrooge, we must first look back into our past, then reassess our present situation, and finally summon the courage to imagine an alternative future.

The first part of journey is, in essence, a search for our ecological roots. Just as awareness of our genealogical roots offers us a sense of heritage and cultural identity, our ecological roots provide a particular appreciation of who we are biologically. It means asking questions about the physical environment we have grown up within, the molecules of which are woven together with the genome we inherited from our ancestors. After all, except for our DNA, with its modest number of genes, all the material that is us—from bone to blood to breast tissue—has come to us from the environment, even including, perhaps, how our genes express themselves.

Going in search of our ecological roots has both intimate and far-flung dimensions. It means learning about the sources of our drinking water (past and present), about the prevailing winds that blow through our communities, and about the agricultural system that provides us food. It involves visiting grain fields, as well as cattle lots, orchards, pastures, and dairy farms. It demands curiosity about how our apartment buildings are, and have been, exterminated, our clothing cleaned, our golf courses maintained. It means asserting our right to know about any and all toxic ingredients in products such as household cleaners, paints, and cosmetics. It requires a determination to find out where the underground storage tanks are located, how the land was used before the subdivision was built over it, what was and is being sprayed along the roadsides and rights-of-way, and what exactly may still be going on behind that barbed-wire fence at the end of the street. Some individuals are taking the right-to-know one step further and are submitting their blood and urine for biomonitoring. Indeed, a new genre of environmental autobiography may be emerging in which the history of the body is read as a history of chemical exposures. With the results of biomonitoring comes the evidence for *toxic trespass*—the involuntary use of one's body as a receptacle for someone else's chemicals.

Acquiring a copy of the Toxics Release Inventory for one's home county, as well as a list of local hazardous waste sites, is a simpler place to begin a right-to-know journey. (See the Afterword for Web sites.) Such information is available for all years back to 1987. These documents often contain clues to the more distant past as well: the toxic chemicals loitering around an abandoned Superfund site, for example, can reveal what kinds of activities occurred there decades earlier.

In full possession of our ecological roots, we can begin to survey our present situation. This requires a human rights perspective. Such a view recognizes that the current system of regulating the use, release, and disposal of known and suspected carcinogens—rather than preventing their generation in the first place—is intolerable. So is the decision to allow untested chemicals free access to our bodies, until which time they are finally assessed for carcinogenic properties. Both practices show reckless disregard for human life.

A human rights view would also recognize that we do not all bear equal risks when carcinogens are allowed to circulate within our environment. Workers who manufacture carcinogens are exposed to higher levels, as are those who live near the chemical graveyards that serve as their final resting place. We know that toxic sites are disproportionately located in poor and minority communities. We know also that disparities in cancer rates exist between U.S. whites and U.S. blacks that cannot be explained by genetic differences. (To its credit, the American Cancer Society points this out.)

Moreover, people are not uniformly vulnerable to the effects of environmental carcinogens. Among those who may be affected more profoundly are infants, whose cellular signaling pathways are still under construction; adolescents, whose bodies are being resculpted by sex hormones; and the elderly, whose detoxifying mechanisms are less efficient. Individuals with genetic predispositions and those with significant prior exposures may also suffer more damage. Cancer may be a lottery, but we do not each of us hold equal chances of winning. When carcinogens are deliberately or accidentally introduced into the environment, some number of vulnerable persons are consigned to death. The impossibility of tabulating an exact body count does not alter this fact. From a human rights standpoint, these deaths must be made visible. Here is one way of doing that. Suppose we assume for the sake of argument that the 1981 estimate concerning the proportion of cancer deaths due to environmental exposures is absolutely accurate. That estimate, put forth by those who seek to dismiss environmental concerns, remember, is 6 percent. (2 for pollution plus 4 for occupational exposures.) Six percent, as the American Cancer Society itself points out, means 33,600 people in the United States expire each year from cancers caused by involuntary exposures to toxic chemicals. All by itself, that would make environmentally caused cancer the eleventh leading cause of death in the United States. 33,600 is greater than the total annual number of homicides in the United States—a figure that is considered a matter of national shame. It exceeds the annual number of suicides—a figure so tragic that phone numbers for suicide prevention hotlines rightly appear on the covers of telephone books. 33,600 is far more than the number of women who die each year from hereditary breast cancer—an issue that launched multi-million-dollar research initiatives. It is more than ten times the number of nonsmokers estimated to die each year of lung cancer caused by exposure to secondhand smoke—a problem so serious it warranted sweeping changes in laws governing air quality in public spaces.

None of these 33,600 Americans will die quick, painless deaths. They will be amputated, irradiated, and dosed with chemotherapy. They will expire privately in hospitals and hospices and be buried quietly, at a rate of ninety-two funerals a day. Some of them will be children. Photographs of their dead bodies will not appear in newspapers. We will not know who most of them are. Their anonymity, however, does not moderate this violence. In 2007, 834,499,071 pounds of known or suspected carcinogens were released into our air, water, and soil by reporting industries. In this light, the 33,600 deaths can be seen as homicides.

After having carefully appraised the risks and losses that we have endured from chemical carcinogens, we can begin to imagine a future in which our right to an environment free of such substances is respected. It is unlikely that we will ever rid our environment of all of them. However, as Rachel Carson herself observed, the elimination of a great number of them would reduce the carcinogenic burden we all bear and thus would prevent considerable suffering and loss of human life. The *precautionary principle* can assist us in this effort. (See the Afterword for the full text of the Wingspread Statement on the Precautionary Principle.)

The idea behind the precautionary principle is that public and private interests should act to prevent harm before it occurs. It dictates that indication of harm, rather than proof of harm, should be the trigger for action—especially if delay may cause irreparable damage. As explained by the European Environment Agency, the point of anticipatory action is to prevent the construction of "machines of unstoppable consequences." Actions taken to mitigate climate change invoke this principle. Central to it is the recognition that we have an obligation to protect human life. By contrast, our current methods of regulation appear governed by what one researcher has called the reactionary approach: anyone may freely introduce new hazards into the environment, and then regulators wait until damage is proven before action is taken. It is a system tantamount to running an uncontrolled experiment using human subjects.

It is time now, we can insist, to run this experiment in reverse. (This is called an intervention study.) Let's declare that the production of carcinogens and suspected carcinogens is the result of outmoded technologies and invest in green chemistry. Let's aim for zero waste, eliminating the need to bury garbage over drinking water or light it on fire inside incinerators. Let's invest in diversified, local, organic farming. This would yield five immediate benefits: decreased amounts of carcinogenic diesel exhaust created from the long-distance transport of food; decreased pesticide residues in our diets; decreased pesticides in our drinking water; decreased dependency on petroleum-based fertilizers, and an increase in access to healthy foods to fight obesity. Let's invest in green energy sources and so reduce the air's load of ultrafine particles, polycyclic aromatic hydrocarbons, and aromatic amines. Let's end the fifty-year-era of petrochemicals and coal. Then let's see what happens to the cancer rates. And what happens to the cost of health care.

Embedded within the precautionary principle is the *principle of reverse onus.* This means that lack of harm, rather than harm, must be demonstrated. The reversal essentially shifts the burden of proof off the shoulders of the public and onto those who produce, import, or use the substance in question. The principle of reverse onus requires that those who seek to introduce chemicals into our environment must first demonstrate that what they propose to do has been tested and no evidence of harm has been shown. This is already the standard we uphold for pharmaceuticals, and yet for most industrial chemicals, no firm requirement for advance demonstration of safety exists. Europe has already moved toward this standard in its own new toxics policy.

Finally, all activities with potential public health consequences should be guided by an *alternatives assessment*, which presumes that toxic substances will not be used as long as there is another way of accomplishing the task. This means choosing the least harmful way of solving problems—whether it is ridding fields of weeds, school cafeterias of cockroaches, dogs of fleas, woolens of stains, or drinking water of pathogens. Alternatives assessment can be aided by *full-cost accounting*, which considers all the costs of each method, including costs that will be borne by future generations. Alternatives assessment moves us away from protracted, unwinnable debates over how to quantify the cancer risks from each individual carcinogen released into the environment and where to set legal maximum limits for their presence in air, food, water, workplace, and consumer goods. It looks toward the day when the availability of safer choices makes the deliberate and routine release of chemical carcinogens into the environment as unthinkable as the practice of slavery.

The story of cancer's ecological roots is a story of disconnections. On the one hand, we live in an era of toxicogenomics that makes possible the simultaneous investigation of all possible biochemical pathways down which toxic exposures can operate, nudging a human cell toward malignant transformation. So much data are generated by these investigations that a whole new field of computational biology has had to be invented—bioinformatics—to analyze it all. On the other hand, we still ignite coal to turn the lights on, blow up mountaintops to get at it, and fill the air with tons of known carcinogens at all stages of the process. We still pour gasoline into combustion engines in order to go places—a hundred-year-old method of transportation—and in so doing, fill the air with more carcinogens. We still pour chlorine into the drinking water—a hundred-year-old method of disinfection—and in so doing, create new carcinogens out of carbon-based chemicals that were not always present in the water supply a century ago.

On the one hand, biomonitoring can measure levels of contaminants in human blood down to parts per trillion. On the other hand, one in every [illegible] children [illegible] within a mile of a [illegible]. On the one hand, we've mapped all three billion base pairs along the human genome. On the other hand, we still don't know how many chemicals in commerce cause breast cancer. (Recently, a massive investigation was required just to pull together a database of all chemicals known to act as mammary gland carcinogens. Before 2007, no one had thought to do that.) On the one hand, the cells within my urine have undergone the latest fluorescence *in situ* hybridization testing to check for biomarkers consistent with bladder cancer. On the other hand, agriculture in my home state is dependent on the use of a fifty-year-old chemical (atrazine) that disturbs pituitary hormones and, quite possibly, breast development.

What's needed now is not more data but what is always needed in a crisis: vision, courage, and the willingness to not be paralyzed by uncertainty. I believe these qualities come easily to cancer patients and their allies. I believe we could speak with a mighty voice. Here is what Bradford Hill, medical statistician and inventor of the randomized clinical trial, told us in 1965, the year I turned six:

> "All scientific work is incomplete—whether it be observational or experimental. All scientific work is liable to be upset or modified by advancing knowledge. That does not confer upon us a freedom to ignore the knowledge we already have or postpone the action that it appears to demand at a given time."

∿∿∿∿∿

Sitting at my desk, I am skimming through a journal article about hormone disruption in young female rats. The study is unusual because the animals were exposed not to a single chemical but to a real-life, low-level mixture of substances derived from the dust, soil, and air from a dioxin-contaminated landfill. After only two days, the test animals exhibited abnormal changes in their livers, reproductive organs, and thyroid glands. Even rats exposed only to air from the landfill experienced significant changes in their development. These results indicate that current methods used for calculating health risks from chemical mixtures may underestimate certain biological effects.

Flipping back to the beginning of the report, my eye catches on a familiar word: *Illinois.* The contaminated dust, soil, and air mixtures in this study were collected from an old, inoperative landfill near my home.

Dust. Soil. Air. The year after my cancer diagnosis, I signed up for a field ecology class and learned to identify plant species in the rarest of rare Illinois habitats: the black soil prairie. Its remnants are almost completely confined to a few old pioneer gravesites. Hunkered down between headstones, I cupped the unfamiliar plants in my hands and tried to will into existence thousands of acres of these grasses and herbs, the sound of animals running, wildfires, birdsong.

As I became ever more enchanted with the Illinois prairie, I found that I was, nevertheless, unable to banish from my heart its remaining enemies—the nonnative invading species. Queen Anne's lace, ox-eye daisy, chicory, foxtail, goat's beard, teasel: all European immigrants, these are the familiar weeds of roadsides and fallow fields. My mother taught me the names of most of them. I am especially fond of teasel. It represents a special threat to prairie plants because mourners brought bouquets of it into the old prairie cemeteries, where it set seed and spread. In the winter, its stiff wands stand in the snow like pinecones on the ends of antennas. I keep a few stalks near my desk to remind me of home. I keep a scientific monograph of prairie plants on the shelf for the same reason.

After finishing the article on the health hazards of trace chemical mixtures, I look at the brown, spiny flowers and then out the window. Dust. Soil. Air. What I see are the contours of home.

afterword

In January 1998, I was invited to join an international group of scientists, lawyers, farmers, government officials, physicians, urban planners, environmental thinkers, and others for a conference on the Precautionary Principle, the subject of *Living Downstream's* final chapter. Snowbound in the elegant confines of Frank Lloyd Wright's Wingspread house in Racine, Wisconsin, we discussed what the principle really means and how its ideals might be realized.

The text of our consensus statement has gone on to have a life of its own—reprinted and referenced as an authoritative description of a promising new standard of ethical decision making.

As a witness to history in the making, I'd like to say I played a role in the crafting of this now-famous statement. In truth, I was newly pregnant with my daughter, Faith, and spent most of that weekend in the bathroom. And yet, however scant my contribution to the various drafts of our collective credo, I was, nevertheless, living its purpose. My body was busy creating new life out of molecules of air, food, and water streaming in from my environment. It was equally preoccupied with the task of turning itself into a habitable environment for a burrowing ball of cells who shared half my chromosomes, whose epigenetic programming was under way, whose developmental journey was already headed down hormone-blazed trails. (And who now sits on the couch in the next room, reading long and complicated novels while twirling her hair into cold knots. If anyone needed the protection of environmental pre-caution that weekend, it was the both of us.

Now that I'm the mother of two school-aged children, the Precau-[illegible]
daily—sometimes hourly—basis. As it does almost every parent I know.
Earlier this evening, I denied my seven-year-old's request for permission to ride his bike around the neighborhood. I almost said yes, but then I remembered that the church at the end of our block holds its weekly youth meeting on Mondays. I don't require proof that my son will suffer bodily harm if he shares the roadway with cars piloted by teenagers late to Bible study. I just require knowledge of an inherently dangerous situation. *Inherently dangerous* is my trigger for action. It was precaution that made me say, "Wait for an hour. Then you can go."

There are, in fact, two churches with active youth groups, one at each end of my street. There is also an art conservatory around the corner where my children take piano lessons. There is a weekly farmers' market by the post office, a public bus service, a library with a marvelous summer reading program, walkable sidewalks to get there, and, just beyond the village limits, a swimmable lake with a waterfall and a seriously fast sledding hill. A lot of people love this community.

There is also talk of an incinerator siting north of town. There is a gas station going in near the grade school. There is a coal-burning power plant on the lake's opposite shore that is one of the state's dirtiest. There are lawn chemicals and bad air days. There are fields of corn planted year after year on the erodable hills above the public drinking water wells.

Because there are skill sets I don't want my children to have to learn—how to schedule cancer check-ups in between college exams is one—I am determined to see the Precautionary Principle implemented in the public sphere as a tool of environmental decision making, and not just within my own household. With an emphasis on better safe than sorry, the Precautionary Principle does not tell us what we should do, but it does serve as a starting point for imagining a future where nontoxic alternatives to inherently dangerous practices are embraced as the commonsense solution. For this reason, I include the entire text of our 1998 statement here. The list of resources that follow may offer further inspiration.

And inspiration is what's required. Now that my children are old enough to ask me what it is I do exactly in my professional life and why, and what they should do with their lives and why, I tell them this. I believe we are musicians in human orchestra. It is time now to play the Save the World Symphony. It is a vast orchestral piece, and you are but one musician. You are not required to play a solo. But you are required to figure out what instrument you hold and play it as well as you can.

In the end, the environment is not just something else to worry about. It is connected to all the things we already worry about—our children, our health, our homeland—and love with all our hearts.

~~~~~~

Wingspread Statement on the Precautionary Principle

The release and use of toxic substances, the exploitation of resources, and physical alterations of the environment have had substantial unintended consequences affecting human health and the environment. Some of these concerns are high rates of learning deficiencies, asthma, cancer, birth defects and species extinctions, along with global climate change, stratospheric ozone depletion and global worldwide contamination with toxic substances and nuclear material.

We believe existing environmental regulations and other decisions, particularly those based on risk assessment, have failed to adequately protect human health and the environment—the larger system of which humans are but a part.

We believe there is compelling evidence that damage to humans and the worldwide environment is of such magnitude and seriousness that new principles for conducting human activities are necessary.

While we realize that human activities may involve hazards, people must proceed more carefully than has been the case in recent history. Corporations, government entities, organizations, communities, scientists, and
~~~~~~

other individuals must adopt a precautionary approach to all human endeavors.

Therefore, it is necessary to implement the Precautionary Principle: When an activity raises threat of harm to human health or the environment, precautionary measures should be taken even if some cause and effect relationships are not fully established scientifically. In this context, the proponent of an activity, rather than the public, should bear the burden of proof.

The process of applying the Precautionary Principle must be open, informed and democratic and must include potentially affected parties. It must also involve an examination of the full range of alternatives, including no action.

further resources

Beyond Pesticides
www.beyondpesticides.org

A nonprofit organization that identifies the risks of pesticides, promotes nonchemical alternatives, and assists individuals and community-based organizations that are struggling with issues involving pesticides—from golf courses and schools to lawn care and food production. Provides databases and fact sheets.

Breast Cancer Action
www.bcaction.org

A membership organization based in San Francisco that views breast cancer as a public health emergency and addresses social injustices, including involuntary chemical exposures, that put women at risk for breast cancer.

Breast Cancer Fund
www.breastcancerfund.org

A prevention-based organization that advocates for the identification and elimination of environmental causes of breast cancer.

Campaign for Safe Cosmetics
www.safecosmetics.org/
www.cosmeticdatabase.org

A coalition of organizations seeking to require the personal care products industry to eliminate the use of chemicals linked to cancer and birth defects. The Environmental Working Group, one of its founding members, has created the searchable Skin Deep database that profiles twenty-five thousand personal care products and their ingredients.

Canadian Cancer Society
www.cancer.ca

A national organization in Canada that supports community right to know and the precautionary principle. The Canadian Cancer Society seeks the identification and elimination of cancer-causing substances in the workplace, home, and environment.

Canadian Partnership for Children's Health and Environment
www.healthyenvironmentforkids.ca

An affiliation of groups that seek to improve children's environmental health in Canada.

Collaborative on Health and the Environment
www.healthandenvironment.org
www.database.healthandenvironment.org/

A network of more than three thousand individuals and organizations working to address growing concerns about the links between human health and environmental factors. The Toxicant and Disease Database summarizes links between chemical contaminants and 180 human diseases.

The Endocrine Disruption Exchange
www.endocrinedisruption.com

A nonprofit organization that compiles scientific evidence on the health and environmental problems caused by exposure to chemicals that interfere with hormones, with a special focus on development.

Environmental Health News
www.environmentalhealthnews.org

A syndication service that is published daily by Environmental Health Sciences. It both publishes its own articles and offers summaries of articles published in referenced journals as well as a daily digest of articles on environmental health topics, including cancer, that are published each day in the world press. I subscribe to its free daily e-letter, Above The Fold.

Environmental Working Group
www.ewg.org

A non-profit research organization that brings to light information on public health and the environment. Provides a database of chemicals found in municipal drinking water supplies throughout the United States.

Health and Environment Alliance
www.env-health.org/
www.chemicalshealthmonitor.org

The European sister organization of the Collaborative on Health and Environment. HEAL, located in Brussels, Belgium, oversees a project called Chemicals Health Monitor that provides a comprehensive compilation of recent information and evidence about the links between chemical contaminants in the environment and human health problems. HEAL provides news on chemical safety policy, especially the EU legislation called REACH, and on human biomonitoring.

International Chemical Secretariat
www.chemsec.org

A nonprofit organization in Sweden that serves as a watchdog for the legislative process and implementation of the European Union's chemical policy REACH, which entered into force in June 2007. It seeks the elimination of hazardous substances and is the standard-bearer of the Precautionary Principle in Europe.

The Land Connection
www.thelandconnection.org

An Illinois-based nonprofit organization that promotes community-based, organic food systems by training new farmers and connecting them with farmland. Thanks to this organization, my family's four-generation farm is, once again, an organic operation.

National Pollutant Release Inventory

www.ec.gc.ca/inrp-npri/

Canada's publicly accessible inventory of pollutant releases to air, water, and land.

Pesticide Action Network, North America

www.panna.org

www.pesticideinfo.org/

A nonprofit organization that promotes the elimination of highly hazardous pesticides and offers a pesticide database on toxicity and regulatory information, including data on pesticide use in California.

Silent Spring Institute

www.silentspring.org

www.silentspring.org/sciencereview

A partnership of scientists, physicians, public health advocates, and community activists that works to identify the links between the environment and women's health, especially breast cancer. Silent Spring Institute provides a database on the 216 different chemicals shown to cause mammary gland cancer in animals, including individual study results, chemical regulatory status, and likely sources of exposure. It provides another searchable database on the 450 primary epidemiologic research articles on breast cancer and environmental pollutants.

The Stockholm Convention on Persistent Organic Pollutants

http://chm.pops.int/

A global treaty to protect human health and the environment from inherently toxic chemicals that remain intact in the environment for a long time and travel far from their points of manufacture and use. Because of their propensity for long-range transport, no one nation state acting alone can protect its citizens from persistent organic pollutants. Administered by the United Nations, the treaty was adopted in 2001, and enacted into international law in 2004. At this writing, the treaty calls for the worldwide phase out from use and production of [illegible] chemicals and requires parties to take measures to eliminate or reduce the release of these substances into the environment. Currently, 164 nations are parties to the Stockholm Convention. At this writing, the United States is not.

Toxics Release Inventory

Environmental Protection Agency: www.epa.gov/triexplorer

Right-to-Know Network: www.rtk.net

There are two ways of acquiring environmental data available to the public under the Emergency Planning and Community Right-to-Know Act (EPCRA). One is through the U.S. Environmental Protection Agency, which is the branch of the federal government charged with administering EPCRA and maintaining the TRI. The other venue is the Right-to-Know Network, which advocates for improved access to government-held information on the environment, health, and safety. RTK Network puts TRI data together with other environmental data—such as hazardous waste and spill and accident reports. Both the EPA and RTK Net host excellent Web sites with mapping features that allow searches by chemical, by facility, or by ZIP code.

Toxipedia

www.toxipedia.org

A wiki–Web site created to bring experts and lay people together in order to provide educational materials on environmental and public health.

Source notes

Abbreviations used in Notes

ACS	American Cancer Society
AEH	*Archives of Environmental Health*
AJE	*American Journal of Epidemiology*
AJPH	*American Journal of Public Health*
ATSDR	Agency for Toxic Substances and Disease Registry
CDC	Centers for Disease Control and Prevention
EDF	Environmental Defense Fund
EHP	*Environmental Health Perspectives*
EPA	U.S. Environmental Protection Agency
FDA	Food and Drug Administration
GAO	General Accounting Office
IARC	International Agency for Research on Cancer
IASS	Illinois Agricultural Statistics Service
IDA	Illinois Department of Agriculture
IDC	Illinois Department of Conservation
IDENR	Illinois Department of Energy and Natural Resources
IDPH	Illinois Department of Public Health
IEPA	Illinois Environmental Protection Agency
IFB	Illinois Farm Bureau
INHS	Illinois Natural History Survey
ISGS	Illinois State Geological Survey
ISGWS	Illinois State Geological and Water Surveys
ISWS	Illinois State Water Survey
JAMA	*Journal of the American Medical Association*

JNCI	*Journal of the National Cancer Institute*
JTEH	*Journal of Toxicology and Environmental Health*
MDPH	Massachusetts Department of Public Health
NCI	National Cancer Institute
NEJM	*New England Journal of Medicine*
NIH	National Institutes of Health
NIOSH	National Institute for Occupational Safety and Health
NRC	National Research Council
NRDC	Natural Resources Defense Council
NTP	National Toxicology Program
OSHA	Occupational Safety and Health Administration
PDT	*Pekin Daily Times*
PJS	*Peoria Journal Star*
SSJR	*Springfield State Journal Register*
USDA	U.S. Department of Agriculture
USDHHS	U.S. Department of Health and Human Services
WHO	World Health Organization

Note: Organized by page number, the citations provided below represent the primary sources I consulted and are not intended to serve as a comprehensive review of the scientific literature. Some of the articles, monographs, and texts cited here are difficult to obtain, and some are highly technical in nature. Whenever I was aware of them, I also provide references to articles appearing in popular publications that, I hope, may be more accessible to lay readers.

vii (epigraph): Living downstream parable adapted from "Population Health Looking Upstream" (editorial), *Lancet* 343 (1994): 429–30.

Foreword to Second Edition

The opening of this foreword appeared, in a slightly different form, in S. Steingraber, "Three Bets on Ecology, Economy, and Human Rights," *Orion Magazine*, May/June 2009.

xii **data going back a hundred years:** This evidence was so convincing that the International Labour Office declared that two types of aromatic amines were carcinogens in 1921. P. Vineis and R. Pirastu, "Aromatic Amines and Cancer," *Cancer Causes and Control* 8 (1997): 346–55.

xiii **regulation of chemicals:** M. Schapiro, *Exposed: The Toxic Chemistry of Everyday Products and What's at Stake for American Power* (White River Junction, VT: Chelsea Green, 2007).

xiii **216 breast cancer carcinogens:** R. A. Rudel et al., "Chemicals Causing Mammary Gland Tumors in Animals Signal New Directions for Epidemiology, Chemicals Testing, and Risk Assessment for Breast Cancer Prevention," *Cancer* 109 (2007): 2635–66.

xviii **cancer causation is complex:** F. Mazzocchi, "Complexity in Biology: Exceeding the Limits of Reductionism and Determinism Using Complexity Theory," *European Molecular Biology Organization Reports* 9 (2008): 10–14.

xix **epigenetics:** B. Sadikovic et al., "Cause and Consequences of Genetic and Epigenetic Alterations in Human Cancer," *Current Genomics* 9 (2008): 394–408.

xix **endocrine disruption:** Endocrine disruptors do not always behave identically to endogenous hormones. They vary in potency, and, even when they activate hormone receptors within the cell, gene expression may vary. A. K. Hotchkiss, "Fifteen Years after 'Wingspread'—Environmental Endocrine Disrupters and Human and Wildlife Health: Where We Are Today and Where We Need to Go," *Toxicology Sciences* 105 (2008): 235–59; A. Kortenkamp, "Low Dose Mixture Effects of Endocrine Disrupters: Implications for Risk Assessments and Epidemiology," *International Journal of Andrology* 31 (2008): 233–40.

xx **the timing makes the poison:** In May 2007, two hundred leading environmental scientists gathered in the Faroe Islands north of Scotland and released a signed declaration, *The Faroes Statement*, which summarized the evidence for the link between low-level exposures to common environmental chemicals during fetal life and infancy and subsequent risks for adult-onset health problems, including cancer. P. Grandjean et al., "The Faroes Statement: Human Health Effects of Developmental Exposures to Chemicals in Our Environment," *Basic & Clinical Pharmacology & Toxicology* 102 (2008): 73–75. See also S. A. Vogel, "From 'The Dose Makes the Poison' to 'The Timing Makes the Poison': Conceptualizing Risk in the Synthetic Age," *Environmental History* 13 (2008): 667–73.

xx **chemicals that alter breast development in early life:** J. L. Rayner et al., "Adverse Effects of Prenatal Exposure to Atrazine During a Critical

Period of Mammary Gland Growth," *Toxicological Sciences* 87 (2005): [illegible]–66; L. S. Birnbaum and S. E. Fenton, "Cancer and Developmental Exposure to Endocrine Disruptors," *EHP* 111 (2003): 389–94.

xx [illegible] Fenton et al., "Prenatal TCDD Exposure Predisposes for Mammary Cancer in Rats," *Reproductive Toxicology* 23 (2007): [illegible]; A. Kortenkamp, "Breast Cancer, Oestrogens and Environmental Pollutants: A Re-evaluation from a Mixture Perspective," *International Journal of Andrology* 29 (2006): 193–98.

xx **mixtures of chemicals with other stressors:** The emergent cancer risks created by mixtures of variables—psychosocial stress plus poor nutrition plus environmental exposures—are largely unexplored. T. Schettler, "Toward an Ecological View of Health: An Imperative for the 21st Century," presentation before the Robert Wood Johnson Foundation, Sept. 2006.

xx **precautionary principle:** A. Stirling, "Risk, Precaution and Science: Towards a More Constructive Policy Debate: Talking Point on the Precautionary Principle," *European Molecular Biology Association Report* 8 (2007): 309–15; D. Gee, "Late Lessons from Early Warnings: Toward Realism and Precaution with Endocrine-Disrupting Substances," *EHP* 114 (2006; S-1): 152–60; J. G. Brody, "Breast Cancer and Environment Studies and the Precautionary Principle," *EHP* 113 (2005): 920–25.

xxi **costs of Appalachian coal mining:** The coal industry generates $8 billion in economic benefits and $42 billion in costs due to premature death of residents living in coal-mining areas. These costs do not include reduced productivity due to illness. M. Hendryx and M. M. Ahern, "Mortality in Appalachian Coal Mining Regions: The Value of Life Lost," *Public Health Reports* 124 (2009): 541–50.

xxi **costs of pollution to workers and children in California:** M. P. Wilson et al., *Green Chemistry: Cornerstone to a Sustainable California* (Berkeley and Los Angeles: University of California Centers for Occupational and Environmental Health, 2008).

xxi **health care costs:** ASPE Issue Brief, *Long Term Growth of Medical Expenditures—Public and Private* (USDHHS, Office of the Assistant Secretary for Planning and Evaluation, May 2005); M. W. Stanton and M. K. Rutherford, *The High Concentration of U.S. Health Care Expenditures*, Research in Action Issue 19, AHRQ Pub. No. 06–0060 (Rockville, MD: Agency for Healthcare Research and Quality, 2005).

xxii **U.S. petroleum industry:** Commission for Environmental Cooperation, *Taking Stock: 2005 North American Pollutant Releases and Transfers* (Montreal, June 2009).

xxii **trends in cancer mortality:** T. R. Frieden et al., "A Public Health Approach to Winning the War Against Cancer," *The Oncologist* 13 (2008), 1306–13.

xxii **trends in cancer incidence:** L. A. G. Ries et al. (eds.), *SEER Cancer Statistics Review, 1975–2004* (Bethesda, MD: NCI, 2007); J. Ahmedin et al., "Annual Report to the Nation on the Status of Cancer, 1975–2005, Featuring Trends in Lung Cancer, Tobacco Use, and Tobacco Control," *JNCI* 100 (2008): 1672–94. See also summary of trends in R. W. Clapp et al., "Environmental and Occupational Causes of Cancer: New Evidence 2005–2007," *Reviews on Environmental Health* 23 (2008): 1–37; R. W. Clapp et al., "Environmental and Occupational Causes of Cancer Re-visited," *Journal of Public Health Policy* 27 (2006): 61–76.

xxiii **projected 45 percent increase in number of people with cancer:** B. D. Smith et al., "Future of Cancer Incidence in the United States: Burdens upon an Aging, Changing Nation," *Journal of Clinical Oncology* 17 (2009): 2758–65.

xxiii **declines in death rate due to smoking cessation:** Frieden et al., "Public Health Approach."

xxiii **declines in lung and colon cancers:** Clapp et al., "Environmental and Occupational Causes of Cancer: New Evidence 2005–2007."

xxiv **proof that smoking causes lung cancer:** Mechanistic proof was demonstrated when researchers identified in 1996 the subcellular pathway by which tobacco causes lung cancer: a smoke-borne chemical called benzo[a]pyrene mutates a gene called p53; it is this alteration that places a lung cell on the pathway to tumor formation. M. F. Denissenko, "Preferential Formation of Benzo[a]pyrene Adducts at Lung Cancer Mutational Hotspots in p53," *Science* 274 (1996): 430–32. Physician Ted Schettler argues that the 1964 decision was only precautionary in the sense that the tobacco companies denied what everyone else knew to be true. From this perspective, the Surgeon General's announcement simply represents the courage and political will to say what was already obvious. Indeed, case-control studies beginning in the 1940s showed clear associations between smoking and lung cancers with dose-response features. This kind of evidence is often impossible to gather for involuntary exposures to other carcinogens. We should not wait for the kind of proof that finally served to denormalize tobacco.

xxiv **President's Cancer Panel:** "Consensus Statement on Cancer and the Environment: Creating a National Strategy to Prevent Environmental Factors in Cancer Causation," submitted by the Collaborative on Health and Environment to the President's Cancer Panel, October 2008.

xxiv **types of cancers rising in incidence:** L. A. G. Ries et al. (eds.), *SEER Cancer Statistics Review, 1975–2004* (Bethesda, MD: NCI, 2007); J. Ahmedin et al., "Annual Report to the Nation on the Status of Cancer, 1975–2005, Featuring Trends in Lung Cancer, Tobacco Use, and Tobacco Control," *JNCI* 100 (2008): 1672–94. See also summary of trends in Clapp et al., "Environmental and Occupational Causes of Cancer: New Evidence 2005–2007."

xxiv **childhood cancer trends:** Clapp et al., "Environmental and Occupational Causes of Cancer: New Evidence 2005–2007."

xxv **"to ignore the scientific evidence":** Clapp et al., "Environmental and Occupational Causes of Cancer Re-visited."

xxv **historical studies bladder cancer and aromatic amines:** These studies are reviewed in Vineis and Pirastu, "Aromatic Amines and Cancer." Aromatic amines are a large class of chemicals that include ingredients in tobacco smoke. In light of their evidence for harm, a few individual aromatic amine substances have been banned or tightly regulated in the workplace. After certain aromatic amines were removed from chemical industry, incidence of bladder cancer among affected workers declined considerably. See also S. P. Lerner et al. (eds.), *Textbook of Bladder Cancer* (London: Taylor and Francis, 2006).

xxv **bladder cancer in farmers:** S. Koutros et al., "Heterocyclic Aromatic Amine Pesticide Use and Human Cancer Risk: Results from the U.S. Agricultural Health Study," *International Journal of Cancer* 124 (2009): 1206–12.

one: trace amounts

2 **Mahomet River:** J. P. Kempton and A. P. Visocky, *Regional Groundwater Resources in Western McLean and Eastern Tazewell Counties with an Emphasis on the Mahomet Bedrock Valley*, Cooperative Groundwater Report 13 (Champaign, IL: ISGWS, 1992); J. P. Kempton et al., "Ma-

homet Bedrock Valley in East-Central Illinois: Topography, Glacial Drift Stratigraphy, and Hydrogeology," in W. Melhorn and J. P. Kempton (eds.), *Geology and Hydrology of the Teays-Mahomet Bedrock Valley System*, Special Report 258 (Boulder, CO: Geological Society of America, 1991); J. P. Gibb et al., *Groundwater Conditions and River-Aquifer Relationships along the Illinois Waterway* (Champaign, IL: ISWS, 1979); M. M. Killey, "Do You Live above an Underground River?" *Geogram* 8 (Urbana, IL: ISGS, 1977).

2 **the ancestral Mississippi River valley:** M. A. Marion and R. J. Schicht, *Groundwater Levels and Pumpage in the Peoria-Pekin Area, Illinois, 1890–1966* (Champaign, IL: ISWS, 1969), 3; S. L. Burch and D. J. Kelly, *Peoria-Pekin Regional Groundwater Quality Assessment*, Research Report 124 (Champaign, IL: ISWS, 1993), 6.

2 **Illinois farm statistics:** IFB, *Farm and Food Facts 2007* (Bloomington, IL: IFB, 2008).

3 **disappearance of the Illinois prairie:** IDENR, *The Changing Illinois Environment: Critical Trends*, summary report and vol. 3, ILENR/RE-EA-94/05 (Springfield, IL: IDENR, 1994); S. L. Post, "Surveying the Illinois Prairie," *The Nature of Illinois* (Winter 1993): 1–8; R. C. Anderson, "Illinois Prairies: A Historical Perspective," in L. M. Page and M. R. Jeffords (eds.), *Our Living Heritage: The Biological Resources of Illinois* (Champaign, IL: INHS, 1991).

4 **pesticide application in Illinois:** Fifty-four million represents pounds of active ingredient. This is a 1995 extrapolation derived from small-scale surveys: L. P. Gianessi and J. E. Anderson, *Pesticide Use in Illinois Crop Production* (Washington, DC: National Center for Food and Agricultural Policy, 1995), table B-2. Other than California and New York, both of which maintain state pesticide registries, no state keeps track of pesticide use unless the pesticide is classified as restricted. See IDENR, *Changing Illinois*, summary report, 81. In recent years, the state of Illinois has not provided compiled estimates of total pesticide use. According to the National Agricultural Statistics Service, 14,143 million pounds of atrazine was applied to Illinois corn in 2005. NASS estimates total insecticide and herbicide use on Illinois corn and soybeans at 46,544 million pounds in 2005. The USDA has not collected agricultural chemical use since 2005. This lack of data complicated the *Peoria Journal Star*'s attempt to investigate trends in the use of fungicide

between 2006 and 2008. S. Tarter, "Illinois Farmers Have Increased the Use of Plane-Sprayed Fungicide," *PJS*, 20 July 2009.

4 **percentage of corn treated with pesticides in 1950:** IDPNR, *Changing Illinois* [illegible].

4 **percentage of corn treated with pesticides in 2005:** [illegible] USDA National Agricultural Statistics Service, www.nass.usda.gov.

5 **atrazine use in Illinois:** D. Coursey, *Illinois Without Atrazine: Who Pays? Economic Implications of an Atrazine Ban in the State of Illinois* (University of Illinois Harris School of Public Policy Working Paper, Feb. 2007).

5 **pesticide drift:** C. M. Benbrook et al., *Pest Management at the Crossroads* (Yonkers, NY: Consumers Union, 1996); C. A. Edwards, "The Impact of Pesticides on the Environment," in D. Pimentel et al. (eds.), *The Pesticide Question: Environment, Economics, and Ethics* (New York: Routledge, 1993); D. E. Glotfelty et al., "Pesticides in Fog," *Nature* 325 (1987): 602–5; C. Howard, "Chemical Drift a Growing Concern for Rural Residents," *PJS*, 25 July 2009; S. M. Miller et al., "Atrazine and Nutrients in Precipitation: Results from the Lake Michigan Mass Balance Study," *Environmental Science & Technology* 34 (2000): 55–61.

5 **pesticides in Illinois surface and ground water:** M. Wu et al., *Poisoning the Well: How EPA Is Ignoring Atrazine Contamination in Surface and Drinking Water in the Central United States* (New York: NRDC, 2009); IDA, Pesticide Monitoring Network (Springfield, IL, 2006); R. B. King, *Pesticides in Surface Water in the Lower Illinois River Basin 1996–98* (U.S. Geological Survey Water Resources Investigations Report 2002-4097, 2003); A. G. Taylor and S. Cook, "Water Quality Update: The Results of Pesticide Monitoring in Illinois' Streams and Public Water Supplies" (paper presented at the 1995 Illinois Agricultural Pesticides Conference, University of Illinois, Urbana, 4–5 Jan. 1995); A. G. Taylor, "The Effects of Agricultural Use on Water Quality in Illinois" (paper presented at the 1993 American Chemical Society Agrochemicals Division Symposium, "Pesticide Management for the Protection of Ground and Surface Water Resources," Chicago, 25–26 Aug. 1993); S. C. Schock et al., *Pilot Study: Agricultural Chemicals in Rural, Private Wells in Illinois*, Cooperative Groundwater Report 14 (Champaign, IL: ISGWS, 1992).

5 **2009 report on drinking water:** Wu et al., *Poisoning the Well.* The two Illinois communities with chronically elevated atrazine levels are Mount Olive and Evansville.

5 **withered vineyards:** C. Howard, "Chemical Drift a Growing Concern for Rural Residents," *PJS*, 25 July [illegible].

5 **DDT most common pesticide in fish:** R. J. Gilliom et al., *The Quality of Our Nation's Waters: Pesticides in the Nation's Streams and Ground Water, 1992–2001* (U.S. Geological Survey Circular 1291, 2006).

6 **DDT residues on kitchen floors:** The same study found chlordane in the kitchen floor dust of [illegible] percent of U.S. homes and chlorpyrifos and diazinon in 70 and [illegible] percent of homes, respectively. [illegible] and permethrin were also commonly detected. The house dust in some homes contained 24 different insecticides. D. M. Stout et al., "American Healthy Homes Survey: A National Study of Residential Pesticides Measured from Floor Wipes," *Environmental Science and Technology* 43 (2009): 4294–4300.

6 **health effects of DDT:** B. Eskenazi et al., "The Pine River Statement: Human Health Consequences of DDT Use," *EHP* 117 (2009): 1359–67.

6 **hormonal effects of atrazine exposure:** For example, H. Shibayama et al., "Collaborative Work on Evaluation of Ovarian Toxicity. 14) Two-or Four-week Repeated-Dose Studies and Fertility Study of Atrazine in Female Rats," *Journal of Toxicological Sciences* 34 (2009, S-1): SP147–55; J. R. Lenkowski et al., "Perturbation of Organogenesis by the Herbicide Atrazine in the Amphibian *Xenopus laevis*," *EHP* 116 (2008): 223–30; M. Suzawa and H. A. Ingraham, "The Herbicide Atrazine Activates Endocrine Gene Networks via Non-Steroidal NR5A Nuclear Receptors in Fish and Mammalian Cells," *PLoS ONE* 3 (2008): e2117; R. L. Cooper et al., "Atrazine and Reproductive Function: Mode and Mechanism of Action Studies," *Birth Defects Research* (Part B), 80 (2007): 98–112; R. R. Enoch et al., "Mammary Gland Development as a Sensitive End Point after Acute Prenatal Exposure to an Atrazine Metabolite Mixture in Female Long-Evans Rats," *EHP* 115 (2007): 541–47; V. M. Rodriguez et al., "Sustained Exposure to the Widely Used Herbicide Atrazine: Altered Function and Loss of Neurons in Brain Monoamine Systems," *EHP* 113 (2005): 708–15; T. Hayes et al., Hermaphroditic Demasculinized Frogs after Exposure to the Herbicide Atrazine at Low Ecologically Relevant Doses," *Proceedings of the National Academy of Sciences* 99 (2002): 5476–80.

6 **toxic releases in Illinois:** Data from the 2007 Toxics Release Inventory for Illinois, retrieved from the Right-to-Know Network database, www.rtknet.org. Ethanol production unites agriculture with industry

in Illinois. IDA, *Facts about Illinois Agriculture* (Springfield, IL: IDA, 2001).

6 **metal degreasers and dry-cleaning fluids:** IDPH, *Chlorinated Solvents in Drinking Water* (Springfield, IL: IDPH, Division of Environmental Health, n.d.).

6 **dry cleaners:** Investigation was prompted by solvent-contaminated drinking water in Crestwood, Illinois. M. Hawthorne, "Dry Cleaners Leave a Toxic Legacy—Despite Cleanup Effort, Chemicals Still Taint Hundreds of Illinois Sites," *Chicago Tribune*, 26 July 2009.

6 **quote from a state assessment:** IDENR, *Changing Illinois*, summary report, 6.

7 **atrazine in U.S. water:** R. J. Gilliom et al., *The Quality of Our Nation's Waters: Pesticides in the Nation's Streams and Ground Water, 1992–2001* (U.S. Geological Survey Circular 1291, 2006).

7 **PCBs in Illinois fish:** C. L. Straub et al., "Trophic Transfer of Polychlorinated Biphenyls in Great Blue Heron (*Ardea Herodias*) at Crab Orchard National Wildlife Refuge, Illinois, United States," *Archives of Environmental Contamination and Toxicology* 52 (2007): 572–79; IDPH, "Illinois Fish Advisory: Illinois River, Contaminant—PCBs" (Springfield, IL: IDPH, 2005).

7 **DDT off the California coast:** J. Gottlieb, "EPA Seeks to Clean Up DDT-tainted Site off Palos Verdes Peninsula," *Los Angeles Times*, 12 June 2009.

7 Archival film clips appear in *Rachel Carson's* Silent Spring, documentary film by Peace River Films, aired on PBS, *The American Experience*, 8 Feb. 1993.

7 Old magazine ads for DDT are reprinted in E. P. Russell III, "'Speaking of Annihilation': Mobilizing for War Against Human and Insect Enemies, 1914–1945," *Journal of American History* 82 (1996): 1505–29; and in J. Curtis et al., *After* Silent Spring: *The Unsolved Problem of Pesticide Use in the United States* (New York: NRDC, 1993), 2.

8 **DDT for polio control:** T. R. Dunlap, *DDT: Scientists, Citizens and Public Policy* (Princeton, NJ: Princeton University Press, 1981), 65.

8 **DDT in paint:** This ad, for Sherwin-Williams, appeared in 1946. See E. C. Helfrick as told to M. Riddle, "Mass Murder Introduces Sherwin-Williams' 'Pestroy,'" *Sales Management*, 15 Oct. 1946, 60–64. See also E. P. Russell III, *War and Nature: Fighting Humans and Insects with*

Chemicals from World War I to Silent Spring (New York: Cambridge University Press, 2001).

8 **[illegible]:** [illegible] of [illegible], MA, and John Gephart of Ithaca, NY.

8 **"the harmless aspect of the familiar":** R. Carson, *Silent Spring* (Boston: Houghton Mifflin, 1962), 20.

8 **"It is not my contention . . .":** Ibid., 12.

9 **Carson on future generations:** Ibid., 13.

9 **"killer of killers," "the atomic bomb of the insect world":** J. Warton, *Before* Silent Spring: *Pesticides and Public Health in Pre-DDT America* (Princeton, NJ: Princeton University Press, 1974), 248–55.

9 **failure of DDT:** Carson, *Silent Spring*, 20–23, 58, 103, 107–9, 112, 113, 120–22, 125, 143–44, 206–7, 225, 267–73; Dunlap, *DDT*, 63–97.

9 **DDT in breast milk:** E. P. Laug et al., "Occurrence of DDT in Human Fat and Milk," *AMA Archives of Industrial Hygiene and Occupational Medicine* 3 (1951): 245–46. DDT remains a common contaminant of human breast milk as well as human blood. It is also found in the bodies of migrating songbirds and in forest soils. USDA, *Pesticide Data Program, Annual Summary Calendar Year 1994* (Washington, DC: USA, Agricultural Marketing Service, 1994), 13; R. G. Harper et al., "Organochlorine Pesticide Contamination in Neotropical Migrant Passerines," *Archives of Environmental Contamination and Toxicology* 31 (1996): 386–90; ATSDR, "DDT, DDE, and DDD" (fact sheet) (Atlanta: ATSDR, 1995); R. G. Lewis et al., "Evaluation of Methods for Monitoring the Potential Exposure of Small Children to Pesticides in the Residential Environment," *Archives of Environmental Contamination and Toxicology* 26 (1996): 37–46; W. H. Smith et al., "Trace Organochlorine Contamination of the Forest Floor of the White Mountain National Forest, New Hampshire," *Environmental Science and Technology* 27 (1993): 2244–46; EPA, *Deposition of Air Pollutants to the Great Lakes: First Report to Congress*, EPA-453/R-93-055 (Washington, DC: EPA, 1994).

10 **history of lindane:** Once used widely in the Christmas tree industry, lindane is now banned in over fifty countries and in the state of California. It is being phased out globally under the Stockholm Convention on Persistent Organic Pollutants, an international treaty negotiated under the custody of the United Nations Environment Programme in Geneva, Switzerland, and to which the United States is not party

(http://chm.pops.int); M. P. Purdue et al., "Occupational Exposure to Organochlorine Insecticides and Cancer Incidence in the Agricultural Health Study," *International Journal of Cancer* 120 (2007): 642–49; EPA, "Lindane: Cancellation Order," *Federal Register* 71 (Dec. 13, 2006): 74905; M. Moses, *Designer Poisons: How to Protect Your Health and Home from Toxic Pesticides* (San Francisco: Pesticide Education Center, 1995); EPA, *Suspended, Cancelled and Restricted Pesticides*, 20T-1002 (Washington, DC: EPA, 1990); Curtis, *After* Silent Spring.

10 **my 1992 discovery of lindane:** Pesticides banned for use in the 1970s and 1980s were nevertheless still exported from the United States to nations with more lenient pesticide restrictions until at least the early 1990s. This practice seems now to have ended. The chemical company in my hometown may have been formulating lindane for the export market, but I was not able to confirm this. In 1992, 600,000 pounds of DDT were shipped out of U.S. ports. Some analysts suspect this cargo may represent a transshipment—cargo imported and then exported again. Poor labeling of pesticide exports make careful tracking very difficult. J. Raloff, "The Pesticide Shuffle," *Science News* 149 (1996): 174–75; Foundation for the Advancement of Science and Education, *Exporting Risk: Pesticide Exports from U.S. Ports* (Los Angeles: Foundation for the Advancement of Science and Education, 1996); J. Wargo, *Our Children's Toxic Legacy: How Science and Law Fail to Protect Us from Pesticides* (New Haven, CT: Yale University Press, 1996), 163–64; D. J. Hanson, "Administration Seeks Tighter Curbs on Exports of Unregistered Pesticides," *Chemical and Engineering News*, 14 Feb. 1994, 16–18; Monica Moore, Pesticide Action Network, personal communication.

10 **aldrin and dieldrin:** J. B. Barnett and K. E. Rodgers, "Pesticides," in J. H. Dean et al. (eds.), *Immunotoxicology and Immunopharmacology*, 2nd ed. (New York: Raven Press, 1994); R. Spear, "Recognized and Possible Exposures to Pesticides," in W. J. Hayes and E. R. Laws Jr. (eds.), *Handbook of Pesticide Toxicology*, vol. 1 (New York: Academic Press, 1991); EPA, 1990, *Suspended*; Carson, *Silent Spring*, 26.

10 **chlordane and heptachlor:** J. J. Spinelli et al., "Organochlorines and Risk of Non-Hodgkin Lymphoma," *International Journal of Cancer* 121 (2007): 2767–75; Spear, "Possible Exposures," 245; P. F. Infante et al., "Blood Dyscrasias and Childhood Tumors and Exposure to Chlordane and Heptachlor," *Scandinavian Journal of Work Environment and Health* 4 (1978): 137–50.

10 [illegible]

11 **women with breast cancer have higher levels of DDE and PCBs in their tumors:** M. Wasserman, "Organochlorine Compounds in Neoplastic and Adjacent Apparently Normal Breast Tissue," *Bulletin of Environmental Contamination and Toxicology* 15 (1976): 478–84.

11 **other studies followed:** H. Mussalo-Rauhamaa et al., "Occurrence of Beta-Hexachlorocyclohexane in Breast Cancer Patients," *Cancer* 66 (1990): 2124–28 (lindane is the gamma isomer of hexachlorocyclohexane); F. Falck Jr. et al., "Pesticides and Polychlorinated Biphenyl Residues in Human Breast Lipids and Their Relation to Breast Cancer," *AEH* 47 (1992): 143–46.

11 **Wolff study:** M. S. Wolff et al., "Blood Levels of Organochlorine Residues and Risk of Breast Cancer," *JNCI* 85 (1993): 648–52; D. J. Hunter and K. T. Kelsey, "Pesticide Residues and Breast Cancer: The Harvest of a Silent Spring?" *JNCI* 85 (1993): 598–99; M. P. Longnecker and S. J. London, "Re: Blood Levels of Organochlorine Residues and Risk of Breast Cancer" (letter and response by M. S. Wolff), *JNCI* 85 (1993): 1696–97.

11 **role of breast cancer activism in redirecting scientific inquiry:** See chapter two in Phil Brown, *Toxic Exposures, Contested Illnesses and the Environmental Movement* (New York: Columbia University Press, 2007).

11 **pesticide use since *Silent Spring*:** Pesticide use doubled between 1964 and 1982, as measured by weight of active pesticidal ingredients. See Wargo, *Toxic Legacy*, 132.

11 **breast cancer among women born between 1947 and 1958:** D. L. Davis et al., "Decreasing Cardiovascular Disease and Increasing Cancer among Whites in the United States from 1973 through 1987: Good News and Bad News," *JAMA* 271 (1994) 431–37.

12 **contradictory studies:** É. Dewailly et al., "High Organochlorine Body Burden in Women with Estrogen Receptor-Positive Breast Cancer," *JNCI* 86 (1994): 232–34. Increasing incidence of receptor-positive breast cancer is largely responsible for the increase in breast cancer rates that occurred between the mid-1970s and the mid-1980s. See A. G. Glassand and R. N. Hoover, "Rising Incidence of Breast Cancer: Relationship to State and Receptor Status," *JNCI* 82 [1990]: 693–96; N. Krieger et al., "Breast Cancer and Serum Organochlorines: A Prospective Study among White, Black and Asian Women," *JNCI* 86 (1994):

(1994): 1255; M. S. Wolff, "Pesticides—How Research Has Succeeded and Failed in Informing Policy: DDT and the Link with Breast Cancer," *EHP* 103, S-6 (1995): 87–91.

12 ***NEJM* study:** D. J. Hunter et al., "Plasma Organochlorine Levels and the Risk of Breast Cancer," *NEJM* 337 (1997): 1303–4.

12 **animal studies and early-life exposure:** L. S. Birnbaum and S. E. Fenton, "Cancer and Developmental Exposure to Endocrine Disruptors," *EHP* 111 (2003): 389–94.

13 **Cohn study:** B. A. Cohn et al., "DDT and Breast Cancer in Young Women: New Data on the Significance of Age at Exposure," *EHP* 115 (2007): 1406–14.

13 **gene-environment interaction in PCB cancer risk:** J. G. Brody et al., "Environmental Pollutants and Breast Cancer: Epidemiologic Studies," *Cancer* 109 (2007; S-12): 2667–2711; Y. Zhang et al., "Serum Polychlorinated Biphenyls, Cytochrome P-450 1A1 Polymorphisms, and Risk of Breast Cancer in Connecticut Women," *AJE* 160 (2004): 1177–83.

14 **atrazine's ranking:** Ranked by pounds of active ingredient, atrazine was the most abundantly applied pesticide in the United States from 1987 until 2001, when glyphosate surpassed it. KRSNetwork, *2005 U.S. Pesticide Industry Report* (Covington, GA, 2006).

14 **atrazine and human cancer:** For example, D. W. Gammon et al., "A Risk Assessment of Atrazine Use in California: Human Health and Ecological Aspects," *Pest Management Science* 61 (2005): 331–55; J. A. Rusiecki et al., "Cancer Incidence Among Pesticide Applicators Exposed to Atrazine in the Agricultural Health Study," *JNCI* 96 (2004): 1375–82; P. A. MacLennan et al., "Cancer Incidence Among Triazine Herbicide Manufacturing Workers," *Journal of Occupational and Environmental Medicine* 45 (2003): 243–44.

14 **human studies needed on early-life exposure to atrazine:** J. R. Roy et al., "Estrogen-like Endocrine-Disrupting Chemicals Affecting Puberty in Humans—A Review," *Medical Science Monitor* 15 (2009):

RA197–45; D. A. Crain et al., "Female Reproductive Disorders: The Role of Endocrine-Disrupting Compounds and Developmental Timing," *Fertility and Sterility* 90 (2008): 911–40.

15 **U.S. regulatory decision on atrazine and aftermath:** T. B. Hayes, "There Is No Denying This: Defusing the Confusion about Atrazine," *BioScience* 54 (2004): 1138–40; J. Huff, "Industry Influence on Occupational and Environmental Public Health," *International Journal of Occupational and Environmental Health* 13 (2007): 107–17; and J. Huff and J. Sass, "Atrazine—A Likely Human Carcinogen?" [letter], *International Journal of Occupational and Environmental Health* 13 (2007): 356–57; EPA, "EPA Begins New Scientific Evaluation of Atrazine," press release, October 7, 2009; Wu et al., *Poisoning the Well.*

15 **atrazine is banned in Europe:** The inability to keep atrazine out of drinking water was the basis for the ban. J. B. Sass and A. Colangelo, "European Union Bans Atrazine, while the United States Negotiates Continued Use," *International Journal of Occupational and Environmental Health* 12 (2006): 260–67.

15 **failure to pursue research on cancer's environmental connections:** This question is brilliantly addressed by Devra Davis, Ph.D. MPH, the director of the Center of Environmental Oncology at the University of Pittsburgh Cancer Institute, in her book, *The Secret History of the War on Cancer* (New York: Basic Books, 2007).

two: silence

I explore more deeply the ways in which my work has been influenced by that of Rachel Carson in essays published in two anthologies: "*Silent Spring*: A Father-Daughter Dance," in Peter Matthiessen (ed.), *Courage for the Earth: Writers, Scientists, and Activists Celebrate the Life and Writing of Rachel Carson* (Boston: Houghton Mifflin, 2007) and "Living Downstream of *Silent Stream*," in Lisa Sideris and Kathleen Dean Moore (eds.), *Rachel Carson: Legacy and Challenge* (Albany: State University of New York Press, 2008). Both commemorate the 100th anniversary of Carson's birth in 1907.

18 **Carson's concern about pesticide debates:** L. J. Lear, "Rachel Carson's *Silent Spring*," *Environmental History Review* 17 (1993): 23–48. See also Lear's definitive biography, *Rachel Carson: Witness for Nature* (New York: Holt, 1997).

19 **Iroquois County:** Rachel Carson, *Silent Spring* (Boston: Houghton Mifflin, 1962), 91–100.

20 **refusal of scientists to send Carson information:** Dr. Linda Lear, personal communication.

20 **threat of defunding:** Carson, *Silent Spring*, 94–95.

20 **"The other day . . . ":** Carson's letter to Freeman, 27 June 1962, reprinted in Freeman, *Always, Rachel*, 408. Abraham Lincoln is likely not the source of this quote. "To sin by silence" is a line from a poem entitled "Protest," by Ella Wheeler Wilcox, which was published in her collection *Poems of Problems* (Chicago: W. B. Conkey, 1914). Wheeler Wilcox was known to be inspired by Lincoln.

20 **Carson's speech:** quoted in Brooks, *House of Life*, 302–4.

23 **twenty years of life lost:** Dr. Devra Lee Davis, personal communication.

24 **Carson's cancer diagnosis and physical ailments:** Carson's letters to Freeman, 1960–1964, in Freeman, *Always, Rachel*; Brooks, *House of Life*; Dr. Linda Lear, personal communication.

24 **Carson's relief at finishing *Silent Spring*:** Carson's letter to Freeman, 6 Jan. 1962, in Freeman, *Always, Rachel*, 391.

24 **two quotes from letters to Freeman:** 3 Nov. 1963 and 9 Jan. 1964, ibid., 490, 515. See also letters dated 6 Jan. 1962; 2 Mar. 1963; and 25 Apr. 1963.

27 **Carson's letters to Freeman that speak openly:** 3 Jan. 1961; 23 Mar. 1961; 25 Mar. 1961; and 18 Sept. 1963, ibid., 326, 364, 365–66, 469.

27 **letters that speak elliptically:** 17 Jan. 1961; 15 Feb. 1961; 25 Oct. 1962; 25 Dec. 1962; and 2 Jan. 1964, ibid., 331, 346, 414, 420, 508.

28 **Freeman's reference to Carson's mastectomy:** Freeman's letter to Carson, 30 Apr. 1960, ibid., 305.

28 **their entreaties and admissions:** See, for example, Freeman's letter to Carson, 6 Mar. 1963, ibid., 441.

28 **the darker story:** Freeman's letters to Carson, 4 and 17 Mar. 1961, ibid., 356, 363.

28 **confessions and recantations:** Carson's letters to Freeman, 23 Jan. 1962; 26 Mar. 1962; 10 Apr. 1962; 14 Feb. 1963; 18 Feb. 1963; 3 Mar. 1963; 14 Jan. 1964, ibid., 395, 399, 404, 434–37, 439–40, 516.

28 **Carson's prohibition of discussions about her health:** M. Spock, "Rachel Carson: A Portrait," *Rachel Carson Council News* 82 (1994): 1–4; Dr. Linda Lear, personal communication.

28 **quotes instructing Dorothy:** Carson's letters to Freeman, 1 Apr. 1962 and 20 May 1962, in Freeman, *Always, Rachel*, 401, 405.

29 **photographs and old film clips:** Beinecke Library archives, Yale University; *Rachel Carson's* Silent Spring, documentary film by Peace River Films, aired on PBS, *The American Experience*, 8 Feb. 1993.

30 **farmers and housewives with cancer:** Carson, *Silent Spring*, 227–30.

30 **first line of evidence:** Ibid., 219–20.

31 **second and third lines of evidence:** Ibid., 221.

31 **"whatever seeds of malignancy . . . ":** Ibid., 226.

31 **death certificates and children's cancers:** Ibid., 221–22.

31 **animals with cancer:** Ibid., 221–22.

31 **cellular mechanisms of carcinogenesis:** Ibid., 231–35.

32 **effect on sex hormones:** Ibid., 235–37.

32 **effect on metabolism:** Ibid., 231–32. Carson was particularly prescient on this point. The ability of certain chemicals to alter metabolic processes by which other chemicals are converted into genotoxic metabolites was acknowledged in a 2009 paper as an important nontraditional pathway to tumor formation. L. G. Hernández et al., "Mechanisms of Non-Genotoxic Carcinogens and the Importance of a Weight of the Evidence Approach," *Mutation Research* 2009 [in press].

32 **Carson's prediction:** Ibid., 232–33.

32 **interspecies differences in susceptibility:** H. C. Pitot III and Y. P. Dragan, "Chemical Carcinogens," in D. Klaassen (ed.), *Casarett and Doull's Toxicology: The Basic Science of Poison*, 5th ed. (New York: McGraw-Hill, 1996); NRC, *Animals as Sentinels of Environmental Health Hazards* (Washington, DC: National Academy Press, 1991).

32 **uncontrolled human experiment:** A lack of unexposed controls makes human studies difficult but not impossible. Theoretically, all that is required for such studies are measurable difference in exposure levels among segments of the human population. For example, all of us are believed to carry detectable levels of dioxin in our tissues. The

question of whether dioxin contributes to human cancers can be addressed by studies that compare cancer incidence rates among those heavily, moderately, and lightly exposed. All other things being equal, a positive trend would indicate a dose-response relationship, which is considered strong evidence by cancer researchers. The wider the spread in exposure levels, the more likely the relationship—if indeed one exists—will reveal itself. As such, researchers interested in conducting human studies often look for "natural experiments" where an unfortunate event—such as a toxic spill of some sort—has exposed an identifiable sector of the population to a heavy dose of the substance in question. Disease rates among this group can then be compared to those of the general population, whose exposures to this substance may be common and ongoing but are occurring at much lower levels.

33 **Olga Owen Huckins, the Committee Against Mass Poisoning, and the *New Yorker*:** L. Lear, *Rachel Carson: The Life of the Author of* Silent Spring (New York: Henry Holt, 1997), 312–38; S. Steingraber, "*Silent Spring*: A Father-Daughter Dance," in Matthiessen, *Courage for the Earth.*

three: time

Unless otherwise stated below, cancer statistics quoted in this chapter are drawn from the following two sources: U.S. Cancer Statistics Working Group. *United States Cancer Statistics: 1999–2005 Incidence and Mortality Web-Based Report.* (Atlanta: USDHHS, CDC and NCI, 2009), available at www.cdc.gov/uscs; Surveillance, Epidemiology and End Results Program, Delay-Adjusted Incidence Database: SEER Incidence Adjusted Rates, 7 Registries, NCI, 2008. Available at: www.seer.cancer.gov/.

36 **number of cancer diagnoses in 2009:** ACS, *Cancer Facts & Figures—2009* (Atlanta: ACS, 2009).

39 **quantifying and correcting problems in data ascertainment:** H. Menck and C. Smart (eds.), *Central Cancer Registries: Design, Management, and Use* (Chur, Switzerland: Harwood Academic Press, 1994); O. M. Jensen et al. (eds.), *Cancer Registration: Principles and Methods*, IARC Scientific Publication 95 (Lyon, France: IARC, 1991). For an excellent history of cancer registration in the United States, see E. R. Greenberg et al., "Measurements of Cancer Incidence in the

United States: Sources and Uses of Data," *JNCI* 68 (1982): 743–49. For an overview of the system of state registries, see CDC/NPCR, *A National Program of Cancer Registries At-a-Glance, 1994–1995* (Atlanta: CDC, 1995), 69–72.

39 **percentage of upsurge in breast cancer attributable to earlier detection:** R. N. Proctor, *Cancer Wars: How Politics Shapes What We Know and Don't Know about Cancer* (New York: Basic Books, 1995), [illegible]; J. M. Liff, "Does Increased Detection Account for the Rising Incidence of Breast Cancer?" *AJPH* 81 (1991): 462–65.

40 **rise in breast cancer predates mammography:** E. J. Feuer and L. M. Wun, "How Much of the Recent Rise in Breast Cancer Incidence Can Be Explained by Increases in Mammography Utilization?" *AJE* 136 (1992): 1423–36; J. R. Harris, "Breast Cancer," *NEJM* 327 (1992): 319–28.

40 **recent drop in breast cancer incidence:** P. M. Ravdin et al., "The Decrease in Breast-Cancer Incidence in 2003 in the United States," *NEJM* 356 (2007):1670–74; S. L. Stewart et al., "Decline in Breast Cancer Incidence—United States, 1999–2003," *Morbidity and Mortality Weekly Report* 56 (2007): 549–53.

40 **2002 warning on estrogen:** Writing Group for the Women's Health Initiative Investigators, "Risks and Benefits of Estrogen Plus Progestin in Healthy Postmenopausal Women: Principal Results from the Women's Health Initiative Randomized Controlled Trial," *JAMA* 288 (2002): 321–33.

40 **decline in hormone replacement in California:** C. A. Clarke et al., "Recent Declines in Hormone Therapy Utilization and Breast Cancer: Clinical and Population-Based Evidence," *Journal of Clinical Oncology* 33 (2006): 349–50.

40 **decline restricted to estrogen-dependent tumors:** J. Gray et al., "State of the Evidence: The Connection Between Breast Cancer and the Environment," *International Journal of Environmental Health* 15 (2009): 43–78.

41 **racial disparities in breast cancer incidence:** Ibid.

41 **declining rates of mammography:** A. Jemal et al., "Annual Report to the Nation on the Status of Cancer, 1975–2005, Featuring Trends in Lung Cancer, Tobacco Cancer, and Tobacco Control," *JNCI* 100 (2008): 1672–94.

41 **possible declining exposure to other causative agents:** Gray et al., "State of the Evidence," 43–78.

43 **data exchange in Illinois:** H. L. Howe et al., *Effect of Interstate Data Exchange on Cancer Rates in Illinois, 1986–1990*, Epidemiological Report Series, 94:1 (Springfield, IL: IDPH, 1994).

43 **seventeen geographic areas:** These are a mixture of state and metropolitan registries. They are Atlanta, Connecticut, Detroit, Hawaii, Iowa, New Mexico, San Francisco–Oakland, Seattle–Puget Sound, Utah, Los Angeles, San Jose–Monterey, rural Georgia, the Alaska Native Tumor Registry, Greater California, Kentucky, Louisiana, and New Jersey. NCI "Seer Data, 1976–2006." Available at www.seer.cancer.gov/data/.

43 **SEER and NPCR:** P. A. Wingo, "Building the Infrastructure for Nationwide Cancer Surveillance and Control—A Comparison Between the National Program of Cancer Registries (NPCR) and the Surveillance, Epidemiology and End Results Program (United States)," *Cancer Causes and Control* 14 (2003): 175–93.

44 **trends in overall cancer incidence rate:** R. W. Clapp et al., "Environmental and Occupational Causes of Cancer Re-visited," *Journal of Public Health Policy* 27 (2006): 61–76.

44 **mortality rates as more reliable:** The renowned biostatistician John Bailar, for example, holds this view. J. C. Bailar III and E. M. Smith, "Progress Against Cancer?" *NEJM* 314 (1986): 1226–32; J. C. Bailar III, "Observations on Some Recent Trends in Cancer," presentation at the President's Cancer Panel Meeting, NIH, Bethesda, MD, 22 Sept. 1993.

44 **mortality has not changed much in 60 years:** Clapp et al., "Environmental and Occupational Causes of Cancer Re-visited," 61–76; D. L. Davis, "The Need to Develop Centers for Environmental Oncology," *Biomedicine and Pharmacotherapy* 61 (2007): 614–622.

44 **trends in cancer mortality:** T. R. Frieden, "A Public Health Approach to Winning the War on Cancer," *The Oncologist* 13 (2008): 1306–13; Davis, "Need to Develop Centers."

45 **childhood cancers:** P. J. Landrigan, "Childhood Cancer and the Environment," testimony before the President's Cancer Panel, East Brunswick,

NJ, Sept. 16, 2008; Clapp et al., "Environmental and Occupational Causes of Cancer Re-visited"; L. A. G. Ries and S. S. Devesa, "Cancer Incidence, Mortality, and Patient Survival in the United States," in D. Schottenfeld and J. F. Fraumeni (eds.), *Cancer Epidemiology and Prevention*, 3rd ed. (New York: Oxford University Press, 2006); L. L. Robison et al., "Assessment of Environmental and Genetic Factors in the Etiology of Childhood Cancers: The Children's Cancer Group Epidemiology Program," *EHP* 103 (1995, S-6): 111–16; S. H. Zahm and S. S. Devesa, "Childhood Cancer: Overview of Incidence Trends and Environmental Carcinogens," *EHP* 103 (1995, S-6): 177–84.

45 **greater exposure of children:** J. Wargo, *Our Children's Toxic Legacy: How Science and Law Fail to Protect Us from Pesticides* (New Haven, CT: Yale University Press, 1996); L. Mott et al., *Handle with Care: Children and Environmental Carcinogens* (New York: NRDC, 1994).

47 **more than 40 percent of Americans will contract cancer:** ACS, *Cancer Statistics—2009.*

47 **cancer as the leading cause of death under age 85:** Clapp et al., "Environmental and Occupational Causes of Cancer Re-visited."

47 **lung cancer trends:** M. R. Spitz et al., "Cancer of the Lung," in Schottenfeld and Fraumeni, *Cancer Epidemiology and Prevention.*

47 **percent of lung cancer deaths due to smoking:** A. Jernal et al., "Annual Report to the Nation on the Status of Cancer, 1975–2005, Featuring Trends in Lung Cancer, Tobacco Use, and Tobacco Control," *JNCI* 100 (2008): 1672–94.

47 **nonsmoking lung cancer deaths:** A. Jernal et al., "Annual Report to the Nation on the Status of Cancer, 1975–2005, Featuring Trends in Lung Cancer, Tobacco Use, and Tobacco Control," *JNCI* 100 (2008): 1672–94.

48 **testicular cancer:** R. W. Clapp et al., "Environmental and Occupational Causes of Cancer: New Evidence," *Reviews on Environmental Health* 23 (2008): 1–37.

48 **types of cancers that are increasing:** Ibid.

48 **thyroid cancer:** L. Enewold, "Rising Rates of Cancer Incidence in the United States by Demographic and Tumor Characteristics, 1980–2005," *Cancer Epidemiology, Biomarkers and Prevention* 18 (2009): 784–91.

48 **quotes by Hueper and Conway:** W. C. Hueper and W. D. Conway, *Chemical Carcinogenesis and Cancers* (Springfield, IL: Charles Thomas, 1964): 17, 158.

and Increasing Cancer Among Whites in the United States from 1973 through 1987: Good News and Bad News," *JAMA* 271 (1994): 431–37. These results have been replicated in Sweden, where researchers, making use of one of the world's oldest and most reliable cancer registries, have shown increasing cancer rates extending into the 1950s birth cohort: H-O. Adami et al., "Increasing Cancer Risk in Younger Birth Cohorts in Sweden," *Lancet* 341 (1993): 773–77.

51 **quote by Davis:** personal communication.

54 **trends in NHL:** P. Hartge et al., "Non-Hodgkin Lymphoma," in Schottenfeld and Fraumeni, *Cancer Epidemiology and Prevention.*

54 **AIDS and NHL:** L. K. Altman, "Lymphomas Are on the Rise in the U.S., and No One Knows Why," *New York Times*, 24 May 1994, C-3; P. Hartge et al., "Hodgkin's and Non-Hodgkin's Lymphomas," in R. Doll et al. (eds.), *Trends in Cancer Incidence and Mortality*, Cancer Surveys 19/20 (Plainview, NY: Cold Spring Harbor Laboratory Press, 1994).

54 **occupations associated with NHL:** Hartge et al., "Non-Hodgkin Lymphoma."

55 **chemicals associated with NHL:** Ibid.

55 **NHL and PCBs:** Exposures to chlordane, the now-banned termite pesticide, also show connections to lymphoma risk. J. S. Colt et al., "Organochlorine Exposure, Immune Gene Variation, and Risk of Non-Hodgkin Lymphoma," *Blood* 113 (2008): 1899–1905; L. S. Engel et al., "Polychlorinated Biphenyls and Non Hodgkin Lymphoma," *Cancer Epidemiology, Biomarkers, and Prevention* 16 (2007): 373–76; K. Hardell et al., "Concentrations of Organohalogen Compounds and Titres of Antibodies to Epstein-Barr Virus Antigens and the Risk for Non-Hodgkin Lymphoma," *Oncology Reports* 21 (2009): 1567–76; J. J. Spinelli et al., "Organochlorines and Risk of Non-Hodgkin Lymphoma," *International Journal of Cancer* 121 (2007): 2767–75.

55 **NHL and pesticides:** Hartge et al., "Non-Hodgkin Lymphoma"; S. H. Zahm and A. Blair, "Pesticides and Non-Hodgkin's Lymphoma," *Can-*

cer Research 52 (1992, S): 5485s–88s; S. H. Zahm, "The Role of Agricultural Pesticide Use in the Development of Non-Hodgkin's Lymphoma in Women," *AEH* 48 (1993): 353–58.

55 **military history of phenoxy herbicides:** D. E. Lilienfeld and M. A. Gallo, "2,4-D, 2,4,5-T, and 2,3,7,8-TCDD: An Overview," *Epidemiologic Reviews* 11 (1989): 28–58.

56 **trade names:** S. A. Briggs, *Basic Guide to Pesticides: Their Characteristics and Hazards* (Bristol, PA: Taylor & Francis, 1992).

56 **evidence for a connection:** Hartge et al., "Non-Hodgkin Lymphoma"; Institute of Medicine, *Veterans and Agent Orange: Health Effects of Herbicides Used in Vietnam* (Washington, DC: National Academy Press, 1994); D. D. Weisenburger, "Epidemiology of Non-Hodgkin's Lymphoma: Recent Findings Regarding an Emerging Epidemic," *Annals of Oncology* 1 (1994, S-5): s19–s24; Zahm and Blair, "Pesticides and Non-Hodgkin's Lymphoma"; S. Zahm et al., "A Case-Control Study of Non-Hodgkin's Lymphoma and the Herbicide 2,4-Dichlorophenoxyacetic Acid (2,4-D) in Eastern Nebraska," *Epidemiology* 1 (1990): 349–56; L. Hardell et al., "Malignant Lymphoma and Exposure to Chemicals, Especially Organic Solvents, Chlorophenols and Phenoxy Acids: A Case-Control Study," *British Journal of Cancer* 43 (1981): 169–76.

56 **lymphoma in golf course superintendents:** B. C. Kross et al., "Proportionate Mortality Study of Golf Course Superintendents," *American Journal of Industrial Medicine* 29 (1996): 501–06.

56 **lymphoma in dogs:** H. M. Hayes et al., "Case-Control Study of Canine Malignant Lymphoma: Positive Association with Dog Owner's Use of 2,4-Dichlorophenoxyacetic Acid Herbicides," *JNCI* 83 (1991): 1226–31.

56 **residential herbicide use:** P. Hartge et al., "Residential Herbicide Use and Risk of Non-Hodgkin Lymphoma," *Cancer Epidemiology, Biomarkers and Prevention* 14 (2005): 934–37.

56 **2006 review of NHL:** Hartge et al., "Non-Hodgkin Lymphoma."

57 **leukemia in southeastern Massachusetts:** M. S. Morris and R. S. Knorr, "Adult Leukemia and Proximity-Based Surrogates for Exposure to Pilgrim Plant's Nuclear Emissions," *AEH* 51 (1996): 266–74; M. S. Morris and R. S. Knorr, *Southeastern Massachusetts Health Study Final Report: Investigation of Leukemia Incidence in 22 Massachusetts Communities, 1978–86* (Boston: MDPH, 1990); L. Tye, "Screening

four: space

59 **history and environmental problems of Normandale:** T. L. Aldous, "Community Dreads Threat of Disease," *PDT*, 14 Sept. 1991, A-2, A-12.

60 **global patterns of cancer incidence:** IARC, *World Cancer Report, 2008* (Lyon, France: IARC, WHO, 2009). For European data on cancer's geography, consult the Eurostat database. Eurostat has been collecting and disseminating health data on EU member states since 1994. Fact sheets on different cancers in different EU countries are provided by the European Cancer Observatory.

61 **worst-polluted places:** D. Biello, "World's Most Polluted Places," *Scientific American*, Sept. 13, 2007; Blacksmith Institute, *The World's Worst Polluted Places: The Top Ten (of the Dirty Thirty)* (New York: Blacksmith Institute, 2007).

61 **China and coal:** The World Bank and State Environmental Protection Administration, P.R. China, *Cost of Pollution in China: Economic Estimates of Physical Damages* (Washington, DC: World Bank, 2007).

61 **reporting by Steven Ribert:** S. Ribert, "Horrors of Hongwei," *The Standard* (Hong Kong), 16 June 2007.

62 **cancer villages in China:** J. Watts, China's Environmental Health Challenges," *Lancet* 372 (2008): 1451–52; J. F. Tremblay, "China's Cancer Villages," *Chemical and Engineering News* 85 (2007): 18–21.

62 **cancer incidence trend in China:** IARC, *World Cancer Report, 2008* (Lyon, France: IARC, WHO, 2009).

62 **migrant studies:** These are reviewed in J. Gray et al., "State of the Evidence: The Connection Between Breast Cancer and the Environment," *International Journal of Occupational and Environmental Health* 15 (2009): 43–78. See also E. M. John et al., "Migration History, Acculturation, and Breast Cancer Risk in Hispanic Women," *Cancer Epidemiology, Biomarkers and Prevention* 14 (2005): 2905–13; E. V. Kliewar and K. R. Smith, "Breast Cancer Mortality Among Immigrants in Australia and

Canada," *JNCI* 87 (1995): 1154–61; N. Angier, "Woman's Move Can Change Her Risk of Breast Cancer," *New York Times*, 2 Aug. 1995, A-17; H. Shimizu et al., "Cancers of the Prostate and Breast Among Japanese and White Immigrants in Los Angeles County," *British Journal of Cancer* 63 (1991): 963–66; L. Tomatis (ed.), *Cancer: Causes, Occurrence and Control* (London: Oxford University Press, 1990); D. B. Thomas and M. R. Karagas, "Cancer in First and Second Generation Americans," *Cancer Research* 47 (1987): 5771–76.

63 **cancer in Normandale:** Aldous, "Community Dreads Threat."

63 **quote from Normandale resident:** Ibid.

64 **geography of cancer:** Interactive maps and graphs of cancer mortality in the United States are found on the NCI's Web site: www.cancer.gov/atlasplus/index.html; L. W. Pickle et al., "The New United Sates Cancer Atlas," *Recent Results in Cancer Research* 14 (1989): 196–207; C. S. Stokes and K. D. Brace, "Agricultural Chemical Use and Cancer Mortality in Selected Rural Counties in the U.S.A.," *Journal of Rural Studies* 4 (1988): 239–47; B. A. Goldman, *The Truth About Where You Live: An Atlas for Action on Toxins and Mortality* (New York: Random House, 1991); S. S. Devesa, "Recent Cancer Patterns Among Men and Women in the United States: Clues for Occupational Research," *Journal of Occupational Medicine* 36 (1994): 832–41; and S. H. Zahm et al., "Pesticides and Multiple Myeloma in Men and Women in Nebraska," in H. H. McDuffie et al. (eds.), *Supplement to Agricultural Health and Safety Workplace, Environment, Sustainability* (Saskatoon, Saskatchewan, Canada: University of Saskatchewan Press, 1995); J. L. Kelsey and P. L. Horn-Ross, "Breast Cancer: Magnitude of the Problem and Descriptive Epidemiology," *Epidemiologic Reviews* 15 (1993): 7–16; Pickle, "New United States."

65 **maps of childhood cancers in the United Kingdom:** E. G. Knox and E. A. Gilman, "Hazard Proximity of Childhood Cancers in Great Britain from 1953–80," *Journal of Epidemiology and Community Health* 51 (1997): 151–59.

65 **workplace carcinogens:** P. R. Infante, "Cancer and Blue-Collar Workers: Who Cares?" *New Solutions* (Winter 1995): 52–57; J. Randal, "Occupation as a Carcinogen: Federal Researcher Suggests Change in Cancer Registries," *JNCI* 86 (1994): 1748–50; P. Landrigan, "Cancer Research in the Workplace," presentation at the President's Cancer Panel meeting, NIH, Bethesda, MD, 22 Sept. 1993.

65 **high cancer rates in farmers:** S. H. Zahm and A. Blair, "Cancer Among Migrant and Seasonal Farmers," *American Journal of Industrial Medicine* 24 [illegible]; A. Blair et al., "Clues to Cancer Etiology from Studies of Farmers," *Scandinavian Journal of Work Environment and Health* 18 [illegible]; D. L. Davis et al., "Agricultural Exposures and Cancer Trends in Developed Countries," *EHP* 100 (1992): 39–44. Excesses in non-Hodgkin lymphoma and brain cancer did not always attain statistical significance.

66 **Agricultural Health Study:** All published papers are available on the AHS Web site: http://aghealth.nci.nih.gov/. Studies cited here are M. C. R. Alavanja et al., "Cancer Incidence in the Agricultural Health Study," *Scandinavian Journal of Work and Environment* 31 (2005; S1): 39–45; G. Andreotti et al., "Agricultural Pesticide Use and Pancreatic Cancer Risk in the Agricultural Health Study Cohort," *International Journal of Cancer* 124 (2009): 2495–2500; A. Blair et al., "Mortality Among Participants in the Agricultural Health Study," *Annals of Epidemiology* 15 (2005): 279–85; B. D. Curwin et al., "Urinary Pesticide Concentrations Among Children, Mothers and Fathers Living in Farm and Non-Farm Households in Iowa," *Annals of Occupational Hygiene* 51 (2007): 53–65; L. S. Engel et al., "Pesticide Use and Breast Cancer Risk Among Farmers' Wives in the Agricultural Health Study," *AJE* 161 (2005): 121–135; S. L. Farr et al., "Pesticide Exposure and Timing of Menopause," *AJE* 163 (2006): 731–42; S. L. Farr et al., Pesticide Use and Menstrual Cycle Characteristics Among Premenopausal Women in the Agricultural Health Study," *AJE* 160 (2004): 1194–1204; K. B. Flower et al., "Cancer Risk and Parental Pesticide Application in Children of Agricultural Health Study Participants," *EHP* 112 (2004): 631–35; J. A. Rusiecki et al., "Cancer Incidence Among Pesticide Applicators Exposed to Permethrin in the Agricultural Health Study," *EHP* 117 (2009): 581–86.

67 **other occupations with high cancer rates:** R. W. Clapp et al., "Environmental and Occupational Causes of Cancer: New Evidence 2005–2007," *Reviews on Environmental Health* 23 (2008): 1–37; R. W. Clapp et al., "Environmental and Occupational Causes of Cancer Re-visited," *Journal of Public Health Policy* 27 (2006): 61–76; E. L. Hall and K. D. Rosenman, "Cancer by Industry: Analysis of a Population-Based Cancer Registry with an Emphasis on Blue-Collar Workers," *American*

Journal of Industrial Medicine 19 (1991): 145–59; J. M. Stellman, "Where Women Work and the Hazards They May Face on the Job," *Journal of Occupational Medicine* 36 (1994): 814–[illegible].

67 **cancer in firefighters:** G. K. LeMasters et al., "Cancer Risk Among Firefighters: A Review and Meta-analysis of 32 Studies," *Journal of Occupational and Environmental Medicine* 48 (2006): 1189–1202. Firefighters' risk of cancer increases with duration of employment. Those with long service also have elevated rates of leukemia as well as cancers of the colon, brain, and kidney. S. Youakim, "Risk of Cancer Among Firefighters: A Quantitative Review of Selected Malignancies," *Archives of Environmental and Occupational Health* 61 (2006): 223–31; G. Tornling et al., "Mortality and Cancer Incidence in Stockholm Firefighters," *American Journal of Industrial Medicine* 25 (1994): 219–28.

67 **cancers in Finnish women workers:** M. L. Lindbohm et al., "Risk of Liver Cancer and Exposure to Organic Solvents and Gasoline Vapors Among Finnish Workers," *International Journal of Cancer* 124 (2009): 2954–59; J. Lohl et al., "Occupational Exposures to Solvents and Gasoline and Risks of Cancers in the Urinary Tract Among Finnish Workers," *American Journal of Industrial Medicine* 51 (2008): 668–72.

67 **Taiwanese electronics workers with breast cancer:** T. I. Sung et al., "Increased Standardized Incidence Ratio of Breast Cancer in Female Electronics Workers," *BMC Public Health* 7 (2007): 102.

67 **concerns about nail salons:** These concerns have been taken up by the California Healthy Nail Salon Collaborative.

67 **professional jobs with high cancer rates:** Clapp et al., "Environmental and Occupational Causes of Cancer: New Evidence," 1–37; Clapp et al., "Environmental and Occupational Causes of Cancer Revisited," 61–76; E. A. Holly, "Intraocular Melanoma Linked to Occupations and Chemical Exposure," *Epidemiology* 7 (1996): 55–61; B. B. Arnetz et al., "Mortality Among Petrochemical Science and Engineering Employees," *AEH* 46 (1991): 237–48.

67 **cancer in dentists, dental assistants, and chemotherapy nurses:** L. M. Pottern et al., "Occupational Cancer Among Women: A Conference Overview," *Journal of Occupational Medicine* 36 (1994): 809–13.

67 **children's cancers related to parental exposures:** L. M. O'Leary et al., "Parental Exposures and Risk of Childhood Cancer: A Review," *American Journal of Industrial Medicine* 20 (1991): 17–35. The gender imbalance in

Pekin (Tazewell County), Illinois" (Springfield, IL: IDPH, 1991); G. Poquette, "Normandale Cancer Study" (memorandum) (Tremont, IL: Tazewell County Health Department, 5 Mar. 1992).

68 **headline:** T. L. Aldous, "Study: Area Cancer Rates Normal," *PDT*, 19 Dec. 1991, A-2, A-12.

69 **Superfund:** Congress enacted legislation on hazardous waste sites in 1980 when it passed the Comprehensive Environmental Response, Compensation and Liability Act (CERCLA), which is generally known as Superfund. The goals of the bill are to inventory hazardous waste sites, to establish priorities for cleanup based on relative danger, to contain dangerous releases, and ultimately to remediate through elimination of unsafe sites. The nomenclature surrounding hazardous waste is confusing. Waste sites appearing on the National Priorities List are called Superfund sites, whereas waste sites inventoried under the program but not on the NPL are commonly referred to as CERCLA sites. The trust fund to clean up NPL sites was created through a corporate "polluter pays" tax, which expired in 1995. Superfund went bankrupt in 2003. For an outstanding history of the program and to locate Superfund sites in your community, see the Center for Public Integrity's 2007 online project, *Wasting Away: Superfund's Toxic Legacy*: http://projects.publicintegrity.org/Superfund/. See also the searchable Superfund database on the EPA's Web site, www.epa.gov/superfund/sites/index.htm. The Web-based public art project Superfund365, launched by digital artist Brooke Singer, is based on travels to 365 different Superfund sites across the nation. www.superfund365.org

69 **children living near Superfund sites:** ATSDR, *Children Living Near Hazardous Waste Sites* (Atlanta: USDHHS, ATSDR, 2003).

69 **750 million tons:** J. Griffith and W. B. Riggan, "Cancer Mortality in U.S. Counties with Hazardous Waste Sites and Ground Water Pollution," *AEH* 44 (1989): 69–74.

69 **cancer in New Jersey:** G. R. Najem et al., "Female Reproductive Organs and Breast Cancer Mortality in New Jersey Counties and the Re-

lationship with Certain Environmental Variables," *Preventive Medicine* 14 (1985): [illegible]; G. R. Najem et al., "Clusters of Cancer Mortality in New Jersey Municipalities, with Special Reference to Chemical Toxic Waste Disposal Sites and Per Capita Income," *International Journal of Epidemiology* 14 (1985): [illegible]; G. R. Najem et al., "Gastrointestinal Cancer Mortality in New Jersey Counties and the Relationship to Environmental Variables," *International Journal of Epidemiology* 12 (1983): 276–89.

70 **cancer in counties with groundwater contamination:** Griffith and Riggan, "Cancer Mortality in U.S. Counties." See also R. Hoover and J. F. Fraumeni Jr., "Cancer Mortality in U.S. Counties with Chemical Industries," *Environmental Research* 9 (1975): 196–207. Bladder cancer is linked to living near toxic waste sites. L. J. Gensburg, "Cancer Incidence among Former Love Canal Residents," *EHP* 117 (2009): 1265–71.

70 **Ecological fallacy:** can also refer to the mistake of applying group attributes to individuals. For example, in a famous study of suicide and religion in nineteenth-century Europe, researchers found that suicide rates rose as the proportion of Protestants living in a given region increased. The obvious conclusion—that Protestants are more likely to kill themselves than Catholics—does not necessarily follow, however. It is entirely possible that all the suicides in Protestant-dominated areas had occurred among Catholics: perhaps people become increasingly vulnerable to suicide as they become an increasingly isolated minority. This latter explanation turned out not to be the case, but an ecological study of groups could not distinguish between these two mutually exclusive conclusions. For a lively discussion of this study, see J. Esteve et al., *Descriptive Epidemiology: Statistical Methods in Cancer Research*, vol. 4 (Lyon, France: IARC, Scientific Pub. No. 128, 1994), 150–54.

72 For an introduction to epidemiological methods and, from a variety of viewpoints, their limitations, see K. J. Rothman and C. Poole, "Causation and Causal Inference," in D. Schottenfeld and J. F. Fraumeni Jr. (eds.), *Cancer Epidemiology and Prevention*, 2nd ed. (Oxford, England: Oxford University Press, 1996); N. Krieger, "Epidemiology and the Web of Causation: Has Anyone Seen the Spider?" *Social Science and Medicine* 39 (1994): 887–903; D. Trichopoulos and E. Petridou, "Epidemiologic Studies and Cancer Etiology in Humans," *Medicine, Exercise, Nutrition, and Health* 3 (1994): 206–25; S. Wing, "Limits of Epidemiology," *Medicine and Global Survival* 1 (1994): 74–86; M. S.

Legator and S. F. Strawn (eds.), *Chemical Alert! Community Action Handbook* (Austin: Texas University Press, 1993).

72 **Mary Wolff's study:** M. S. Wolff et al., "Blood Levels of Organochlo- [illegible]

73 **the vexations of cluster studies:** See the supplemental issue of the *AJE* 132 (1990), which contains the proceedings of the National Conference on Clustering of Health Events, held in Atlanta, 16–17 Feb. 1989. See also G. Taubes, "Epidemiology Faces Its Limits," *Science* 269 (1995): 164–69; Legator and Strawn, *Chemical Alert*; CDC, "Guidelines for Investigating Clusters of Health Events," *Morbidity and Mortality Weekly Report* 39/RR-11 (1990): 1–23; and K. J. Rothman, "Clustering of Disease," *AJPH* 77 (1987): 13–15.

74 **overtly dismissive:** For an example, M. J. Thun and T. Sinks, "Understanding Cancer Clusters," *CA* 54 (2004): 273–80. See also J. D. Besley et al., "Local Newspaper Coverage of Health Authority Fairness During Cancer Cluster Investigations," *Science Communication* 29 (2008): 498–21.

74 **Limited power:** Power and significance can be described in several ways. One of the most commonly used measures in epidemiology is the confidence interval—a computed range with a given probability (95 percent) that the true value of the variable lies within it. Further explanations of epidemiological statistics for laypersons can be found in M. J. Scott and B. L. Harper, "Lots of Information: What to Do with It: Statistics for Nonstatisticians," in Legator and Strawn, *Chemical Alert*.

74 **eight to twenty times higher:** R. R. Neutra, "Counterpoint from a Cluster Buster," *AJE* 132 (1990): 1–8.

75 **TCE:** ATSDR, *Case Studies in Environmental Medicine: Trichloroethylene Toxicity* (Atlanta: ATSDR, 1992).

75 **quote by nurse:** D. Robinson, "Letter in Response to 'Cancer Clusters: Findings vs Feelings,'" *Medscape General Medicine* 4 (2002): 4.

76 **the Eleven Blue Men:** B. Roueche, *Eleven Blue Men and Other Narratives of Medical Detection* (Boston: Little, Brown, 1954). The significance of this case study for cancer clusters is discussed by Neutra, "Counterpoint."

77 **GIS and exposure assessment:** B. S. Kingsley et al., "An Update on Cancer Cluster Activities at the Centers for Disease Control and Prevention," *EHP* 115 (2008): 165–71.

77 **statistics to test for randomness:** M. Kulldorff et al., "Cancer Map Patterns: Are They Random or Not?" *American Journal of Preventive Medicine* 2006; F. Wang, "Spatial Clusters of Cancers in Illinois, 1986–2000," *Journal of Medical Systems* 28 (2004): [illegible].

77 **limits to GIS:** D. C. Wheeler "A Comparison of Spatial Clustering and Cluster Detection Techniques for Childhood Leukemia Incidence in Ohio, 1996–2003," *International Journal of Health Geographics* 6 (2007): 10.

77 **problems with ZIP codes:** T. H. Grubesic and T. C. Matisziw, "On the Use of ZIP Code and ZIP Code Tabulation Areas (ZCTAs) for the Spatial Analysis of Epidemiological Data," *International Journal of Health Geographics* 5 (2006): 58.

77 **the biggest hindrance is ignorance:** Pew Environmental Health Commission, Environmental Health Project Team, *America's Environmental Health Gap: Why the Country Needs a Nationwide Health Tracking Network* (Baltimore: Johns Hopkins School of Hygiene and Public Health, 2000); N. S. Juzych et al., "Adequacy of State Capacity to Address Noncommunicable Disease Clusters in the Era of Environmental Health Tracking, *American Journal of Public Health* 97 (2007, S-1): S163–69.

78 **cluster studies:** A. M. Nieder et al., "Bladder Cancer Clusters in Florida: Identifying Populations at Risk," *Journal of Urology* 182 (2009): 46–51; J. Dahlgren et al., "Cluster of Hodgkin's Lymphoma in Residents Near a Non-Operational Petroleum Refinery," *Toxicology and Industrial Health* 24 (2008): 683–92; R. W. Clapp and K. Hoffman, "Cancer Mortality in IBM Endicott Plant Workers, 1969–2001: An Update on a NY Production Plant," *Environmental Health* 7 (2008):13; M. Gilbert, "Cancer Cluster is Confirmed in Clyde," *Toledo Blade*, 30 May 2009.

79 **two excesses attained statistical significance:** This reanalysis was conducted by Richard Clapp, an epidemiologist at Boston University and the former director of the Massachusetts Cancer Registry.

79 **breast cancer and chemical plants on Long Island:** E. L. Lewis-Michl et al., "Breast Cancer Risk and Residence Near Industry or Traffic in Nassau and Suffolk Counties, Long Island, New York," *AEH* 51 (1996): 255–65; J. Melius et al., "Residence Near Industries and High Traffic Areas and the Risk of Breast Cancer on Long Island" (Albany: New York State Dept. of Health, 1994).

79 **earlier Long Island study:** New York State Department of Health, Department of Community and Preventative Medicine at the State University of New York at Stony Brook, Nassau County Department of Health, and Suffolk County Department of Health Services, *The Long Island Breast Cancer Study*, Reports [illegible]

79 Gina G. Kolata, "Long Island Breast Cancer Called Explainable by U.S.," *New York Times*, 19 Dec. 1992, A-9.

80 **Long Island Breast Cancer Study Project:** M. D. Gammon et al., "Environmental Toxins and Breast Cancer on Long Island: II. Organochlorine Compound Levels in Blood," *Cancer Epidemiology, Biomarkers and Prevention* 11 (2002): 686–97.

80 **breast cancer in Long Island women and pesticide use:** S. L. Teitelbaum et al., "Reported Residential Pesticide Use and Breast Cancer Risk on Long Island, New York," *AJE* 165 (2007): 643–51.

81 **breast cancer in Long Island women with signs of DNA damage:** These are known as DNA adducts. M. D. Gammon et al., "Environmental Toxins and Breast Cancer on Long Island: I. Polycyclic Aromatic Hydrocarbon DNA Adducts," *Cancer Epidemiology, Biomarkers and Prevention* 11 (2002): 677–85; J. G. Brody and R. A. Rudell, "Environmental Pollutants and Breast Cancer: The Evidence from Animal and Human Studies," *Breast Diseases: A Year Book Quarterly* 19 (2008): 17–19.

81 **history of the Upper Cape:** S. Rolbein, *The Enemy Within: The Struggle to Clean Up Cape Cod's Military Superfund Site* (Orleans, MA: Association for the Preservation of Cape Cod, 1995).

81 **cancer rates in the Upper Cape:** MDPH, *Cancer Incidence in Massachusetts, 1982–90* (Boston: MDPH, 1993).

81 **1991 study:** A. Aschengrau and D. M. Ozonoff, *Upper Cape Cancer Incidence Study. Final Report* (Boston: Mass. Depts. of Public Health and Environmental Protection, 1991).

82 **Silent Spring Institute:** J. G. Brody et al., "Mapping Out a Search for Environmental Causes of Breast Cancer," *Public Health Reports* 111 (1996): 495–507; "Cape Cod Breast Cancer and Environment Study Overview" (Newton, MA: Silent Spring Institute, July 12, 1995).

82 **Cape Cod studies:** V. Viera et al., "Spatial Analysis of Lung, Colorectal, and Breast Cancer on Cape Cod: An Application of Generalized Additive Models to Case-Control Data," *Environmental Health* 4 (2005):

11; W. McKelvey et al., "Association between Residence on Cape Cod, Massachusetts, and Breast Cancer," *Annals of Epidemiology* 14 (2004): 89–94; L. D. Standley et al., "Wastewater-Contaminated Groundwater as a Source of Endogenous Hormones and Pharmaceuticals to Surface Water Ecosystems," *Environmental Toxicology and Chemistry* 27 (2008): 2457–68; J. Brody et al., "Breast Cancer Risk and Drinking Water Contaminated by Wastewater: A Case-Control Study," *Environmental Health* 5 (2006): 28.

83 **quote from 1991 study:** Aschengrau and Ozonoff, *Upper Cape*, ix.

83 **Cape Cod water pipes:** A. Aschengrau et al., "Cancer Risk and Tetrachloroethylene-Contaminated Drinking Water in Massachusetts," *AEH* 48 (1993): 284–92; T. Webster and H. S. Brown, "Exposure to Tetrachloroethylene via Contaminated Drinking Water Pipes in Massachusetts: A Predictive Model," *AEH* 48 (1993): 293–97.

84 **dry cleaners:** N. S. Weiss, "Cancer in Relation to Occupational Exposure to Perchloroethylene," *Cancer Causes and Control* 6 (1995): 257–66.

84 **1983 study:** C. D. Larsen et al., "Tetrachloroethylene Leached from Lined Asbestos-Cement Pipe into Drinking Water," *Journal of the American Water Works Association* 75 (1983): 184–88.

85 **quote from 1993 study:** Aschengrau, "Tetrachloroethylene-Contaminated," 291.

85 **Normandale:** T. L. Aldous, "State to Probe Cancer in Normandale," *PDT*, 4 Oct. 1991, A-2; T. L. Aldous, "Study: No Cancer Cluster," *PDT*, 6 Mar. 1992, A-1, A-12.

86 **newspaper investigation of death certificates:** Ibid.

87 **quote from Normandale widower:** Aldous, "Area Cancer Rates Normal."

five: war

90 **World War II in *Silent Spring*:** R. Carson, *Silent Spring* (Boston: Houghton Mifflin, 1962). (See especially Chapters 2 and 3.)

91 **all life was caught in the crossfire:** Ibid., 8.

91 **trends in chemical production:** Graphs from International Trade Commission, Washington, DC.; R. C. Thompson et al., "Our Plastic Age," *Philosophical Transactions of the Royal Society B* 364 (2009): 1973–76.

May/June 2005, 16–27.

94 **collect in tissues high in fat:** J. D. Sherman, *Chemical Exposure and Disease: Diagnostic and Investigative Techniques* (Princeton, NJ: Princeton Scientific Publishing, 1994); L. S. Welch, "Organic Solvents," in M. Paul (ed.), *Occupational and Environmental Reproductive Hazards: A Guide for Clinicians* (Baltimore: Williams & Wilkins, 1993).

95 **chloroform:** ATSDR, *Toxicological Profile for Chloroform* (Atlanta: US-DHHS, ATSDR, 1997).

95 **DDT in World War II:** E. P. Russell III, "'Speaking of Annihilation': Mobilizing for War Against Human and Insect Enemies, 1914–1945," *Journal of American History* 82 (1996): 1505–29; T. R. Dunlap, DT: *Scientists, Citizens, Public Policy* (Princeton, NJ: Princeton University Press, 1981), 61–62; J. Whorton, *Before* Silent Spring: *Pesticides and Public Health in Pre-DDT America* (Princeton, NJ: Princeton University Press, 1974), 248–55.

96 **Hitler's head:** This ad appeared in the trade magazine *Soap and Sanitary Chemicals* in April 1944 and is reprinted in Russell, "'Speaking of Annihilation.'"

96 **phenoxy herbicides:** D. E. Lilienfeld and M. A. Gallo, "2,4-D, 2,4,5-T, and 2,3,7,8-TCDD: An Overview," *Epidemiologic Reviews* 11 (1989): 28–58.

96 **parathion and other organophosphates:** Sherman, *Chemical Exposure and Disease*, 24; H. W. Chambers, "Organophosphorous Compounds: An Overview," in J. E. Chambers and P. E. Levi (eds.), *Organophosphates: Chemistry, Fate, and Effects* (San Diego: Academic Press, 1992).

96 **mechanisms of action:** L. J. Fuortes et al., "Cholinesterase-Inhibiting Insecticide Toxicity," *American Family Physician* 47 (1993): 1613–20; F. Matsumura, *Toxicology of Insecticides*, 2nd ed. (New York: Plenum, 1985), 111–202.

96 **organophosphates as German nerve gas:** Sherman, *Chemical Exposure and Disease*, 161; J. Borkin, *The Crime and Punishment of I. G. Farben* (New York: Harper & Row, 1978), 722–23.

96 **phenoxy herbicides in war:** P. F. Cecil, *Herbicidal Warfare: The Ranch Hand Project in Vietnam* (New York: Praeger, 1986); A. Ihde, *The Development of Modern Chemistry* (New York: Harper & Row, 1964), 722–23.

97 **by 1960, 2,4-D accounted for half:** Lilienfeld and Gallo, "2,4-D, 2,4,5-T."

97 For more on the rise of herbicide use in the United States, see NRC, *Pesticides in the Diets of Infants and Children* (Washington, DC: National Academy Press, 1993), 15.

97 **graphs of pesticide use:** W. J. Hayes Jr. and E. R. Laws (eds.), *Handbook of Pesticide Toxicology*, vol.1, *General Principles* (New York: Academic Press, 1991), 22.

97 **capturing 90 percent of the market:** NRC, *Pesticides in the Diets*, 15.

97 **trends in pesticide use:** About half of pesticides used in U.S. agriculture are herbicides. Insecticides and fungicides and others make up the remainder. Some indirect evidence suggests that herbicide use has peaked and insecticide use is declining, but exact figures are not available because, at this writing, pesticide use data have not been released by the EPA for eight years. S. K. Ritter, "Pinpointing Trends in Pesticide Use—Limited Data Indicate that Pesticide Use Has Dropped Since the 1970s," *Chemical and Engineering News* 87 (2009).

97 **household pesticide use:** T. Kiely et al., *Pesticides—Industry Sales and Usage, 2000 and 2001 Market Estimates* (Washington, DC: U.S. EPA Office of Prevention, Pesticides, and Toxic Substances, 2004).

97 **pesticides in carpet fibers:** R. G. Lewis et al., "Evaluation of Methods for Monitoring the Potential Exposure of Small Children to Pesticides in the Residential Environment," *Archives of Environmental Contamination and Toxicology* 26 (1994): 37–46; M. Moses et al., "Environmental Equity and Pesticide Exposure," *Toxicology and Industrial Health* 9 (1993): 913–59.

97 **HUD study of kitchen floors:** D. M. Stout et al., "American Healthy Homes Survey: A National Study of Residential Pesticides Measured from Floor Wipes," *Environmental Science and Technology* 43 (2009): 4294–4300.

98 **childhood cancers and household pesticide use:** C. Infante-Rivard and S. Weichenthal, "Pesticide and Childhood Cancer: An Update of Zahm and Ward's 1998 Review," *Journal of Toxicology and Environmental Health, Part B* 10 (2007): 81–99; X. Ma et al., "Critical Windows of

Exposure to Household Painting and Risk of Childhood Leukemia," *EHP* 110 (2002): 955–60; C. Metayer and P. A. Buffler, "Residential Exposures to Pesticides and Childhood Leukemia," *Radiation Protection Dosimetry* 132 (2008): 212–19; A. L. Rosso et al., "A Case-Control Study of Childhood Brain Tumors and Fathers' Hobbies: A Children's Oncology Group Study," *Cancer Causes and Control* 19 (2008): 1201–7; J. Rudant et al., "Household Exposure to Pesticides and Risk of Hematopoietic Malignancies: The ESCALE Study (SFCE)," *EHP* 115 (2007): 1787–93; O. P. Soldin et al., "Pediatric Acute Lymphoblastic Leukemia and Exposure to Pesticides," *Therapeutic Drug Monitoring* 32 (2009): 495–501.

98 **rise of petrochemicals:** R. F. Sawyer, "Trends in Auto Emissions and Gasoline Composition," *EHP* 101 (1993, S-6): 5–12; Ihde, *Modern Chemistry.*

99 **Germany's artificial fertilizer:** Ihde, *Modern Chemistry*, 680–81.

99 **chlorine gas and chlorinated solvents:** International Programme on Chemical Safety, WHO, "Chlorine and Hydrogen Chloride," *Environmental Health Criteria* 21 (1982): 54–60; Dr. Edmund Russell III, personal communication.

99 **after the war ended:** A. Thackary et al., *Chemistry in America, 1876–1976* (Dordecht, Netherlands: Reidel, 1985).

99 **by the 1930s:** Ihde, *Modern Chemistry.*

99 **all-out assaults of World War II:** Ibid.

99 **fear of national leaders:** Dr. Edmund Russell III, personal communication.

99 **quote by Aaron Ihde:** Ihde, *Modern Chemistry*, 674.

99 **transformation from a carbohydrate-based economy to a petrochemical-based one:** D. Morris and I. Ahmed, *The Carbohydrate Economy: Making Chemicals and Industrial Materials from Plant Matter* (Washington, DC: Institute for Local Self-Reliance, 1992). For an entertaining history of plant-derived plastics and their replacement by petrochemical plastics, see S. Fenichell, *Plastic: The Making of a Synthetic Century* (New York: Harper-Business, 1996).

100 **plastics consume 8 percent of world oil:** 4 percent is used for feedstock and the other 4 percent to provide energy for their manufacture. About one-third of the annual production of plastic is for disposable packaging. Plastics account for 10 percent by weight of the municipal waste stream. Annual growth of plastic production is about 9 percent.

R. C. Thompson et al., "Our Plastic Age," *Philosophical Transactions of the Royal Society* 364 (2009): 1973–76; R. C. Thompson et al., "Plastics, the Environment, and Human Health: Current Consensus and Future Trends," *Philosophical Transactions of the Royal Society* 364 (2009): 2153–66.

100 **formaldehyde:** L. E. Beane Freeman et al., "Mortality from Lymphohematopoietic Malignancies Among Workers in Formaldehyde Industries: The National Cancer Institute Cohort," *JNCI*, 101 (2009): 751–61; NCI, "Fact Sheet: Formaldehyde and Cancer Risk," May 2009; NTP, "Formaldehyde," *Report on Carcinogens*, 11th ed. (USDHHS, Public Health Service, 2005).

100 **formaldehyde in foam insulation:** IDPH, "Urea Formaldehyde Foam Insulation" (pamphlet) (Springfield, IL: IDPH, 1992).

100 **formaldehyde as an indoor air pollutant:** M. C. Marbury and R. A. Krieger, "Formaldehyde," in J. M. Samet and J. D. Spengler (eds.), *Indoor Air Pollution: A Health Perspective* (Baltimore, MD: Johns Hopkins University Press, 1991).

100 **headlines about formaldehyde in trailers:** For example, M. Engel, "Fuming over Formaldehyde," *Los Angeles Times*, 7 Oct. 2008.

100 **embalmers:** "Formaldehyde," NTP, *Report on Carcinogens*, 11th ed. (USDHHS, Public Health Service, 2005).

101 **soybeans as a formaldehyde predecessor:** Morris and Ahmed, *Carbohydrate Economy*.

101 **other oil-based plants:** Ibid.

101 **synthetic cutting fluids in machine shops:** Y. T. Fan, "N-Nitrosodiethanolamine in Synthetic Cutting Fluids: A Part-per-Hundred Impurity," *Science* 196 (1977): 70–71.

101 **contaminants in cutting fluids:** NTP, *Seventh Annual Report*, 282.

101 **quote from cutting-fluid study:** Fan, "N-Nitrosodiethanolamine," 71.

101 **fall 2009 announcement:** K. Zito, "EPA Wants More Oversight on Chemicals," *San Francisco Chronicle*, 30 Sept. 2009.

101 **flaws of TSCA:** My analysis is based on the following sources: A. Daemmrich, "Risk Frameworks and Biomonitoring: Distributed Regulation of Synthetic Chemicals in Humans," *Environmental History* 13 (2008): 684–95; M. Schapiro, *Exposed: The Toxic Chemistry of Everyday Products and What's at Stake for American Power* (White River Junction, VT: Chelsea Green Publishing, 2007); U.S. GAO, Chemical Regulation: Options Exist to Improve EPA's Ability to Assess Health

Risks and Manage Its Chemical Review Program (U.S. Government Accountability Office, GAO-05-458, June 2005); M. P. Wilson et al., *Green Chemistry in California: A Framework for Leadership in Chemicals Policy and Innovation* (University of California, California Policy Research Center, 2006); M. P. Wilson and M. R. Schwarzman, *Green Chemistry: Cornerstone to a Sustainable California* (Centers for Occupational and Environmental Health, University of California, Jan. 2008).

103 **FFDCA and FIFRA:** For a thoughtful discussion of the loopholes and shortcomings of both of these laws, see J. Wargo, *Our Children's Toxic Legacy: How Science and Law Fail to Protect Us from Pesticides* (New Haven, CT: Yale University Press, 1996); and GAO, *Food Safety: Changes Needed to Minimize Unsafe Chemicals in Food*, Report to the Chairman, Human Resources and Intergovernmental Relations Subcommittee, Committee on Government Operations, House of Representatives, GAP/RCED-94-192, Sept. 1994.

103 **issuing everyone a driver's license:** D. Ozonoff, "Taking the Handle off the Chlorine Pump" (presentation at the public health forum "Environmental and Occupational Health Problems Posed by Chlorinated Organic Chemicals," Boston University School of Public Health, 5 Oct. 1993).

103 **history of right-to-know laws:** The EPCRA legislation was a response to citizen activism at the state and local levels, as well as a direct reaction to the 1984 chemical disaster in Bhopal, India, which occurred when a feedstock for pesticide manufacture escaped from a Union Carbide plant and killed many thousands of sleeping residents in their homes. Emergency medical efforts were frustrated by the fact that no one knew what the chemical was. A similar chemical release occurred at a sister plant in West Virginia. Shortly thereafter, Congress voted EPCRA into law. Key parts of this legislation passed by a one-vote margin. B. A. Goldman, "Is TRI Useful in the Environmental Justice Movement?" (presentation to the Toxics Release Inventory Data Use Conference, Boston, 6 Dec. 1994), reprinted in *EPA Proceedings: Toxics Release Inventory (TRI) Data Use Conference, Building TRI and Pollution Prevention Partnerships*, EPA 749-R-95–001 (Washington, DC: EPA, 1995), 133–37; and Paul Orum, Working Group on Community-Right-to-Know, personal communication.

104 **description of TRI:** Commission for Environmental Cooperation, *Taking Stock: 2005 North American Pollutant Releases and Transfers* (Montreal: CEC, 2009).

104 **disappearance of right-to-know data under homeland security:** [illegible] Allen, "Environment, Health, and Missing Information," *Environmental History* 13 (2008): 659–66.

104 **falling number of facilities reporting:** Commission for Environmental Cooperation, *Taking Stock.*

104 **reversal of changes:** OMB Watch, "The Toxics Release Inventory Is Back," press release, 24 Mar. 2009.

104 **phantom reductions:** Working Group on Community Right-to-Know, "New Toxics Data Show Little Progress in Source Reduction," press release, Washington, DC, 27 Mar. 1995.

105 **declines in releases not tethered to decline in production:** Inform, Inc., *Toxics Watch* 1995 (New York: Inform, Inc., 1995).

105 **impact of the TRI report:** J. H. Cushman, "Efficient Pollution Rule under Attack," *New York Times*, 28 June 1995, A-16; K. Schneider, "For Communities, Knowledge of Polluters Is Power," *New York Times*, 24 Mar. 1991, A-5.

105 **quotes from chemical industry representatives:** Reprinted in *Working Notes on Community-Right-to-Know* (Washington, DC: Working Group on Community Right-to-Know, May–June 1995), 3.

105 **most recent TRI figures:** Carcinogens are as defined by OSHA. TRI Explorer, "Releases—Chemical Report for Release Year 2007" (EPA, 2009). Available at www.epa.gov/triexplorer/chemical.htm.

106 **data for the whole continent:** Commission for Environmental Cooperation, *Taking Stock: 2005 North American Pollutant Releases and Transfers* (Montreal: CEC, 2009).

106 **hazardous waste landfill in Peoria:** T. Bibo, "Peoria County Climbs Toxic Rankings," *PJS*, 26 April 2009; Commission for Environmental Cooperation, *Taking Stock.*

108 **hydrologist's description:** L. Hoburg et al., *Groundwater in the Peoria Region*, Cooperative Research Bulletin 39 (Urbana, IL: ISGWS, 1950), 53.

108 **history of Pekin:** *Pekin, Illinois, Sesquicentennial (1824–1974): A History* (Pekin, IL: Pekin Chamber of Commerce, 1974).

108 **one of Pekin's distilleries:** Midwest Grain Products, *1994 Annual Report.*

100 **"Cats" in the war:** P. A. Letourneau (ed.), *Caterpillar Military Tractors,* vol.1, (Minneapolis: Iconografix, 1994).

108 **sugar beet fields and starch explosion:** *Pekin, Illinois, Sesquicentennial,* 68.

108 **TRI releases from Aventine:** Carcinogens are as defined by OSHA. TRI Explorer, "Releases: Chemical Report for Release Year 2007" (EPA, 2009). Available at www.epa.gov/triexplorer/chemical.htm.

109 **pollution from Powerton:** "Pekin Edison Plant Named Worst Pollutor," *Bloomington Daily Pantagraph*, 10 Aug. 1974; J. Simpson, "Conservationist Blasts Pekin Energy Plant," *Bloomington Daily Pantagraph*, 30 July 1971.

110 **Keystone:** E. Hopkins, "Keystone Plans Costly Cleanup," *PJS*, 3 July 1993, A-1; E. Hopkins, "Region Awash in Toxic Chemicals: Study," *PJS*, 25 July 1993, A-2.

111 **quote from toxicologist and newspaper's conclusion:** E. Hopkins, "Emissions List Ranks Region 13th," *PJS*, 19 Mar. 1995, A-1, A-22.

111 **statistics on toxic emissions in the Pekin-Peoria:** From TRI. See also Hopkins, "Region Awash."

111 **Captan:** EPA, *Suspended, Cancelled, and Restricted Pesticides*, 20T-1002 (Washington, DC: EPA, 1990).

111 **documents:** From the Right-to-Know Network's copies of EPA's TRI, PCS, and FINDS databases. These searches were conducted by Kathy Grandfield on 1 Jan. 1995. Additional data for Tazewell County were provided by Joe Goodner, TRI coordinator at the IEPA in Springfield.

112 **Tazewell doubled the amount of hazardous waste:** IEPA, *Summary of Annual Reports on Hazardous Waste in Illinois for 1991 and 1992: Generation, Treatment, Storage, Disposal, and Recovery*, IEPA/BOL/94-155 (Springfield, IL: IEPA, 1994), 61.

112 **received four times more waste than it produced:** IEPA, *Illinois Nonhazardous Special Waste Annual Report for 1991* (Springfield, IL: IEPA, 1993), table K.

112 **spill report:** The report is part of the Tazewell County, Illinois, Area Report taken from the Right-to-Know Network's copy of EPA's ERNS database.

112 **methyl chloride:** ATSDR, *Toxicological Profile for Chloromethane* (Atlanta: USDHHS, ATSDR, 1998).

113 **estrogenicity of postwar chemicals:** D. M. Klotz et al., "Identification of Environmental Chemicals with Estrogenic Activity Using a Combi-

nation of *In Vitro* Assays," *EHP* 104 (1996): 1084–89; "Masculinity at Risk" (editorial), *Nature* 375 (1995): 522; R. M. Sharpe, "Another DDT Connection," *Nature* 375 (1995): 538–39; "Male Reproductive Health and Environmental Oestrogens" (editorial), *Lancet* 345 (1995): 933–35; Institute for Environment and Health, *Environmental Oestrogens: Consequences to Human Health and Wildlife* (Leicester, England: University of Leicester, 1995); J. Raloff, "Beyond Estrogens: Why Unmasking Hormone-Mimicking Pollutants Proves So Challenging," *Science News* 148 (1995): 44–46.

113 **DDE:** W. R. Kelce et al., "Persistent DDT Metabolite *p,p'*-DDE is a Potent Androgen Receptor Antagonist," *Nature* 375 (1995): 581–85.

113 **testicular dysgenesis syndrome:** N. E. Skakkebaek et al., "Testicular Dysgenesis Syndrome: An Increasingly Common Developmental Disorder with Environmental Aspects," *Human Reproduction* 16 (2001): 972–78; S. E. Talsness et al., "Components of Plastic: Experimental Studies in Animals and Relevance for Human Health," *Philosophical Transactions of the Royal Society* 364 (2009): 2079–96.

114 **dangers of phthalates to male development:** National Toxicology Program Center for the Evaluation of Risks to Human Reproduction, "Expert Panel Review of Phthalates," final report, National Toxicology Center, 2000.

114 **baby study:** S. H. Swan et al., "Decrease in Anogenital Distance Among Male Infants with Prenatal Phthalate Exposure," *EHP* 113 (2005): 1056–61.

114 **adult study:** S. M. Duty, "The Relationship Between Environmental Exposures to Phthalates and DNA Damage in Human Sperm," *EHP* 111 (2003): 1164–69.

114 **phthalates in people:** CDC, *Second Annual Report on Human Exposure to Environmental Chemicals* (NCEH Pub. No. 02-0716, 2003); B. C. Blount et al., "Levels of Seven Urinary Phthalate Metabolites in a Human Reference Population," *EHP* 108 (2000): 979–82.

114 **description of organochlorines:** J. Thornton, *Pandora's Poison: Chlorine, Health, and a New Environmental Strategy* (Cambridge, MA: MIT, 2000).

115 **PCBs reducing testosterone:** A. Goncharov et al., "Lower Serum Testosterone Associated with Elevated Polychlorinated Biphenyl Concentrations in Native American Men," *EHP* 117 (2009): 1454–60.

117 **Canadian bans on cosmetic use of pesticides:** Hudson, Quebec, became the first Canadian town to pass a bylaw disallowing nonagricultural

pesticide use on public and private lands. In 2001, the Supreme Court of Canada upheld Hudson's law and, in 2005, dismissed a chemical industry challenge to Toronto's bylaw. L. Armstrong et al., *Cancer: 101 Solutions to a Preventable Epidemic* (Gabriola Island, British Columbia: New Society Press, 2007); CBC News, "Cancer Society Pushes for B.C. Pesticide Ban," 5 April 2005; "Canadian Activists Win Pesticide Bylaws," *Global Pesticide Campaigner* 14 (2004); N. Arya, "Pesticides and Human Health: Why Public Health Officials Should Support a Ban on Non-Essential Residential Use," *Canadian Journal of Public Health* 96 (2005): 89–92.

118 **Ontario College of Physicians:** K. L. Bassil et al., "Cancer Health Effects of Pesticides, Systematic Review," *Canadian Family Physician* 53 (2007): 1704–11.

118 **tourism magazine quote:** G. Deacon, "Green Dream," *Toronto*, 2009.

119 **REACH:** For example, H. Foth and A. Hayes, "Concept of REACH and Impact on Evaluation of Chemicals," *Human and Experimental Toxicology* 27 (2008): 5–21. For a highly readable lay summary comparing and contrasting U.S. and E.U. toxics policies, see M. Schapiro, *Exposed: The Toxic Chemistry of Everyday Products and What's at Stake for American Power* (White River Junction, VT: Chelsea Green Publishing, 2007).

119 **Stockholm Convention on Persistent Organic Pollutants:** United Nations Environment Programme, "Ridding the World of POPs: A Guide to the Stockholm Convention on Persistent Organic Pollutants," April 2005. Available at http://chm.pops.int/. At this writing, the United States is not a party to this treaty. The citizen watchdog group for REACH is the International Chemical Secretariat in Sweden: www.chemsec.org. See also C. Hogue, "Persistent Organic Pollutants—Treaty Now Includes PFOS and Brominated Flame Retardants," *Chemical and Engineering News* 87 (2009): 9.

120 **number of cancer survivors:** ACS, *Cancer Facts and Figures—2008* (Atlanta: ACS, 2008).

120 **green chemistry:** P. Anastas and J. Warner, *Green Chemistry: Theory and Practice* (New York: Oxford University Press, 1998); Grossman, *Chasing Molecules*; W. McDonough and M. Braungart, *Cradle to Cradle: Remaking the Way We Make Things* (New York: North Point Press, 2002).

120 **twelve basic principles of green chemistry:** Elizabeth Grossman's book *Chasing Molecules* describes each in clear detail and provides wonderful

examples of green chemistry's principles in action. Here they are as an unannotated list: prevent waste; design safer chemicals and products; design less hazardous chemical syntheses; use renewable feedstocks; use catalysts, not stoichiometric reagents; avoid chemical derivatives; maximize atom efficiency; use safer solvents and reaction conditions; increase energy efficiency; design chemicals and products to degrade after use; analyze in real time to prevent pollution; minimize the potential for accidents (from P. Anastas and J. Warner, *Green Chemistry: Theory and Practice* [New York: Oxford University Press, 1998]).

121 **soy-based adhesive wins award:** Grossman, *Chasing Molecules.*

121 **obstacles to the mainstreaming of green chemistry:** Ibid.; Joseph Guth, personal communication; M. P. Wilson et al., *Green Chemistry in California: A Framework for Leadership in Chemicals Policy and Innovation* (University of California, California Policy Research Center, 2006); M. P. Wilson and M. R. Schwartzman, *Green Chemistry: Cornerstone to a Sustainable California* (Centers for Occupational and Environmental Health, University of California, Jan. 2008).

six: animals

124 **additive effect of estrogen mimics:** A. M. Soto et al., "The E-SCREEN Assay as a Tool to Identify Estrogens: An Update on Estrogenic Environmental Pollutants," *EHP* 103 (1995, S-7): 113–22. See also A. M. Soto et al., "The Pesticides Endosulfan, Toxaphene, and Dieldrin Have Estrogenic Effects on Human Estrogen-Sensitive Cells," *EHP* 102 (1994): 380–83.

124 **endosulfan:** ATSDR, *Toxicological Profile for Endosulfan* (USDHHS, 2000); J. Sass et al., "We Call on the U.S. Environmental Protection Agency to Ban Endosulfan" (open letter to Stephen Johnson, Administrator, U.S. Environmental Protection Agency, 19 May 2008); US EPA Endosulfan Updated Risk Assessment, *Federal Register* 72 (16 Nov. 2007), docket ID HQ-OPP-2002-0262-0067; V. Wilson et al., "Endosulfan Elevates Testosterone Biotransformation and Clearance in CD-1 Mice," *Toxicology and Applied Pharmacology* 148 (1998): 158–68.

125 **endosulfan in California:** T. M. Ole et al., *Water Woes: An Analysis of Pesticide Concentrations in California Surface Water* (San Francisco, CA: California Public Interest Research Group and the Pesticide Action Network Regional Center, 2000).

communication.

126 **origins of MCF-7:** J. Ricci, "One Nun's Living Legacy," *Detroit Free Press*, 30 Sept. 1984, F-1, F-4; H. D. Soule, "A Human Cell Line from a Pleural Effusion Derived from a Breast Carcinoma," *JNCI* 51 (1973): 1409–16.

127 **description of rodent bioassays and work of IARC and NTP:** S. M. Snedeker, "Perspectives on Approaches to Identify Cancer Hazards," *The Ribbon* [newsletter of the Cornell University Program on Breast Cancer and Environmental Risk Factors] 13 (2008): 1–2.

128 **estimates of carcinogens in commerce:** V. A. Fung et al., "The Carcinogenesis Bioassay in Perspective: Application in Identifying Human Cancer Hazards," *EHP* 103 (1995): 680–83.

128 **scientists asking to move on:** J. Huff et al., "The Limits of the Two-Year Bioassay Exposure Regimens for Identifying Chemical Carcinogens," *EHP* 116 (2008): 1439–42.

129 **calls for high-throughput assays:** B. E. Erickson, "Next-Generation Risk Assessment—EPA's Plan to Adopt In Vitro Methods for Toxicity Testing Gets Mixed Reviews from Stakeholders," *Chemical and Engineering News* 87 (2009): 30–33; National Research Council, *Toxicity Testing in the 21st Century: A Vision and a Strategy* (Washington, DC: National Academies Press, 2007).

130 **1938 dog studies:** W. C. Hueper et al., "Experimental Production of Bladder Tumors in Dogs by Administration of beta-Naphthylamine," *Journal of Industrial Hygiene and Toxicology* 20 (1938): 46–84. As a result of this and other studies, Wilhelm Hueper endured industry harassment, firings, and attempts to defund his research. The story of his dog experiments and their political aftermath is brilliantly narrated by Devra Davis in *The Secret History of the War on Cancer* (New York: Basic Books, 2007). Science historian Robert Proctor also provides an excellent overview of Wilhelm Hueper's struggles in *Cancer Wars: How Politics Shapes What We Know and Don't Know about Cancer* (New York: Basic Books, 1995), 36–48.

130 **coincident rise of synthetic dyes and bladder cancer among textile workers:** E. K. Weisburger, "General Principles of Chemical Carcinogenesis," in M. P. Waalkes and J. M. Ward (eds.), *Carcinogenesis* (New York: Raven Press, 1994); NIOSH, *Special Occupational Hazard Review for Benzidine-Based Dyes*, DHEW (NIOSH) Pub. 80-109 (Cincinnati: NIOSH, 1980).

131 **bladder cancer among workers in rubber and metal industries:** P. Vineis and S. Di Prima, "Cutting Oils and Bladder Cancer," *Scandinavian Journal of Work Environment and Health* 9 (1983): 449–50; R. R. Monson and K. Nakano, "Mortality among Rubber Workers: I. White Male Union Employees in Akron, Ohio," *AJE* 103 (1976): 284–96; P. Cole et al., "Occupation and Cancer of the Lower Urinary Tract," *Cancer* 29 (1972): 1250–60.

131 **cancer in dogs:** L. Marconato et al., "Association Between Waste Management and Cancer in Companion Animals," *Journal of Veterinary Internal Medicine* 23 (2009): 564–69; L. T. Glickman et al., "Herbicide Exposure and the Risk of Transitional Cell Carcinoma of the Urinary Bladder in Scottish Terriers," *Journal of the American Veterinary Medical Association* 224 (2004): 1290–97; L. T. Glickman et al., "Epidemiologic Study of Insecticide Exposures, Obesity, and Risk of Bladder Cancer in Household Dogs," *JTEH* 28 (1989): 407–14; H. M. Hayes, "Bladder Cancer in Pet Dogs: A Sentinel for Environmental Cancer?" *AJE* 114 (1981): 229–33.

132 **description of breast development:** C. W. Daniel and G. B. Silverstein, "Postnatal Development of the Rodent Mammary Gland," in M. C. Neville and C. W. Daniel (eds.), *The Mammary Gland: Development, Regulation, and Function* (New York: Plenum, 1987); J. Russo and I. H. Russo, "Development of the Human Mammary Gland," in ibid.; S. Z. Haslam, "Role of Sex Steroid Hormones in Normal Mammary Gland Function," in ibid.

132 **terminal end buds:** L. S. Birnbaum and S. E. Fenton, "Cancer and Developmental Exposure to Endocrine Disrupters," *EHP* 111 (2003): 389–94; S. E. Fenton, "The Mammary Gland: A Tissue Sensitive to Environmental Exposures," presentation before the President's Cancer Panel, Indianapolis, IN, 21 Oct. 2008; S. E. Fenton, "Endocrine Disrupting Compounds and Mammary Gland Development: Early Exposure and Later Life Consequences," *Endocrinology* 147 (supplement): S18–S24; A. Kortenkamp, "Breast Cancer, Oestrogens, and Environmental

[illegible] Point After Acute Prenatal Exposure to an Atrazine Metabolite Mixture in Female Long-Evans Rats," *EHP* 115 (2007): 541–47; J. L. Rayner et al., "Adverse Effects of Prenatal Exposure to Atrazine During a Critical Period of Mammary Gland Growth," *Toxicological Science* 87 (2005): 255–66.

133 **relevance to humans of atrazine and breast cancer studies in rats:** Reviewed in R. A. Rudel et al., "Chemicals Causing Mammary Gland Tumors in Animals Signal New Directions for Epidemiology, Chemicals Testing, and Risk Assessment for Breast Cancer Prevention," *Cancer* 109 (2007): 2635–66. See also California Breast Cancer Research Program, *Identifying Gaps in Breast Cancer Research: Addressing Disparities and the Roles of the Physical and Social Environment* (Oakland, CA: University of California Office of the President, California Breast Cancer Research Program, 2007), draft report.

133 **decision of the EPA to allow atrazine:** The rationale was, "It is unlikely that the mechanism by which atrazine induces mammary gland tumors in female SD rats could be operational in humans." EPA, *Decision Documents for Atrazine: Atrazine IRED* (January 2003).

133 **bladder cancer in whales and smelter workers:** P. Béland, "About Carcinogens and Tumors," *Canadian Journal of Fisheries and Aquatic Sciences* 45 (1988): 1855–56; D. Martineau et al., "Transitional Cell Carcinoma of the Urinary Bladder in a Beluga Whale (*Delphinapterus leucas*)," *Journal of Wildlife Diseases* 22 (1985): 289–94.

134 **1988 study:** D. Martineau et al., "Pathology of Stranded Beluga Whales (*Delphinapterus leucas*) from the St. Lawrence Estuary, Québec, Canada," *Journal of Comparative Pathology* 98 (1988): 287–311.

134 **1994 autopsy reports:** S. de Guise et al., "Tumors in St. Lawrence Beluga Whales," *Veterinary Pathology* 31 (1994): 444–49.

135 **cancers in the beluga to date:** C. Cirard et al., "Adenocarcinoma of the Salivary Gland in a Beluga Whale (*Delphinapterus leucas*)," *Journal of Veterinary Diagnostic Investigation* 3 (1991): 264–65; D. Martineau et al., "Cancer in Wildlife, a Case Study: Beluga from the St. Lawrence

Estuary, Québec, Canada," *EHP* 110 (2002): 285–92; D. Martineau, "Intestinal Adenocarcinomas in Two Beluga Whales (*Delphinapterus leucas*) from the Estuary of the St. Lawrence River," *Canadian Veterinary Journal* 36 (1995): 563–65; D. McAloose and A. L. Newton, "Wildlife Cancer: A Conservation Perspective," *Nature Reviews* 9 (2009): 517–26; D. E. Sargent and W. Hoek, "An Update of the Status of White Whales *Delphinapterus leucas* in the St. Lawrence Estuary, Canada," in J. Prescott and M. Gauquelin (eds.), *Proceedings of the International Forum for the Future of the Beluga* (Sillery, Québec: Presses de l'Université du Québec, 1990).

135 **St. Lawrence whales have shorter life spans:** McAloose and Newton, "Wildlife Cancer."

135 **quote about belugas:** Martineau et al., "Cancer in Wildlife."

135 **belugas not reproducing:** Martineau et al., "Cancer in Wildlife." See also S. de Guise et al., "Possible Mechanisms of Action of Environmental Contaminants on St. Lawrence Beluga Whales (*Delphinapterus leucas*)," *EHP* 103, S-4 (1995): 73–77; A. Motluk, "Deadlier Than the Harpoon?" *New Scientist*, 1 July 1995, 12–13; D. Martineau et al., "Levels of Organochlorine Chemicals in Tissues of Beluga Whales (*Delphinapterus leucas*) from the St. Lawrence Estuary, Québec, Canada," *Archives of Environmental Contamination and Toxicology* 16 (1987): 137–47; R. Masse et al., "Concentrations and Chromatographic Profile of DDT Metabolites and Polychlorobiphenyl (PCB) Residues in Stranded Beluga Whales (*Delphinapterus leucas*) from the St. Lawrence Estuary, Canada," *Archives of Environmental Contamination and Toxicology* 15 (1986): 567–79.

135 **free-swimming live whales:** K. E. Hobbs et al., "PCBs and Organochlorine Pesticides in Blubber Biopsies from Free-Ranging St. Lawrence River Estuary Beluga Whales (*Delphinapterus leucas*), 1994–1998," *Environmental Pollution* 122 (2003): 291–302.

135 **airborne deposition of chlordane and toxaphene:** D. Muir, "Levels and Possible Effects of PCBs and Other Organochlorine Contaminants and St. Lawrence Belugas," in Prescott and Gauquelin, *Future of the Beluga.*

135 **eels, whales, and mirex:** P. Béland et al., "Toxic Compounds and Health and Reproductive Effects in St. Lawrence Beluga Whales," *Journal of Great Lakes Research* 19 (1993): 766–75; T. Colborn, *Great Lakes,*

137 **benzo[a]pyrene and St. Lawrence belugas:** P. Béland, "The Beluga Whales of the St. Lawrence River," *Scientific American*, May 1996, 74–81; D. Martineau et al., "St Lawrence Beluga Whales, the River Sweepers?" *EHP* 110 (2002): A562–A64; McAloose and Newton, "Wildlife Cancer."

137 **chemistry and carcinogenicity of benzo[a]pyrene:** NTP, *Report on Carcinogens*, 11th ed. (USDHHS, Public Health Service, 2005)—Polycyclic Aromatic Hydrocarbons.

138 **mechanism of action:** M. E. Hahn and J. J. Stegeman, "The Role of Biotransformation in the Toxicity of Marine Pollutants," in Prescott and Gauquelin, *Future of the Beluga.*

138 **whale and DNA adducts:** D. Martineau et al., "Pathology and Toxicology of Beluga Whales from the St. Lawrence Estuary, Québec, Canada: Past, Present and Future," *Science of the Total Environment* 154 (1994): 201–15; L. R. Shugart and C. Theodorakis, "Environmental Toxicology: Probing the Underlying Mechanisms," *EHP* 102, S-12 (1994): 13–17; L. R. Shugart et al., "Detection and Quantitation of Benzo[a]pyrene-DNA Adducts in Brain and Liver Tissues of Beluga Whales (*Delphinapterus leucas*) from the St. Lawrence and Mackenzie Estuaries," in Prescott and Gauquelin, *Future of the Beluga.*

139 **quote by Leone Pippard:** L. Pippard, "Ailing Whales, Water and Marine Management Systems: An Urgency for Fresh, New Approaches," in Prescott and Gauquelin, *Future of the Beluga.*

139 **Clyde Dawe's discovery:** J. C. Harshbarger, "Introduction to Session on Pathology and Epizootiology," *EHP* 90 (1991): 5.

139 **Registry of Tumors in Lower Animals:** J. C. Harshbarger, "Role of the Registry of Tumors in Lower Animals in the Study of Environmental Carcinogenesis in Aquatic Animals," *Annals of the New York Academy of Sciences* 298 (1977): 280–89; J. C. Harshbarger, "The Registry of Tumors in Lower Animals," in *Neoplasia and Related Disorders in Invertebrates and Lower Vertebrate Animals*," *NCI Monograph* 31 (1969); J. C.

Wolf et al., "Updating the Registry of Tumors in Lower Animals," 28th Annual Eastern Fish Health Workshop, April 2003.

140 **tumors associated with contaminated sediments:** McAloose and Newton, "Wildlife Cancer"; M. J. Moore and M. S. Myers, "Pathobiology of Chemical-Associated Neoplasia in Fish," in D. C. Malins and G. K. Ostrander (eds.), *Aquatic Toxicology: Molecular, Biochemical and Cellular Perspectives* (Boca Raton, FL: Lewis, 1994).

140 **laboratory experiments with contaminated sediments:** J. C. Harshbarger and J. B. Clark, "Epizootiology of Neoplasms in Bony Fish of North America," *Science of the Total Environment* 94 (1990): 1–32; Dr. William Hawkins, Gulf Coast Research Laboratory, personal communication.

141 **Fox River in Illinois:** J. A. Couch and J. C. Harshbarger, "Effects of Carcinogenic Agents on Aquatic Animals: An Environmental and Experimental Overview," *Environmental Carcinogenesis Reviews* 3 (1985): 63–105; E. R. Brown et al., "Frequency of Fish Tumors Found in a Polluted Watershed as Compared to Nonpolluted Canadian Waters," *Canada Research* 33 (1973): 189–98.

141 **earth mounds on Buffalo Rock:** D. C. McGill, *Michael Heizer: Effigy Tumuli: The Reemergence of Ancient Mound Building* (New York: Abrams, 1990). Susan Post of the INHS informs me that hikers are no longer allowed to walk on the mounds, which are eroding.

seven: earth

144 **number of farms in Illinois:** 1960: 159,000; 2008: 75,900. IFB, *Farm and Food Facts, 2008*. Available at www.ilfb.org.

144 **number of cows and chickens:** Ibid.

145 **changes in farming:** J. Bender, *Future Harvest: Pesticide-Free Farming* (Lincoln: University of Nebraska Press, 1994), 2; IDENR, *The Changing Illinois Environment: Critical Trends, Summary Report*, IDENR/RE-EA-94/05 (Springfield, IL: IDENR, 1994), 54–55.

145 **corn and beans are 90 percent of cash receipts:** IFB, *Farm and Food Facts, 2008*.

146 **postwar changes in agricultural economy:** F. Kirschenmann, "Scale—Does It Matter?" in A. Kimbrell (ed.), *Fatal Harvest: The Tragedy of Industrial Agriculture* (Washington, DC: Foundation for

146 **alfalfa:** Ibid.

146 **natural history of soybeans:** American Soybean Association, *Soy Stats: A Reference Guide to Important Soybean Facts and Figures* (St. Louis: American Soybean Association, 1994); S. L. Post, "Miracle Bean," *The Nature of Illinois* (Fall 1993), 1, 3; Illinois Soybean Association, *Soybeans: The Gold That Grows* (pamphlet) (Bloomington, IL: Illinois Soybean Association, n.d.).

147 **natural history of corn:** *The Nature of Corn* (pamphlet) (Springfield: Illinois State Board of Education, 1996).

149 **Illinois is a protein production machine:** L. L. Jackson, "Who 'Designs' the Agricultural Landscape?" *Landscape Journal* 27 (2008): 23–40.

150 **years of peak livestock production:** IDA, Facts About Illinois Agriculture—Economic History. Available at www.agr.state.il.us/about.

150 **livestock raised time zones away:** 99 percent of all U.S. hogs, turkeys, and cattle are produced in confinement facilities. Most beef cattle spend some time on pasture. Jackson, "Who 'Designs' the Agricultural Landscape?" 23–40.

150 **industrial food system:** See ibid.

150 **loud critics:** My personal favorites are Wendell Berry, Terra Brockman, Kamyar Enshayan, Wes Jackson, Laura Jackson, Barbara Kingsolver, Fred Kirschenmann, and Michael Pollan.

151 **2006 spinach scare:** D. G. Maki, "Don't Eat the Spinach—Controlling Foodborne Infectious Disease," *NEJM* 355 (2006): 1952–55.

151 **mustard greens:** IFB, *Farm and Food Facts, 2008*.

151 **Terra Brockman in the *Chicago Tribune*:** B. Mahany, "Dirty Stories," *Chicago Tribune*, 24 Sept. 2006. Reiterated in her book, *The Seasons on Henry's Farm: A Year of Food and Life on a Sustainable Farm* (Evanston, IL: Agate Surrey, 2009).

151 **the public health argument:** D. S. Ludwig and H. A. Pollack. "Obesity and the Economy: From Crisis to Opportunity," *JAMA* 301 (2009): 533–35.

151 **statistics on obesity trends and 2009 report on obesity:** Robert Wood Johnson Foundation and the Trust for America's Health, *F Is for Fat: How Obesity Policies Are Failing in America*, July 2009.

153 **how obesity contributes to cancer:** [illegible] et al., "Epidemiological and Molecular Mechanisms Aspects Linking Obesity and Cancer," *Arquivos Brasileiros de Endocrinologia e Metabologia* 53 (2009): 219–26.

153 **ethanol:** Illinois leads the nation in ethanol production: IFB, *Farm and Food Facts, 2008*. My conclusions on the net energy balance of ethanol follow those of environmental engineer Kamyar Enshayan at Northern Iowa University. K. Enshayan, *Living Within Our Means: Beyond the Fossil Fuel Credit Card* (Cedar Falls, IA: UNI Local Food Project, 2005).

154 **pesticides used on corn and soy:** R. Gilliom et al., *Pesticides in the Nation's Streams and Ground Water, 1992–2001* (USGS, Circular 1291, 2006).

154 **Illinois weeds:** IDENR, *The Changing Illinois Environment: Critical Trends*, vol. 3, IDENR/RE-EA-94/05 (Springfield, IL: IDENR, 1994), 84; R. L. Zimdahl, *Fundamentals of Weed Science* (San Diego: Academic Press, 1993); M. J. Chrispeels and D. Sadava, *Plants, Food, and People* (San Francisco: Freeman, 1977), 163–64.

155 **density of seedbank:** F. Forcella et al., "Weed Seedbanks of the U.S. Corn Belt: Magnitude, Variation, Emergence, and Application," *Weed Science* 40 (1992): 636–44.

155 **direction of weed control research:** D. D. Buhler et al., "Integrated Weed Management Techniques to Reduce Herbicide Inputs in Soybeans," *Agronomy Journal* 84 (1992): 973–78.

155 **GMO corn:** D. Coursey, *Illinois Without Atrazine: Who Pays? Economic Implications of an Atrazine Ban in the State of Illinois* (University of Chicago, Harris School of Public Policy Working Paper, 2007).

155 **poisoning mechanisms of herbicides:** A. Cobb, *Herbicides and Plant Physiology* (New York: Chapman & Hall, 1992).

155 **2,4-D in Illinois:** IASS, *Agricultural Fertilizer and Chemical Usage: Corn—1993* and *Agricultural Fertilizer and Chemical Usage: Soybeans—1993*.

156 **Weed Science Society of America:** www.weedscience.org/In.asp

156 **herbicide-resistant weeds:** C. A. Edwards, "Impact of Pesticides on the Environment," in D. Pimentel and H. Lehman (eds.), *The Pesticide*

Physiology of Herbicide Action (Englewood Cliffs, NJ: Prentice Hall, 1993), 113–40.

157 **maps of atrazine:** For example, R. Gilliom et al., *Pesticides in the Nation's Streams and Ground Water, 1992–2001* (USGS, Circular 1291, 2006).

157 **atrazine in water:** Gilliom et al., *The Quality of Our Nation's Water*; Wu et al., *Poisoning the Well: How the EPA Is Ignoring Atrazine Contamination in Surface and Drinking Water in the Central United States* (New York: NRDC, 2009); EPA, *The Triazine Herbicides, Atrazine, Simazine, and Cyanazine: Position Document 1, Initiation of Special Review*, OPP-30000-60, FRL-4919-5 (Washington, DC: EPA, 1994).

157 **atrazine a proven endocrine disruptor:** R. L. Cooper et al., "Atrazine and Reproductive Function: Mode and Mechanism of Action Studies," *Birth Defects Research, Part B* 80 (2007): 98–112.

157 **atrazine and ovulation:** R. L. Cooper et al., "Atrazine Disrupts the Hypothalamic Control of Pituitary-Ovarian Function," *Toxicological Sciences* 53 (2000): 297–307.

157 **atrazine and frogs:** T. B. Hayes, "There Is No Denying This: Defusing the Confusion About Atrazine," *Bioscience* 54 (2004): 1138–49.

157 **atrazine and human cancers:** D. A. Crain, et al., "Female Reproductive Disorders: The Roles of Endocrine-Disrupting Compounds and Developmental Timing," *Fertility and Sterility* 90 (2008): 911–40; A. Donna et al., "Triazine Herbicides and Ovarian Cancer Neoplasms," *Scandinavian Journal of Work and Environmental Health* 15 (1989): 47–53; J. A. Rusiecki et al., "Cancer Incidence Among Pesticide Applicators Exposed to Atrazine in the Agricultural Health Study," *JNCI* 96 (2004): 1375–82; H. A. Young et al., "Triazine Herbicides and Epithelial Ovarian Cancer Risk in Central California," *Journal of Occupational and Environmental Medicine* 47 (2005): 1148–56.

158 **atrazine in the urine of farmers:** B. Bakke et al., "Exposure to Atrazine and Selected Non-Persistent Pesticides Among Corn Farmers During a Growing Season," *Journal of Exposure Science and Environmental Epidemiology* 19 (2009): 544–54.

158 **atrazine's intracellular effects:** L. Albanito et al., "G-Protein-Coupled Receptor 30 and Estrogen Receptor α Are Involved in the Proliferative Effects Induced by Atrazine in Ovarian Cancer Cells," *EHP* 116 (2008): 1648–55; W. Fan et al., "Atrazine-Induced Aromatase Expression Is SF-1 Dependent: Implications for Endocrine Disruption in Wildlife and Reproductive Cancers in Humans," *EHP* (2007): 720–27; T. Hayes, "The One Stop Shop: Chemical Causes and Cures for Cancer," white paper submitted to the President's Cancer Panel, Indianapolis, IN, 21 Oct. 2008; M. Suzawa and H. A. Ingraham, "The Herbicide Atrazine Activates Endocrine Gene Networks via Non-Steroidal NR5A Nuclear Receptors in Fish and Mammalian Cells," *PLoS ONE* 3 (2008): e2117.

159 **atrazine and obesity:** S. Lim et al., "Chronic Exposure to the Herbicide, Atrazine, Causes Mitochondrial Dysfunction and Insulin Resistance," *PLoS ONE* 4 (2009): e5186.

159 **ecological effects of nitrogen:** S. Fields, "Global Nitrogen: Cycling out of Control," *EHP* 112 (2004): A556–63.

160 **anhydrous as fertilizer:** J. M. Shutske, "Using Anhydrous Ammonia Safely on the Farm," University of Minnesota Extension, FO-02326, 2005.

160 **anhydrous and methamphetamine:** S. Simstad and D. Jeppsen, "Preventing Theft of Anhydrous Ammonia," Ohio State University: OSU Extension Fact Sheet, AEX-594.1

160 **nitrogen fixation by lightning:** R. D. Hill et al., "Atmospheric Nitrogen Fixation by Lightning," *Journal of the Atmospheric Sciences* 37 (1980): 179–92.

161 **nitrosomo compounds in urine:** S. S. Mirvish et al., "N-nitrosoproline Excretion by Rural Nebraskans Drinking Water of Varied Nitrate Content," *Cancer Epidemiology, Biomarkers and Prevention* 1 (1992): 455–61.

161 **IARC on nitrates:** Y. Grosse et al., "Carcinogenicity of Nitrate, Nitrite, and Cyanobacterial Peptide Toxins," *Lancet Oncology* 7 (2008): 628–29.

161 **mixed results of human studies:** M. H. Ward, "Too Much of a Good Thing? Nitrate from Nitrogen Fertilizers and Cancer," presentation before the President's Cancer Panel, Indianapolis, IN, 21 Oct. 2008.

161 **nitrates and bladder cancer:** M. H. Ward et al., "Nitrate in Public Water Supplies and Risk of Bladder Cancer," *Epidemiology* 14 (2003): 183–90.

161 **yields on organic farms:** L. G. Malkina-Pykh and Y. A. Pykh, *Sustainable Food and Agriculture* (Southampton, UK: WIT Press, 2003), 205–07; P. Mader et al., "Soil Fertility and Biodiversity in Organic Farming," *Science* 296 (2002): 1694–97.

161 **Rodale field trials:** D. Pimentel et al., "Environmental, Energetic, and Economic Comparisons of Organic and Conventional Farming Systems," *Bioscience* 55 (2005): 573–82.

162 **other studies:** T. Gomiero et al., "Energy and Environmental Issues in Organic and Conventional Agriculture," *Critical Reviews in Plant Sciences* 27 (2008): 239–54; R. Welsh, *The Economics of Organic Grain and Soybean Production in the Midwestern United States* (Greenbelt, MD: Henry A. Wallace Institute, 1999).

163 **externalized costs of pesticides:** D. Pimentel, "Environmental and Economic Costs of the Application of Pesticides Primarily in the United States," *Environment, Development and Sustainability* 7 (2005): 229–52.

164 **carcinogenic pesticides used in California:** Peggy Reynolds, "Agricultural Exposures and Children's Cancer," presentation before the President's Cancer Panel, Indianapolis, IN, 21 Oct. 2008.

164 **pesticide drift:** L. Lu et al., "Pesticide Exposure of Children in an Agricultural Community: Evidence of Household Proximity to Farmland and Take Home Exposure Pathways," *Environmental Research* 84 (2000): 290–302.

164 **farmworkers exposed to pesticides:** A. Bradman et al., Pesticides and Their Metabolites in the Homes and Urine of Farmworker Children Living in the Salinas Valley, CA," *Journal of Exposure Science and Environmental Epidemiology* 2006. Three-quarters of U.S. farmworkers are born in Mexico. U.S. Department of Labor, *Findings from the National Agricultural Workers Survey (NAWS), 2001–2002: A Demographic and Employment Profile of United States Farm Workers* (USDL, May 2005).

164 **pesticide registry:** California Department of Pesticide Registration, Pesticide Use Reporting: An Overview of California's Unique Full Reporting System (Sacramento, CA: California Department of Pesticide Regulation, 2000). Available at www.cdpr.ca.gov/docs/pur/purmain.htm.

164 **cancers in high-pesticide areas in California:** Reynolds, "Agricultural Exposures and Children's Cancer."

164 **California breast cancers not related to pesticide use:** P. Reynolds et al., "Residential Proximity to Agricultural Pesticide Use and Incidence of Breast Cancer in California, 1988–1997," *EHP* 113 (2005): 993–1000.

165 **breast cancer among women farmworkers:** P. K. Mills, "Breast Cancer Risk in Hispanic Agricultural Workers in California," *International Journal of Occupational and Environmental Health* 11 (2005): 123–31.

165 **quote by Joan Flocks:** "Pesticide Policy and Farmworker Health," presentation before the President's Cancer Panel, Indianapolis, IN, 21 Oct. 2008.

166 **Carson's remark:** R. Carson, *Silent Spring* (Boston: Houghton Mifflin, 1962).

166 **The Land Connection:** Described in Brockman, *The Seasons on Henry's Farm*. See also www.thelandconnection.org.

166 **Black Hawk County:** K. Enshayan, Northern Iowa University, personal communication.

167 **Illinois acreage under organic production:** IFB, *Farm and Food Facts, 2008.*

167 **Laura Jackson:** Jackson, "Who 'Designs' the Agricultural Landscape?" 23–40.

167 **2007 study forecasts dire consequences:** Coursey, *Illinois Without Atrazine.*

eight: air

172 **Paracelsus:** M P. Hall, *The Secret Teaching of All Ages* (Los Angeles: Philosophical Research Society, 1988), 107–108.

173 **DDT and PCBs in Hubbard Brook:** W. H. Smith et al., "Trace Organochlorine Contamination of the Forest Floor of the White Mountain National Forest, New Hampshire," *Environmental Science and Technology* 27 (1993): 2244–46; "DDT and PCBs, Long Banned in the U.S., Found in Remote Forest, Suggesting Global Distribution via the Atmosphere" (Yale University press release, 14 Dec. 1993).

173 **rain-fed bogs:** R. A. Rapaport et al., "'New' DDT Inputs to North America: Atmospheric Deposition," *Chemosphere* 14 (1985): 1167–73.

173 **world's trees:** S. L. Simonich and R. A. Hites, "Global Distribution of Persistent Organochlorine Compounds," *Science* 269 (1995); 1851–54.

of Low Volatility Organochlorine Compounds in Polar Regions," *Ambio* 22 (1993): 10–18.

174 **Lake Laberge:** K. A. Kidd et al., "High Concentrations of Toxaphene in Fishes from a Subarctic Lake," *Science* 269 (1995): 240–42. See also J. Raloff, "Fishy Clues to a Toxaphene Puzzle," *Science News* 148 (1995): 38–39.

174 **carcinogens into air:** EPA, *2007 Toxics Release Inventory (TRI) Public Data Release Report* (EPA 260-R-09-001, Mar. 2009) and www.rtk.net.org/db/tri.

175 **airborne carcinogens:** A. Pintér et al., "Mutagenicity of Emission and Immission Samples around Industrial Areas," in H. Vainio et al. (eds.), *Complex Mixtures and Cancer Risk*, IARC Scientific Pub. 104 (Lyon, France: IARC, 1990).

175 **60 percent live with bad air:** 61.7 percent of Americans are exposed to either ozone or particulate matter above levels considered healthful. American Lung Association, *State of the Air*, 2009 (Washington, DC: American Lung Association, 2009).

175 **actual contribution elusive:** G. Pershagen, "Air Pollution and Cancer," in Vainio, *Complex Mixtures.*

175 **fluidity of air:** F. E. Speizer and J. M. Samet, "Air Pollution and Lung Cancer," in J. M. Samet (ed.), *Epidemiology of Lung Cancer* (New York: Marcel Dekker, 1994); K. Hemminki, "Measurement and Monitoring of Individual Exposures," in L. Tomatis (ed.), *Air Pollution and Human Cancer* (New York: Springer-Verlag, 1990).

176 **ultrafine particles:** American Lung Association, *State of the Air, 2009* (Washington, DC: ALA, 2009); J. Raloff, "Bad Breath: Studies are Homing In on Which Particles Polluting the Air Are Most Sickening—And Why," *Science News* 176 (2009): 26.

176 **transmutational quality of air:** L. Fishbein, "Sources, Nature, and Levels of Air Pollutants," in Tomatis, *Air Pollution*; L. Lewtas, "Experimental Evidence for Carcinogenicity of Air Pollutants," in ibid.

177 **air pollutants that may morph into carcinogens:** "OSU to Study Air Pollutant's Impact on Chinese, U.S. Health," press release, Oregon State University, Corvallis, OR, 28 Apr. 2009.

177 **ozone:** K. Breslin, "The Impact of Ozone," *EHP* 103 (1995): 660–64; G. J. Jakab et al., "The Effects of Ozone on Immune Function," *EHP* 103, S-2 (1995): 77–89.

178 **National Air Toxics Assessment:** In June 2009, the EPA released the results of its 2002 assessment. EPA, *2002 National-Scale Air Toxics Assessment* (Washington, DC: EPA, 2009). Available at www.epa.gov/nata2002/.

178 **five-year survival rate:** M. P. Spitz et al., "Cancer of the Lung," in D. Schottenfeld and J. F. Fraumeni (eds.), *Cancer Epidemiology and Prevention*, 3rd ed. (New York: Oxford University Press, 2006).

178 **guilt and blame:** "Lung Cancer: Dying in Disgrace?" *Harvard Health Letter* 20 (1995): 4–6.

179 **primacy of tobacco:** Spitz et al., "Cancer of the Lung."

179 **lung cancer among nonsmokers the sixth most common cancer death:** Between seventeen thousand and twenty-six thousand deaths each year are caused by lung cancer in people who are not active cigarette smokers. Of these, fifteen thousand deaths occur among lifelong nonsmokers. M. J. Thun et al., "Lung Cancer Deaths in Lifelong Nonsmokers," *JNCI* 98 (2006): 691–99.

179 **lung cancer among nonsmokers:** T. Reynolds, "EPA Finds Passive Smoking Causes Lung Cancer," *JNCI* 85 (1993): 179–80; Spitz et al., "Cancer of the Lung."

179 **unavoidable and interactive effects of air pollution:** K. Hemminki and G. Pershagen, "Cancer Risk of Air Pollution: Epidemiological Evidence," *EHP* 102 (1994): 187–92.

179 **adenocarcinoma:** A. Charioux et al., "The Increasing Incidence of Lung Adenocarcinoma: Reality or Artefact? A Review of the Epidemiology of Lung Adenocarcinoma," *International Journal of Epidemiology* 26 (1997): 14–23.

179 **urban factors in lung cancer:** A. J. Cohen, "Outdoor Air Pollution and Lung Cancer," *EHP* 108 (2000, S-4): 743–50; R. W. Clapp et al., "Environmental and Occupational Causes of Cancer: New Evidence 2005–2007," *Reviews on Environmental Health* 23 (2008): 1–36; P. Vineis and K. Husgafvel-Pursiainen, "Air Pollution and Cancer: Biomarker Studies in Human Populations," *Carcinogenesis* 26 (2005): 1846–55.

179 **other epidemiological studies:** American Lung Association, *State of the Air, 2009*; W. J. Blot and J. F. Fraumeni Jr., "Geographic Patterns of Lung Cancer: Industrial Correlations," *AJE* 103 (1976): 539–50; D. W. Dockery, "An Association Between Air Pollution and Mortality in Six U.S. Cities," *NEJM* 329 (1993): 1753–59; P. Gustavsson et al., "Excess Mortality Among Swedish Chimney Sweeps," *British Journal of Industrial Medicine* 44 (1987): 738–43; Hemminki and Pershagen, "Cancer Risk of Air Pollution," 187–92; D. Krewski et al., *Extended Follow-Up and Spatial Analysis of the American Cancer Society Study Linking Particulate Air Pollution and Mortality* (Boston: Health Effects Institute, Report 140, 2009); J. M. Samet and A. J. Cohen, "Air Pollution," in Schottenfeld and Fraumeni, *Cancer Epidemiology and Prevention*.

180 **breast cancer and air pollution:** J. G. Brody et al., "Environmental Pollutants and Breast Cancer: Epidemiologic Studies," *Cancer* 109 (2007, S12): 2667–2711; N. M. Perry et al., "Exposure to Traffic Emissions Throughout Life and Risk of Breast Cancer," *Cancer Causes and Control* 19 (2008): 435; J. M. Melius et al., *Residence near Industries and High Traffic Areas and the Risk of Breast Cancer on Long Island* (Albany: New York State Dept. of Health, 1994); J. E. Vena, "Lung, Breast, Bladder and Rectal Cancer: Indoor and Outdoor Air Pollution and Water Pollution," presentation before the President's Cancer Panel, Charleston, SC, 4 Dec. 2008. See also M. Spencer, "Overlooking Evidence: Media Ignore Environmental Connections to Breast Cancer," *Extra!* (Fairness and Accuracy in Reporting, Feb. 2009).

180 **benzo[a]pyrene and breast cancer:** J. J. Morris and E. Seifter, "The Role of Aromatic Hydrocarbons in the Genesis of Breast Cancer," *Medical Hypotheses* 38 (1992): 177–84.

180 **bladder cancer and air pollution:** C. C. Liu et al., "Ambient Exposure to Criteria Air Pollutants and Risk of Death from Bladder Cancer in Taiwan," *Inhalation Toxicology* 21 (2009): 48–54; B. J. Pan et al., "Excess Cancer Mortality Among Children and Adolescents in Residential Districts Polluted by Petrochemical Manufacturing Plants in Taiwan," *JTEH* 43 (1994): 117–29; D. Trichopoulos and F. Petridou, "Epidemiologic Studies and Cancer Etiology in Humans," *Medicine, Exercise, Nutrition, and Health* 3 (1994): 206–25; S. S. Tsai et al., "Association of Bladder Cancer with Residential Exposure to Petrochemical Air Pollutant Emissions in Taiwan," *Journal of Toxicology and Environmental Health. Part A* 72 (2009): 53–59.

181 **nitrogen dioxide and lung tumors:** K. A. Fackelmann, "Air Pollution Boosts Cancer Spread," *Science News* 137 (1990): 221; A. Richters, "Effects of Nitrogen Oxide and Ozone on Blood-Borne Cancer Cell Colonization of the Lungs," *JTEH* 25 (1988): 383–90.

181 **spread of cancer to lungs:** E. Ruoslahti, "How Cancer Spreads," *Scientific American*, Sept. 1996, 72–77.

181 **quote by Richters:** Fackelmann, "Air Pollution Boosts," 221.

182 Quotations are from, in the order of presentation, the following sources: A. Biggeri, "Air Pollution and Lung Cancer in Trieste, Italy: Spatial Analysis of Risk as a Function of Distance from Sources," *EHP* 104 (1996) 750–54; G. Pershagen and L. Simonato, "Epidemiological Evidence on Air Pollution and Cancer," in Tomatis, *Air Pollution*; ibid.; C. W. Sweet and S. J. Vermette, *Toxic Volatile Organic Chemicals in Urban Air in Illinois*, HWRIC RR-057 (Champaign, IL: Hazardous Waste Research and Information Center, 1991), 1; Hemminki and Pershagen, "Cancer Risk of Air Pollution"; and Lewtas, "Experimental Evidence."

183 **the miasma theory:** S. N. Tesh, *Hidden Arguments: Political Ideology and Disease Prevention Policy* (New Brunswick, NJ: Rutgers University Press, 1988), 825–32.

nine: water

187 **photograph of mussel gatherers:** L. M. Talkington, *The Illinois River: Working for Our State*, Misc. Pub. 128 (Champaign, IL: ISWS, 1991), 11.

187 **demise of button factories:** Ibid., 10–11.

188 **demise of diving ducks:** H. B. Mills, *Man's Effect on the Fish and Wildlife of the Illinois River*, Biological Notes 57 (Urbana, IL: INHS, 1966).

188 **demise of scaups and fingernail clams:** F. C. Bellrose et al., *Waterfowl Populations and the Changing Environment of the Illinois River Valley*, Bulletin 32 (Urbana, IL: INHS, 1979); Mills, *Man's Effect.*

188 **quotes on bird identification:** From D. L. Stokes, *Stokes Field Guide to Birds, Eastern Region* (Boston: Little, Brown, 1996); National Geographic Society, *Field Guide to the Birds of North America*, 2nd ed. (Washington, DC: National Geographic Society, 1987); C. S. Robbins et al., *Birds of North America* (New York: Golden Press, 1966).

188 **demise of dabbling ducks and aquatic plants:** E. Hopkins, "Pollution Keeps Preying on Plants in Illinois River," *PJS*, 25 July 1993, A-2; Mills, *Man's Effect.*

188 **fishing industry on the Illinois River:** P. Ross and R. Sparks, *Identification of Toxic Substances in the Upper Illinois River*, Report 193 (Urbana, IL: INHS, 1989).

189 **geology and ecology of the Illinois River:** M. Runkle, "Plight of the Illinois: A River in Transition," *Illinois Audubon* 236 (1991): 2–7; Talkington, *Illinois River*; Bellrose et al., *Waterfowl Populations*; Blodgett, INHS, personal communication.

189 **S&S Canal:** Talkington, *Illinois River*; Bellrose et al., *Waterfowl Populations.*

189 **photograph of Illinois fish:** Mills, *Man's Effect.*

189 **improvement after 1972:** IDENR, *The Changing Illinois Environment: Critical Trends*, summary report, ILENR/RE/-EA-94/05 (SR) 20M (Springfield, IL: IDENR, 1994), 16–17.

190 **advisories:** *Illinois 1994 Fishing Information* (Springfield, IL: IDC, 1994).

190 **impact of barges and tugs:** T. A. Butts and D. B. Shackleford, *Impacts of Commercial Navigation on Water Quality in the Illinois River Channel*, Research Report 122 (Champaign, IL: ISWS, 1992); R. M. Sparks, "River Watch: The Surveys Look After Illinois' Aquatic Resources," *The Nature of Illinois* (Winter 1992): 1–4; W. J. Tucker, *An Intensive Survey of the Illinois River and Its Tributaries: A Comparison Study of the 1967 and 1978 Stream Conditions* (Springfield, IL: IEPA, n.d.); Runkle, "Plight of the Illinois."

190 **toxic spills:** M. Demissie and L. Keefer, *Preliminary Evaluation of the Risk of Accidental Spills of Hazardous Materials in Illinois Waterways*, HWRIC RR-055 (Champaign, IL: Hazardous Waste Research and Information Center, 1991); Talkington, *Illinois River*, 14; "The Illinois River: Its History, Its Uses, Its Problems," *Currents* 5 (Champaign, IL: ISWS, Jan.–Feb. 1993), 1–12; Blodgett, personal communication.

190 **routine industrial discharges:** E. Hopkins, "New Rules, Industry Initiatives May Cut Toxic Dumping in River," *PJS*, 19 Mar. 1995, A-23.

191 **disappearance of fish, amphibians, crayfish, mussels:** IDENR, *The Changing Illinois Environment*, 19–22; J. H. Cushman, "Freshwater Mussels Facing Mass Extinction," *New York Times*, 3 Oct. 1995, C-1, C-7.

191 **poem by Robert Frost:** "The Oven Bird," in E. C. Lathem (ed.), *The Poetry of Robert Frost* (New York: Henry Holt, 1969).

191 **drinking water contaminants:** U.S. EPA, *Ground Water and Drinking Water: Frequently Asked Questions* [Web page]. Available at www.epa

gov/safewater/faq/faq.html. USEPA, *Drinking Water Contaminants* [Web page]. Available at www.epa.gov/safewater/contaminants/index.html. See also U.S. EPA, *Water on Tap: What You Need to Know* (Washington, DC: EPA, 2003, 816-K-03-2007).

193 **12 of 216 mammary carcinogens in water:** The 12 mammary gland carcinogens regulated under the Safe Drinking Water Act are acrylamide, the triazine herbicides atrazine and symazine, DBCP, 1,2-dibromoethane, 1,2-dichloropropane, 1,2-dichloroethane, benzene, carbon tetrachloride, 3,3-dimethyoxybenzidine, styrene, and vinyl chloride. R. A. Rudel et al., "Chemicals Causing Mammary Gland Tumors in Animals Signal New Directions for Epidemiology, Chemicals Testing, and Risk Assessment for Breast Cancer Prevention," *Cancer* 109 (2007, S-12): 2635–66.

193 **personal care products unregulated:** D. W. Kolpin et al., "Pharmaceuticals, Hormones, and Other Organic Wastewater Contaminants in U.S. Streams, 1999–2000: A National Reconnaissance," *Environmental Science and Technology* 36 (2002): 1202–11; H. M. Kuch and K. Ballschmiter, "Determination of Endocrine-disrupting Phenolic Compounds and Estrogens in Surface and Drinking Water by HRGC-(NCI)-MS in the Picogram Per Liter Range," *Environmental Science and Technology* 35 (2001): 3201–06; P. E. Stackelberg et al., "Persistence of Pharmaceutical Compounds and Other Organic Wastewater Contaminants in a Conventional Drinking-Water Treatment Plant," *Science of the Total Environment* 329 (2004): 99–113.

193 **regulation of nitrates:** M. H. Ward et al., Workgroup Report: Drinking-water Nitrate and Health—Recent Findings and Research Needs," *EHP* 113 (2005): 1607–14.

193 **nitrates and cancer:** evidence reviewed in K. P. Cantor et al., "Water Contaminants," in D. Schottenfeld and J. F. Fraumeni (eds.), *Cancer Prevention and Epidemiology*, 3rd ed. (New York: Oxford University Press, 2006).

193 **1995 study of herbicides in drinking water:** B. Cohen et al., *Weed Killers by the Glass: A Citizens' Tap Water Monitoring Project in 29 Cities* (Washington, DC: Environmental Working Group, 1995).

194 **2009 study of atrazine in drinking water:** M. Wu et al., *Poisoning the Well: How the EPA Is Ignoring Atrazine Contamination in Surface and Drinking Water in the Central United States* (New York: NRDC, 2009).

194 **dermal and inhalation routes of exposure:** S. M. Gordon et al., "Changes in Breath Trihalomethane Levels Resulting from Household Water-Use Activities," *EHP* 114 (2006): 514–21; C. Howard and R. L. Corsi, "Volatilization of Chemicals from Drinking Water to Indoor Air: The Role of Residential Washing Machines," *Journal of Air and Waste Management Association* (1998): 907–14; L. M. Kerger et al., "Influence of Tap Water Quality and Household Water Use Activities on Indoor Air and Internal Dose Level of Trihalomethanes," *EHP* 113 (2005): 863–70; S. D. Richardson, "Water Analysis: Emerging Contaminants and Current Issues," *Analytical Chemistry* 79 (2007): 4295–4323.

195 **danger to women and infants:** C. W. Forrest and R. Olshansky, *Groundwater Protection by Local Government* (Springfield, IL: IDENR and IEPA, 1993), 16.

195 **bathing and showering:** S. M. Gordon et al., "Changes in Breath Trihalomethane Levels Resulting from Household Water-Use Activities, *EHP* 114 (2006): 514–21; C. P. Weisel and W. K. Jo, "Ingestion, Inhalation, and Dermal Exposures to Chloroform and Trichloroethene from Tap Water," *EHP* 104 (1996): 48–51.

195 **southeastern Rockford Superfund site:** J. E. Keller and S. W. Metcalf, *Exposure Study of Volatile Organic Compounds in Southeast Rockford* (Springfield, IL: IDPH, Division of Epidemiologic Studies, 1991, Epidemiologic Report Series 91:3).

196 **bladder cancer cluster in Rockford:** K. Mallin, "Investigation of a Bladder Cancer Cluster in Northwestern Illinois," *AJE* 132 (1990, S-1): S96–106.

196 **the 1918 survey form:** IEPA, *Pilot Groundwater Protection Program Needs Assessment for Pekin Public Water Supply Facility Number 1795040* (Springfield, IL: IEPA, Division of Public Water Supplies, 1992), appendix C.

197 **lengthy report on Pekin's groundwater:** IEPA, *Pilot Groundwater Protection.*

197 **city's response:** T. L. Aldous, "Committee Examines Aquifer Protection," *PDT*, 11 Dec. 1993, A-1, A-12.

197 **quote by mayor:** Ibid.

198 **review by Kenneth Cantor:** Cantor et al., "Water Contaminants."

198 **Camp Lejeune:** "Key Events in Camp Lejeune's Water Contamination," Associated Press, 20 June 2009; "Hagan Wants a Conclusion to

the Ongoing Camp Lejeune Water Contamination Issue," press release, Kay R. Hagan, U.S. Senator, North Carolina, 16 June 2009; A. Aschengrau et al., "Statement in Response to the National Research Council Report on Camp Lejeune," letter by scientists to the U.S. Agency on Toxic Substances and Disease Registry, June 2009 [full disclosure: I am one of the signatories of this letter]; R. Beamish, "False Comfort: US Pulls Report That Minimized Cancer Risk from Toxic Water at Marine Base," Associated Press, 20 June 2009; W. R. Levesque, "More Vets Report Cancer," *St. Petersburg Times*, 3 July 2009; NRC, Committee on Contaminated Drinking Water at Camp Lejeune, *Contaminated Drinking Water at Camp Lejeune—Assessing Potential Health Effects* (Washington, DC: National Academies Press, 2009); M. Quillan, "Marine Battles over Contaminated LeJeune Water, *The News and Observer*, 31 May 2009. A Web site entitled *The Few, The Proud, The Forgotten* has been established by former Marines to serve as a clearinghouse of information about water contamination at the base and as a registry for former Marines and their family members who lived on the base during the three-decade period of water contamination: www.tftptf.com.

199 **ecologic studies of drinking water and cancer:** L. D. Budnick et al., "Cancer and Birth Defects near the Drake Superfund Site, Pennsylvania," *AEH* 39 (1984): 409–13; A. Aschengrau et al., "Cancer Risk and Tetrachloroethylene-Contaminated Drinking Water in Massachusetts," *AEH* 48 (1993): 284–92; Cantor et al., "Water Contaminants"; J. Fagliano et al., "Drinking Water Contamination and the Incidence of Leukemia: An Ecologic Study," *AJPH* 80 (1990): 1209–12; J. Griffith et al., "Cancer Mortality in U.S. Counties with Hazardous Waste Sites and Ground Water Pollution," *AEH* 44 (1989): 69–74; W. Hoffmann et al., "Radium-226-Contaminated Drinking Water: Hypothesis on an Exposure Pathway in a Population with Elevated Childhood Leukemias," *EHP* 101 (1993, S-3): 113–15; S. W. Lagakos et al., "An Analysis of Contaminated Well Water and Health Effects in Woburn, Massachusetts," *Journal of the American Statistical Association* 395 (1986): 583–96; P. Lampi et al., "Cancer Incidence following Chlorophenol Exposure in a Community in Southern Finland," *AEH* 47 (1992): 167–75; J. S. Osborne et al., "Epidemiologic Analysis of a Reported Cancer Cluster in a Small Rural Population," *AJE* 132 (1990, S-1): 87–95.

201 **History of water chlorination:** R. D. Morris et al., "Chlorination, Chlorination By-Products, and Cancer: A Meta-analysis," *AJPH* 82 (1992): 955; R. L. Zierler, "Bladder Cancer in Massachusetts Related to Chlorinated and Chloraminated Drinking Water: A Case-Control Study," *AEH* 43 (1988): 195–200; R. L. Jolley et al. (eds.), *Water Chlorination: Chemistry, Environmental Impact, and Health Effects*, vol. 5 (Chelsea, MI: Lewis, 1985).

201 **link between water chlorination and bladder and other cancers:** Cantor et al., "Water Contaminants"; K. P. Cantor, "Water Chlorination, Mutagenicity, and Cancer Epidemiology" (editorial), *AJPH* 84 (1994): 1211–13.

202 **six hundred different disinfection by-products:** This number refers to all products of disinfection, not just the ones produced by chlorination. H. S. Weinberg et al., *Disinfection By-Products (DBPs) of Health Concern in Drinking Water: Results of a Nationwide DBP Occurrence Study* (Athens, GA: EPA Office of Research and Development, National Exposure Research Laboratory, 2002, EPA/600/R-02/068). For a comprehensive discussion about the evidence linking disinfection by-products in drinking water to breast cancer risk, see California Breast Cancer Research Program, *Identifying Gaps in Breast Cancer Research: Addressing Disparities and the Roles of the Physical and Social Environment* (Oakland, CA: CBCRP, 2007), draft report.

202 **MX:** T. A. McDonald et al., "Carcinogenicity of the Chlorination Disinfection By-Product MX," *Journal of Environmental Science and Health C: Environmental Carcinogenicity and Ecotoxicology Review* 23 (2005): 163–214; R. A. Rudel et al., "Chemicals Causing Mammary Gland Tumors in Animals Signal New Directions for Epidemiology, Chemicals Testing, and Risk Assessment for Breast Cancer Prevention," *Cancer* 109 (2007, S-12): 2635–66.

202 **two groups of regulated disinfection byproducts:** Trihalomethanes were first identified in drinking water in the 1970s. In the 1990s, researchers discovered the second group of volatile disinfection by-products: the haloacetic acids (Ronnie Levin, EPA, personal communication). For a fascinating history of the EPA's decision-making process regarding the regulation of disinfection byproducts, see R. Morris, *The Blue Death: Disease, Disaster and the Water We Drink* (New York: Harper Collins, 2007), 163–77.

202 **EPA table of drinking-water standards:** EPA, "List of Drinking Water Contaminants and Their MCLs." Available at www.epa.gov/safewater/contaminants/

203 **Kenneth Cantor's study:** K. P. Cantor et al., "Bladder Cancer, Drinking Water Source, and Tap Water Consumption: A Case-Control Study," *JNCI* 79 (1987): 1269–79

203 **corroborating studies on bladder cancer and disinfection by-products:** V. Bhardwaj, "Disinfection By-Products, *Journal of Environmental Health* 68 (2006): 61–63; Cantor et al., "Water Contaminants"; C. M. Villanueva et al., "Meta-analysis of Studies on Individual Consumption of Chlorinated Drinking Water and Bladder Cancer," *Journal of Epidemiology and Community Health* 57 (2003): 166–73.

203 **alternatives to chlorination:** None of these alternatives alone provides a perfect solution. Moving chlorination to the end of the process, for example, decreases its killing time and so may increase the numbers of microorganisms in finished drinking water. There is no technological substitute for watershed protection. Cantor et al., "Water Contaminants"; B. A. Cohen and E. D. Olsen, *Victorian Water Treatment Enters the 21st Century: Public Health Threats from Water Utilities' Ancient Treatment and Distribution Systems* (New York: NRDC, 1994).

203 **ozonation more effective against *Cryptosporidium*:** J. E. Simmons, "Development of a Research Strategy for Integrated Technology-Based Toxicological and Chemical Evaluation of Complex Mixtures of Drinking Water Disinfection Byproducts," *EHP* 110 (2002, S-6): 1013–24.

203 **Susan Richardson:** See, for example, S. D. Richardon et al., "Integrated Disinfection Byproducts Mixtures Research: Comprehensive Characterization of Water Concentrates Prepared from Chlorinated and Ozonated/Postchlorinated Drinking Water," *Journal of Toxicology and Environmental Health, Part A* 71 (2008): 1165–86; S. D. Richardson, "Water Analysis: Emerging Contaminants and Current Issues," *Analytical Chemistry* 79 (2007): 4295–24.

204 **quote by Kenneth Cantor:** Cantor et al., "Water Contaminants." This sentiment is also echoed by Robert Morris, another leading researcher in the field of drinking water and health: "If we do not protect and preserve our waterways, the most advanced filter will not protect us. . . . If we do not take on the stewardship of our planet with evangelistic fervor,

we will accumulate an ecological budget deficit that future generations can never repay." Morris, *The Blue Death*, 292.

205 **New Jersey [illegible] examiner:** K. P. Cantor, "Water Chlorination, Mutagenicity, and Cancer Epidemiology [editorial]," *AJPH* 84 (1994): 1211–13.

206 **Sankoty Aquifer:** W. H. Walker et al., *Preliminary Report on the Groundwater Resources of the Havana Region in West-Central Illinois*, Cooperative Groundwater Report 3 (Urbana, IL: ISGWS, 1965); L. Horberg et al., *Groundwater in the Peoria Region*, Cooperative Bulletin 39 (Urbana, IL: ISGWS, 1950).

207 **types of aquifers:** Horberg et al., *Groundwater in the Peoria Region*, 16; "Surveying Groundwater," *The Nature of Illinois* (Winter 1992): 9–12.

207 **1989 survey:** IEPA, *Illinois American Water Company, Pekin, Facility Number 1795040 Well Site Survey Report* (Springfield, IL: IEPA, 1989).

207 **Creve Coeur advisory:** S. L. Burch and D. J. Kelly, *Peoria-Pekin Regional Ground-Water Quality Assessment*, Research Report 124 (Champaign, IL: ISWS, 1993).

207 **contaminants in North Pekin wells:** Ibid.

208 **quote from 1993 assessment:** Burch and Kelly, *Peoria-Pekin Regional Ground-Water*, 56.

208 **recharge areas:** IEPA, *A Primer Regarding Certain Provisions of the Illinois Groundwater Protection Act* (Springfield, IL: IEPA, 1988); ISGS, *Ground-Water Contamination: Problems and Remedial Action*, Environmental Geology Notes 81 (Champaign, IL: ISGS, 1977).

208 **difficulty of remediation:** W. T. Piver, "Contamination and Restoration of Groundwater Aquifers," *EHP* 100 (1992): 237–47; IDPH, *Chlorinated Solvents in Drinking Water* (Springfield, IL: IDPH, n.d.); ISGS, *Ground-Water Contamination.*

209 **Pekin's ordinance:** City of Pekin Groundwater Protection Area Ordinance.

209 **growing public awareness:** D. Rheingold, "Pekin Readies Water Watch," *PDT*, 17 Jan. 1994, A-1, A-12.

209 **quote by gas station owner:** Ibid.

209 **observation by superintendent:** Kevin W. Caveny, personal communication.

210 **ongoing detections in Pekin's wells:** Ibid.; Interagency Coordinating Committee on Groundwater, *Illinois Groundwater Protection Program*, vols. 1 and 2, *Biennial Technical Appendices Report* (Springfield, IL:

IEPA, 1994). Since 2004, twenty-seven chemicals have been detected in Pekin's drinking water supply and fourteen of these exceeded health-based limits. Two pollutants (perchloroethylene and nitrates) have exceeded legal limits. To investigate drinking water contaminates in your zip code, see the Drinking Water Quality Report at www.ewg.org.

210 **sooner or later:** contaminants in groundwater typically move between one inch and one foot each year. The contaminants we drink now may be those spilled onto the ground decades ago.

210 **40 percent drink water from aquifers:** S. S. Hutson et al., *Estimated Use of Water in the United States in 2000* (Denver, CO: USGS Information Services, 2004, circular 1268).

210 **observation by Rachel Carson:** R. Carson, *Silent Spring* (Boston: Houghton Mifflin, 1962), 42.

210 **Illinois groundwater report:** Interagency Coordinating Committee on Groundwater, *Illinois Groundwater Protection Program Biennial Comprehensive Status and Self-Assessment Report* (Springfield, IL: Illinois EPA, 2008, IEPA/BOW/08-001).

210 **USGS report:** L. A. DeSimone et al., *Quality of Water from Domestic Wells in Principal Aquifers of the United States, 1991–2004—Overview of Major Findings* (USGS Circular 1332, 2009).

211 My imaginary description of subterranean Illinois is inspired by old Tazewell County drilling logs, as well as by geological background information provided in M. A. Marino and R. J. Shicts, *Groundwater Levels and Pumpage in the Peoria-Pekin Area, Illinois, 1890–1966* (Urbana, IL: ISWS, 1969), and in Horberg et al., *Groundwater in the Peoria Region.*

ten: fire

213 Epigraph is from John Knoepfle, *Poems from the Sangamon* (Urbana: University of Illinois Press, 1985).

214 **the plan:** T. L. Aldous, "Developer Proposes a Site for Burner," *PDT*, 22 July 1992, A-1, A-12.

215 **recycling competes with incinerators:** K. Schneider, "Burning Trash for Energy: Is It an Endangered Species?" *New York Times*, 11 Oct. 1994, C-18.

215 **Columbus incinerator:** Ibid.; S. Powers, "From Trash Burner to Cash Burner," *Columbus Dispatch*, 4 Sept. 1994, B-6; S. Powers, "Board Votes to Close Trash Plant," *Columbus Dispatch*, 2 Nov. 1994, A-1.

Dioxins and Health (New York: Plenum, 1994).

215 **dioxin harmful in trace amounts:** T. Webster and B. Commoner, "Overview: The Dioxin Debate," in Schecter, *Dioxins and Health.*

215 **draft reassessment:** EPA, *Estimating Exposure to Dioxin-Like Compounds,* vols. 1–3, EPA/600/6-88/005Ca,b,c (Washington, DC: EPA, 1994); EPA, *Health Assessment Document for 2,3,7,8-Tetrachlorodibenzo-p-dioxin (TCDD) and Related Compounds,* vol. 1–3, EPA/600/BP-92/001a,b,c (Washington, DC: EPA, 1994).

215 **case against dioxin has been strengthened:** IARC, *IARC Monographs on the Evaluation of Carcinogenic Risks to Humans—Polychlorinated Dibenzo-para-Dioxins and Polychlorinated Dibenzofurans,* vol. 69 (Lyon, France: WHO, IARC, 1997); NTP, *Report on Carcinogens,* 11th ed. (Washington, DC: USDHHS Public Health Service, 2005).

216 **dioxin the most potent carcinogen:** Based on its ability to induce cancer in animals at vanishingly low concentrations. S. Jenkins et al., "Prenatal TCDD Exposure Predisposes for Mammary Cancer in Rats," *Reproductive Toxicology* 23 (2007): 391–96.

216 **suggested rewrites:** NAS, *Health Risks from Dioxin and Related Compounds: Evaluation of the EPA Reassessment* (Washington, DC: National Academies Press, 2006). Available at www.nap.edu/catalog.php?record_id=11688.

216 **waxing and waning of incinerator popularity:** Schneider, "Burning Trash for Energy."

217 **incinerator in Rutland, Vermont:** S. Hemingway, "Report: Trash-to-Energy Plants Pose Environmental Hazard," *Burlington Free Press,* 15 June 2009.

217 **incinerator in Detroit:** D. Ciplet, *An Industry Blowing Smoke: 10 Reasons Why Gasification, Pyrolysis, and Plasma Incineration are Not "Green Solutions"* (Berkeley, CA: Global Alliance for Incinerator Alternatives and Global Anti-Incinerator Alliance, June 2009).

217 **incinerators repackaged as a source of renewable energy:** Ibid.

217 **trends in dioxin emissions:** Dioxin was first reportable to the Toxics Release Inventory in 2000. Available at www.epa.gov/triexplorer/.

218 **if 1,000 tons go in, [illegible] tons come out:** Actually, the final mass exceeds the mass of the material the incinerator is stoked with. Because oxygen combines with fuel in the process of burning, total combustion emissions—ash plus smoke plus vapors—are somewhat heavier than the solid ingredients fed into the incinerator initially.

218 **John Kirby's demonstration:** T. L. Aldous, "Hearing Has Havana Humming," *PDT*, 23 Oct. 1993, A-1, A-10.

218 **toxicity of incinerator ash:** P. Connett and E. Connett, "Municipal Waste Incineration: Wrong Question, Wrong Answer," *The Ecologist* 24 (1994): 14–20; K. Schneider, "In the Humble Ashes of a Lone Incinerator, the Makings of a Law," *New York Times*, 18 Mar. 1994, A-22.

218 **18 boxcars produce 10 truckloads:** The 18:10 ratio was part of the Pekin incinerator proposal.

219 **formation of fly ash:** Connett and Connett, "Municipal Waste Incineration"; T. G. Brna and J. D. Kilgore, "The Impact of Particulate Emissions Control on the Control of Other MWC Air Emissions," *Journal of Air and Waste Management Association* 40 (1990): 1324–29.

219 **types of dioxins and furans:** M. J. Devito and L. S. Birnbaum, "Toxicology of Dioxins and Related Compounds," in Schecter, *Dioxins and Health.*

219 **TCDD:** "NTP Technical Report on the Toxicology and Carcinogenesis Studies of 2,3,7,8-Tetrachlorodibenzo-p-dioxin (TCDD) (CAS No. 1746-01-6) in Female Harlan Sprague-Dawley Rats (Gavage Studies)," *National Toxicology Program Report Service* 521 (2006): 4–232.

220 **incineration not the only source:** Municipal water incinerators are now ranked third. Backyard burn barrels are now presumed the leading source. EPA, *The Inventory of Sources and Environmental Releases of Dioxin-like Compounds in the United States: The Year 2000 Update* (Washington, DC: EPA, National Center for Environmental Assessment; EPA/600/P-03/002A, 2005). Available from National Technical Information Service, Springfield, VA, and online at http://epa.gov/ncea.

220 **PVC plastic:** Polyvinyl chloride, 59 percent chlorine by weight, is the dominant source of organically bound chlorine in hospital waste, where it takes the form of IV-bags, gloves, bedpans, tubing, and packaging. Much of this waste is incinerated. Indeed, medical waste incineration

222 **excerpt from testimony:** C. West-Williams, "State Board Decision on Hold for Now," *PDT*, 7 Apr. 1994, A-1, A-12.

223 **food as a source of dioxin:** ATSDR, TOXFAQs for Chlorinated Dibenzo-p-dioxins (CDDs), Feb. 1999. Available at www.atsdr.coc.gov/tfacts104.html.

223 **dioxin in cow's milk near incinerators:** A. K. D. Liem et al., "Occurrence of Dioxin in Cow's Milk in the Vicinity of Municipal Waste Incinerators and a Metal Reclamation Plant in the Netherlands, *Chemosphere* 23 (1991): 1675–84; P. Connett and T. Webster, "An Estimation of the Relative Human Exposure to 2,3,7,8-TCDD Emissions via Inhalation and Ingestion of Cow's Milk," *Chemosphere* 16 (1987): 2079–84.

223 **dioxin in rivers, fish, soil, and crops:** B. Paigen, "What Is Dioxin?" in Gibbs, *Dying from Dioxin*.

224 **public health threats of incinerators:** NRC, Committee on the Health Effects of Waste Incineration, *Waste Incineration and Public Health* (Washington, DC: National Academies Press, 2000). See also D. Porta et al., "Systemic Review of Epidemiological Studies on Health Effects Associated with Management of Municipal Solid Waste," *Environmental Health* 8 (2009): 60.

224 **clues from animals:** M. J. DeVito et al., "Comparisons of Estimated Human Body Burdens of Dioxinlike Chemicals and TCDD Body Burdens in Experimentally Exposed Animals," *EHP* 103 (1995): 820–31.

224 **quote from James Huff:** J. Huff, "Dioxins and Mammalian Carcinogenesis," in Schecter, *Dioxins and Health*.

224 **dioxin and liver cancer:** A. M. Tritscher et al., "Dose-Response Relationships from Chronic Exposure to 2,3,7,8-Tetrachlorodibenzo-*p*-dioxin in a Rat Tumor Promotion Model: Quantification and Immunolocalization of CYP1A1 and CYP1A2 in the Liver," *Cancer Research* 52 (1992): 3436–42.

224 **dioxin and lung cancer:** G. W. Lucier et al., "Receptor Mechanisms and Dose-Response Models for the Effects of Dioxin," *EHP* 101 (1993): 36–44.

224 **dioxin's effects on hormones and growth factors:** M. La Merrill, "Mouse Breast Cancer Model-Dependent Changes in Metabolic Syndrome-Associated Phenotypes Caused by Maternal Dioxin Exposure and Dietary Fat," *American Journal of Physiology, Endocrinology and Metabolism* 296 (2009): E203–10; A. Schecter et al., "Dioxins: An Overview," *Environmental Research* 101 (2006): 419–28.

224 **dioxin as a developmental toxicant:** L. S. Birnbaum and S. E. Fenton, "Cancer and Developmental Exposure to Endocrine Disruptors," *EHP* 111 (2003): 389–94; S. Jenkins et al., "Prenatal TCDD Exposure Predisposes for Mammary Cancer in Rats," *Reproductive Toxicology* 23 (2007): 391–96; B. J. Lew et al., "Activiation of the Aryl Hydrocarbon Receptor (AhR) during Different Critical Windows in Pregnancy Alters Mammary Epithelial Cell Proliferation and Differentiation," *Toxicological Sciences* 111 (2009): 151–62; B. A. Vorderstrasse et al., "A Novel Effect of Dioxin: Exposure During Pregnancy Severely Impairs Mammary Gland Differentiation," *Toxicological Science* 78 (2004): 248–57.

225 **human studies:** Summarized in NRC, Committee on the Health Effects of Waste Incineration, *Waste Incineration and Public Health* (Washington, DC: National Academies Press, 2000). See also H. Becher et al., "Cancer Mortality in German Male Workers Exposed to Phenoxy Herbicides and Dioxin," *Cancer Causes and Control* 7 (1996): 312–21; L. Hardell et al., "Cancer Epidemiology," in Schecter, *Dioxins and Health*; M. A. Fingerhut et al., "Cancer Mortality in Workers Exposed to 2,3,7,8-Tetrachlorodibenzo-*p*-diozin," *NEJM* 324 (1991): 212–18; J. H. Leem et al., "Risk Factors Affecting Blood PCDD's and PCDF's in Residents Living Near an Industrial Incinerator in Korea," *Archives of Environmental Contamination and Toxicology* 51 (2006): 478–84; A. Manz et al., "Cancer Mortality Among Workers in a Chemical Plant Contaminated with Dioxin," *Lancet* 338 (1991): 959–64; A. Zober et al., "Thirty-four-year Mortality Follow-up of BASF Employees Exposed to 2,3,7,8-TCDD after the 1953 Accident," *International Archives of Occupational and Environmental Health* 62 (1990): 139–57.

226 **early studies from Seveso:** P. A. Bertazzi and A. di Domenico, "Chemical, Environmental, and Health Aspects of the Seveso, Italy, Accident," in Schecter, *Dioxins and Health*; P. A. Bertazzi et al., "Cancer Incidence in a Population Accidentally Exposed to 2,3,7,8-tetrachlorodibenzoparadioxin," *Epidemiology* 4 (1993): 398–406; R. Stone, "New Seveso Findings Point to Cancer," *Science* 261 (1993): 1383.

226 **Seveso updates:** A. Baccarelli et al., "Immunologic Effects of Dioxin: New Results from Seveso and Comparison with Other Studies," *EHP* 110 [illegible] [illegible] Follow-Up," [illegible] M. Warner et al., "Serum Dioxin Concentrations and Breast Cancer Risk in the Seveso Women's Health Study," *EHP* 110 (2002): 625–28.

227 **Michigan study:** The half-life of TCDD is 9–15 years in surface soil and 25–100 years in subsurface soil. D. Dai and T. J. Oyana, "Spatial Variations in the Incidence of Breast Cancer and Potential Risks Associated with Soil Dioxin Contamination in Midland, Saginaw, and Bay Counties, Michigan, USA," *Environmental Health* 7 (2008): 49.

227 **Havana's feasibility study:** T. L. Aldous, "Study: Trash Burner a Boon," *PDT*, 20 May 1992, A-1, A-12.

228 **the first rebuttal:** T. Webster, "Comments on 'A Feasibility Study of Operating a Waste-to-Energy Facility in Mason County Near Havana, Illinois'" (unpub. Ms., 7 Oct. 1992, 4 pp.).

228 **second rebuttal:** T. L. Aldous, "Farm Bureau Members Oppose New Incinerator," *PDT*, 24 July 1992, A-1; S. Iyengar, "Farm Bureau: SIU Study Skewed," *PDT*, 8 Oct. 1992, A-1, A-12.

228 **popcorn threat:** Dr. Dorothy Anderson, personal communication.

228 **Fourth of July:** K. McDermott, "Havana Incinerator Backers Hot about 'Devil Burns' Parade Float," *SSJR*, 1 July 1992, 1.

228 **letter to the editor in Havana:** A. Robertson, *Mason City Banner Times*, 10 June 1992, 11.

229 **letter to the editor in Forrest:** C. Kaisner, "Suddenly in Forrest, Greed Has Become No. 1 Attitude," *Bloomington Daily Pantagraph*, 6 Aug. 1994.

229 **letter about Kirby's smoking habits:** R. Hankins, letter to the editor, *Mason County Democrat*, 3 June 1992, 2.

229 **endorsement of risk:** "Editorial," *Fairbury Blade*, 20 July 1994, 2.

230 **condemnation of risk:** "Dioxin Findings Raise New Fears" (editorial), *Jacksonville Journal-Courier*, 15 Sept. 1994, 10.

230 **P450 enzymes and Ah receptors:** Webster and Commoner, "Dioxin Debate"; G. Lucier et al., "Receptor Model and Dose-Response Model for the Effects of Dioxin," *EHP* 101 (1993): 36–44; T. R. Sutter et al., "Targets for Dioxin: Genes for Plasminogen Activator Inhibitor-2 and Interleukein-1B," *Science* 254 (1991): 415–18.

230 **[illegible]** K. Steenland et al., "Dioxin Revisited: Developments Since the 1997 IARC Classification of Dioxin as a Human Carcinogen," *EHP* 112 (2004): 1265–68.

231 **John Kirby's career:** T. L. Aldous, "Kirby Sees Havana Opportunity, Opposition," *RDT*, 22 Oct. 1993, A-1, A-12; A. Lindstrom, "Sherman Horse Track Sure Bet—Promoters," *SSJR*, 9 Oct. 1973; J. O'Dell, "Hens with Glasses a Barnyard Spectacle," *SSJR*, 27 Aug. 1973; K. Watson, "John Kirby Eyes Candidacy," *SSJR*, 8 Aug. 1968; K. Watson, "Page Names Kirby," *SSJR*, 7 Jan. 1963.

232 **quotes by Kirby:** Aldous, "Kirby sees Havana."

235 **trends in percent of waste incinerated:** In 2007, on average, each person in the United States generated each day 4.62 pounds of garbage. About 1.54 pounds of that was recycled, and .58 pounds per person per day was incinerated. US EPA, *Municipal Solid Waste Generation, Recycling, and Disposal in the United States: Facts and Figures* (Washington, DC: EPA, 2008). Available at www.epa.gov/epawaste/nonhaz/municipal/pubs/msw07-rpt.pdf.

235 **Zero Waste:** Ciplet, *An Industry Blowing Smoke*. See also Zero Waste International Alliance. www.zwia.org.

235 **fish of the Vermilion:** IDEPA, *Illinois Water Quality Report, 1992–93*, vol. 1, IEPA/WPC/94-160 (Springfield, IL: IEPA, 1994).

236 **quote by John Kirby:** J. Knauer, "Incinerator's Future Smoldering after 'No' Vote," *Fairbury Blade*, 16 Nov. 1994, 1, 3.

236 **appellate court decision:** E. Hopkins, "Court Backs Pollution Board's Incinerator Ruling," *PJS*, 13 Sept. 1995, B-5.

236 **repeal of retail rate law:** R. B. Dold, "Clearing the Air," *Chicago Tribune*, 12 Jan. 1996.

236 Malignant mesothelioma is a cancer of the membranes surrounding the lungs. Mesothelioma is caused almost exclusively by exposure to asbestos.

237 **The Land Connection:** This organization is headquartered in Evanston, IL. www.thelandconnection.org.

eleven: our bodies, inscribed

239 **tree-ring analysis:** R. Phipps and M. Bolin, "Tree Rings—Nature's Signposts to the Past," *Illinois Steward* (Summer 1993): 18–21.

At this writing the most recent report was released in 2005: CDC, *Third National Report on Human Exposure to Environmental Chemicals* (Washington, DC: CDC, 2005). The CDC ran an earlier biomonitoring program from 1967 to 1990 called the National Human Biomonitoring Program, which included measuring organochlorine chemicals in fat samples as part of the National Human Adipose Tissue Survey. Cornell University's Program on Breast Cancer and Environmental Risk Factors, "Questions and Answers: Biomonitoring and Environmental Monitoring," Oct. 2005. Available at http://envirocancer.cornell.edu/learning/biomonitor/biomonfaq.cfm.

242 **flame retardants:** A. Sjodin et al., "Concentrations of Polybrominated Diphenyl Ethers (PBDEs) and Polybrominated Biphenyl (PBB) in the United States Population: 2003–2004," *Environmental Science and Technology* 42 (2008): 1377–84.

242 **bisphenol A:** A. M. Calafat et al., "Exposure of the U.S. Population to Bisphenol A and 4-tertiary-Octylphenol: 2003–2004," *EHP* 116 (2008): 39–44.

242 **falling levels of POPs in pregnant women:** R. Y. Wang et al., "Serum Concentrations of Selected Persistent Organic Pollutants in a Sample of Pregnant Females and Changes in Their Concentrations During Gestation," *EHP* 117 (2009): 1244–49. POPs are also now falling in Swedish breast milk: S. Lignell et al., "Persistent Organochlorine and Organobromine Compounds in Mother's Milk from Sweden 1996–2006: Compound-Specific Temporal Trends," *Environmental Research* 109 (2009): 760–67.

242 **shift to monitoring people:** R. Morello-Forsch et al., "Toxic Ignorance and Right-to-Know in Biomonitoring Results Communication: A Survey of Scientists and Study Participants," *Environmental Health* 8 (2009): 6–18.

242 **Silent Spring Institute:** Ibid.

243 **state-based biomonitoring programs:** J. W. Nelson et al., "A New Spin on Research Translation: The Boston Consensus Conference on Human Biomonitoring," *EHP* 117 (2009): 495–99.

243 **limitations of biomonitoring:** NRC, *Human Biomonitoring for Environmental Chemicals* (Washington, DC: National Academies Press, 2006).

243 **bisphenol A:** Studies conducted in nursery schools found bisphenol A in most food items eaten by preschoolers. Research from laboratories revealed that bisphenol A leaches from polycarbonate baby bottles as well as from the epoxy lining of food cans. Together, these studies point to food as an important route of exposure to this particular contaminant. In other laboratories, bisphenol A exposure induced precancerous lesions to form in the mammary glands of young female rats. In young male rats, exposure provoked similar change in their prostate glands. Together, these studies suggest that bisphenol A may imprint hormonal-sensitive tissues in early life in ways that raise the risk for subsequent cancers. M. Durando et al., "Prenatal Bisphenol A Exposure Induces Preneoplastic Lesions in the Mammary Gland in Wistar Rats," *EHP* 115 (2007): 80–86; S.-M. Ho et al., "Developmental Exposure to Estradiol and Bisphenol A Increases Susceptibility to Prostate Carcinogenesis and Epigenetically Regulates Phosphodiesterase Type 4 Variant 4," *Cancer Research* 66 (2006): 5624–32; S. Snedeker, "Environmental Estrogens: Effects on Puberty and Cancer Risk," *The Ribbon* [newsletter of Cornell University's Program on Breast Cancer and Environmental Risk Factors] 12 (2007): 5–7; N. K. Wilson et al., "An Observational Study of the Potential Exposures of Preschool Children to Pentachlorophenol, Bisphenol-A, and nonylphenol at Home and Daycare," *Environmental Research* 103 (2007): 9–20.

243 **Korean study:** M. S. Lee et al., "Seasonal and Regional Contributors of 1-Hydroypyrene Among Children Near a Steel Mill," *Cancer, Epidemiology, Biomarkers and Prevention* 18 (2009): 96–101.

245 **purposefulness:** Robert Millikan, personal communication. See also S. B. Nuland, *How We Die: Reflections on Life's Final Chapter* (New York: Random House, 1993), 202–21.

245 **behavior of cancer cells:** A. E. Erson and E. M. Petty, "Molecular and Genetic Events in Neoplastic Transformation," in D. Schottenfeld and J. F. Fraumeni (eds.), *Cancer Epidemiology and Prevention*, 3rd ed. (New York: Oxford University Press, 2006), 47–64; D. Hanahan and R. A. Weinberg, "The Hallmarks of Cancer," *Cell* 100 (2000): 57–70; J. E. Klaunig and L. M. Kamendulis, "Chemical Carcinogenesis," in C. D. Klaassen (ed.), *Casarett & Doull's Toxicology*, 7th ed. (New York:

"Uncovering New Clues to Cancer Risk," *Scientific American*, May 1996, 54–62.

247 **benzo[a]pyrene and DNA adducts:** Ibid.

248 **description of carcinogenesis:** R. W. Clapp et al., "Environmental and Occupational Causes of Cancer: New Evidence 2005–2007," *Reviews on Environmental Health* 23 (2008): 1–37; J. E. Klaunig and L. M. Kamendulis, "Chemical Carcinogenesis," in C. D. Klaassen (ed.), *Casarett & Doull's Toxicology*, 7th ed. (New York: McGraw Hill, 2008); R. A. Weinberg, *The Biology of Cancer* (London: Garland Science, 2006).

249 **tissue architecture:** See, for example, C. Sonnenschein and A. M. Soto, "Somatic Mutation Theory of Carcinogenesis: Why It Should Be Dropped and Replaced," *Molecular Carcinogenesis* 29 (2000): 205–11.

250 **cancer and developmental toxicants:** L. S. Birnbaum and S. E. Fenton, "Cancer and Developmental Exposure to Endocrine Disruptors," *EHP* 111 (2003): 389–94.

251 **cancer and inflammation:** O. Bottasso et al., "Chronic Inflammation as a Manifestation of Defects in Immunoregulatory Networks: Implications for Novel Therapies Based on Microbial Products," *Inflammopharmacology* 17 (2009): 193–203; R. E. Harris, "Cyclooxygenase-2 (COX-2) and the Inflammogenesis of Cancer," in R. E. Harris (ed.), *Inflammation in the Pathogenesis of Chronic Diseases* (New York: Springer, 2007); G. Stix, "A Malignant Flame," *Scientific American* 297 (2007): 60–67; R. A. Weinberg, *The Biology of Cancer*.

252 **quote on epigenetics:** J. Qiu, "Epigenetics: Unfinished Symphony," *Nature* 441 (2006): 143–45.

253 **environmental epigenetics:** I. Amato, "Orchestrating Genetic Expression," *Chemical and Engineering News* 87 (2009); D. L. Foley, "Prospects for Epigenetic Epidemiology," *AJE* 15 (2009): 389–400; Klaunig and Kamendulis, "Chemical Carcinogenesis"; Qiu, "Epigenetics: Unfinished Symphony"; S. M. Reamon-Buettner and J. Borlak, "A New Paradigm in Toxicology and Teratology: Altering Gene Activity in the Absence of DNA Sequence Variation," *Reproductive Toxicology* 24 (2007) 2–30.

254 **biological markers:** S. Anderson et al., "Genetic and Molecular Ecotoxicology: A Research Framework," *EHP* 102 (1994, S-12): 3–8; M. Eubanks, "Biological Markers: The Clues to Genetic Susceptibility," *EHP* 102 (1994): 50–56; F. Veglia et al., "DNA Adducts and Cancer Risk in Prospective Studies: A Pooled Analysis and a Meta-Analysis," *Carcinogenesis* 29 (2008): 932–36.

254 **epigenetic biomarkers:** P. Vineis and F. Perera, "Molecular Epidemiology and Biomarkers in Etiologic Cancer Research: The New in Light of the Old," *Cancer, Epidemiology, Biomarkers, and Prevention* 16 (2007): 1954–65.

255 **Spanish study on pesticides and gene expression in breast cancer:** P. F. Valerón et al., "Differential Effects Exerted on Human Mammary Epithelial Cells by Environmental Relevant Organochlorine Pesticides Either Individually or in Combination," *Chemico-Biological Interactions* 180 (2009): 485–91.

255 **decreased methylation among Greenlandic Inuit:** J. A. Rusiecki et al., "Global DNA Hypomethylation is Associated with High Serum Persistent Organic Pollutants in Greenlandic Inuit," *EHP* 116 (2008): 1547–52.

255 **Polish study:** Perera, "Uncovering New Clues to Cancer Risk"; F. P. Perera et al., "Molecular and Genetic Damage in Humans from Environmental Pollution in Poland," *Nature* 360 (1992): 256–58S. See also Øvrebø et al., "Biological Monitoring of Polycyclic Aromatic Hydrocarbon Exposure in a Highly Polluted Area of Poland," *EHP* 103 (1995): 838–43; K. Hemminki et al., "DNA Adducts in Humans Environmentally Exposed to Aromatic Compounds in an Industrial Area of Poland," *Carcinogenesis* 11 (1990): 1229–31.

257 **cancer among adoptees:** T. I. A. Sørensen et al., "Genetic and Environmental Influences on Premature Death in Adult Adoptees," *NEJM* 318 (1988): 727–32.

258 **1974 breast cancer blip:** ACS, *Breast Cancer Facts and Figures 1996* (Atlanta: ACS, 1995), fig. 2.

twelve: ecological roots

261 **twin study:** M. F. Fraga et al., "Epigenetic Differences Arise During the Lifetime of Monozygotic Twins," *Proceedings of the National Academy of Sciences* 102 (2005): 10604–609. This study is beautifully described

"Genetic Concepts and Methods in Epidemiologic Research," in D. Schottenfeld and J. F. Fraumeni eds., *Cancer Epidemiology and Prevention*, 3rd ed. (New York: Oxford University Press, 2006).

262 **twin study:** Some cancers had stronger heritable influences than others. Stomach, colon, lung, breast, and prostate cancer showed the strongest concordance, but for none did genetics explain more than 42 percent of the risk. P. Lichenstein, "Environmental and Heritable Factors in the Causation of Cancer—Analysis of Cohorts of Twins from Sweden, Denmark, and Finland," *NEJM* 343 (2000): 78–85.

263 **Human Genome Project:** www.genomics.energy.gov.

263 **complexity of cancer causation:** A. E. Erson and E. M. Petty, "Molecular and Genetic Events in Neoplastic Transformation," in Schottenfeld and Fraumeni, *Cancer Epidemiology and Prevention*; J. Qiu, "Unfinished Symphony," *Nature* 44 (2006): 143–45; J. R. Weidman et al., "Cancer Susceptibility: Epigenetic Manifestation of Environmental Exposure," *Cancer Journal* 13 (2007): 9–16.

264 **familial cancers in Sweden:** K. Hemminki et al., "How Common Is Familial Cancer?" *Annals of Oncology* 19 (2008): 163–67.

264 **BRCA 1 and 2 carriers:** M. C. King et al., "Breast and Ovarian Cancer Risk Due to Inherited Mutations in BRCA 1 and BRCA 2," *Science* 302 (2003): 643–46; A. Kortenkamp, "Breast Cancer and Exposure to Hormonally Active Chemicals: An Appraisal of the Scientific Evidence," a briefing paper for CHEM Trust, Jan. 2008; Risch and Whittemore, "Genetic Concepts and Methods in Epidemiologic Research."

264 **pancreatic cancer:** T. P. Yeo, et al., "Assessment of 'Gene-Environment' Interaction in Cases of Familial and Sporadic Pancreatic Cancer," *Journal of Gastrointestinal Surgery* 13 (2009): 1487–94.

266 **1983 study:** R. A. Weinberg, "A Molecular Basis of Cancer," *Scientific American*, Nov. 1983, 126–42.

267 **genetic changes involved in bladder cancer:** I. Orlow et al., "Deletion of the p16 and p15 Genes in Human Bladder Tumors," *JNCI* 87 (1995):

1324–33; S. H. Kroft and R. Oyasu, "Urinary Bladder Cancer: Mechanisms of Development and Progression," *Laboratory Investigation* 71 (1994): 158–74; P. Lipponen and M. Eskelinen, "Expression of Epidermal Growth Factor Receptor in Bladder Cancer as Related to Established Prognostic Factors, Oncoprotein Expression and Long-Term Prognosis," *British Journal of Cancer* 69 (1994): 1120–25.

268 **aromatic amines and DNA adducts:** D. Lin et al., "Analysis of 4-Aminobiphenyl-DNA Adducts in Human Urinary Bladder and Lung by Alkaline Hydrolysis and Negative Ion Gas Chromatography-Mass Spectrometry," *EHP* 102 (1994, S-1): 11–16; P. L. Skipper and S. R. Tannenbaum, "Molecular Dosimetry of Aromatic Amines in Human Populations," *EHP* 102 (1994, S-6): 17–21; S. M. Cohen and L. B. Ellwein, *EHP* 101 (1994, S-5): 111–14.

268 **new knowledge about bladder cancer:** Variations in genes are called polymorphisms. A. S. Andrew et al., "DNA Repair Genotype Interacts with Arsenic Exposure to Increase Bladder Cancer Risk," *Toxicology Letters* 187 (2009): 10–14; J. D. Figueroa et al., "Genetic Variation in the Base Excision Repair Pathway and Bladder Cancer," *Human Genetics* 121 (2007): 233–42; M. Franekova et al., "Gene Polymorphisms in Bladder Cancer," *Urologic Oncology* 26 (2008): 1–8; P. Greenwald and B. K. Dunn, "Landmarks in the History of Cancer Epidemiology," *Cancer Research* 69 (2009); R. J. Hung et al., "GST, NAT, SULT1A1, CYP1B1 Genetic Polymorphisms, Interactions with Environmental Exposures and Bladder Cancer Risk in a High-Risk Population," *International Journal of Cancer* 110 (2004): 598–604; A. E. Kilte, "Molecular Epidemiology of DNA Repair Genes in Bladder Cancer," *Methods in Molecular Biology* 472 (2009): 281–306; C. Li et al., "DNA Repair Phenotype and Cancer Susceptibility—A Mini Review," *International Journal of Cancer* 124 (2009): 999–1007; P. D. Negraes et al., "DNA Methylation Patterns in Bladder Cancer and Washing Cell Sediments: A Perspective for Tumor Recurrence Detection," *BMC Cancer* 8 (2008): 238; X. Wu et al., "Bladder Cancer Predisposition: A Multigenic Approach to DNA-Repair and Cell-Cycle—Control Genes," *American Journal of Human Genetics* 78 (2006): 464–79; X. Wu et al., "Genetic Polymorphisms in Bladder Cancer," *Frontiers in Bioscience* 12 (2007): 192–213; Y. Ye et al., "Genetic Variants in Cell Cycle Control Pathway Confer Susceptibility to Bladder Cancer," *Cancer* 112 (2008): 2467–74.

268 **slow and fast acetylators:** P. Vineis and G. Ronco, "Interindividual Variation in Carcinogen Metabolism and Bladder Cancer Risk," *EHP* [illegible]

269 **trends in bladder cancer incidence and mortality:** M. J. Horner et al. (eds.), *SEER Cancer Statistics Review, 1975–2006* (Bethesda, MD: NCI, 2009). Available at http://seer.cancer.gov/

269 ***o*-toluidine releases in 1992:** EPA, *1992 Toxics Release Inventory: Public Data Release*, EPA 745-R-001 (Washington, DC: EPA, 1994), 79.

269 ***o*-toluidine releases in 2007:** EPA, *Toxics Release Inventory—2007* data. Available at www.epa.gov.triexplorer.

270 **bladder carcinogens continue to be made and used:** One researcher offers the following reflection on the bladder cancer situation in England: "The continued use of known carcinogenic substances in British industry for many years after their identification, the wide range of industries with a known or suspected increased risk of bladder cancer, and our ignorance of the carcinogenic potential of many materials used in current manufacturing should be a cause for continuing concern." R. R. Hall, "Superficial Bladder Cancer," *British Medical Journal* 308 (1994): 910–13.

270 **rankings and recurrence of bladder cancer:** S. P. Lerner et al., eds., *Textbook of Bladder Cancer* (London: Taylor and Francis, 2006).

270 **bladder cancer expensive to treat:** B. K. Hollenbeck et al., "Provider Treatment Intensity and Outcomes for Patients with Early-Stage Bladder Cancer," *JNCI* 101 (2009): 571–80; E. B. Avritscher et al., "Clinical Model of Lifetime Cost of Treating Bladder Cancer and Associated Complication," *Urology* 68 (2006): 549–53.

270 **2008 incidence and mortality statistics on bladder cancer:** American Cancer Society, *Cancer Facts and Figures—2008.*

271 **pink and blue brochure:** "Cancer Prevention" (pamphlet) (Bethesda, MD: USDHHS, n.d.).

271 **genetics textbook:** G. Edlin, *Human Genetics: A Modern Synthesis*, 2nd ed. (Boston: Jones & Bartlett, 1990). Quotes are from 184–204.

273 **lifestyle factors and cholera:** C. E. Rosenberg, *The Cholera Years: The United States in 1832, 1849, and 1866* (Chicago: University of Chicago Press, 1962), 1–60.

274 **quotes from American Cancer Society materials:** From the ACS Cancer Reference Information: www.cancer.org. The ACS did not dis-

[illegible] environmental factors in its 1995 report on cancer prevention, ACS, *Cancer Risk Report: Prevention and Control, 1995*. The ACS devotes 1.5 pages to the environment in its most recent, 72-page report *Cancer Facts and Figures—2008*.

274 **removal of pesticides from the shelf:** C. Porter, "Bugs, [illegible] Replace Pesticides Today," *Toronto Star*, 22 April 2009.

275 **Paris Appeal:** available on the Web site of the Association pour la Recherche Therapeutique Anti Cancereuse: www.[illegible].

275 **pie chart:** R. Doll and R. Peto, "The Causes of Cancer: Quantitative Estimates of Avoidable Risks of Cancer in the United States Today," *JNCI* 66 (1981): 1191–1308. Harvard Center for Cancer Prevention, "Harvard Report on Cancer Prevention," *Cancer Causes and Control* 7 (1996-S1): 3–59. Contemporary critique of the pie chart is provided in Collaborative on Health and Environment, "Consensus Statement on Cancer and the Environment: Creating a National Strategy to Prevent Environmental Factors in Cancer Causation," Oct. 2008. The pie chart was not without its critics when it first was published. See, for example, S. S. Epstein and J. B. Swartz, "Fallacies of Lifestyle Cancer Theories," *Nature* 289 (1981): 127–30.

276 **quote from Illinois cancer report:** IDPH, *Cancer Incidence in Illinois by County, 1985–87*, Supplemental Report (Springfield, IL: IDPH, 1990), 7–8.

277 **critique by Dominique Belpomme and Richard Clapp:** P. Irigaray et al., "Lifestyle-Related Factors and Environmental Agents Causing Cancer: An Overview," *Biomedicine and Pharmacotherapy* 61 (2007): 640–58; R. W. Clapp, "Industrial Carcinogens: A Need for Action," presentation before the President's Cancer Panel, East Brunswick, NJ, 16 Sept. 2008.

277 **complex web of causation:** F. Mazzocchi, "Complexity in Biology," *European Molecular Biology Organization Reports* 9 (2008): 10–14; Collaborative on Health and the Environment, "Consensus Statement on Cancer and the Environment: Creating a National Strategy to Prevent Environmental Factors in Cancer Causation," submitted to the President's Cancer Panel, Oct. 2008.

278 **Rachel Carson on environmental human rights:** Senate testimony hearings before the Subcommittee on Reorganization and International Organizations of the Committee on Government Operations,

Environmental Working Group has also conducted biomonitoring on individuals, including (through the use of umbilical cord blood) U.S. infants. See www.ewg.org.

280 **we do not all bear equal risks:** National Environmental Justice Advisory Council, Cumulative Risks/Impacts Work Group, "Ensuring Risk Reduction in Communities with Multiple Stressors: Environmental Justice and Cumulative Risks/Impacts," report to the U.S. EPA, December 2004; ACS, *Cancer Facts and Figures—2008.*

280 **6 percent is 33,600:** ACS, *Cancer Facts and Figures—2008.*

281 **2007 releases of carcinogens:** EPA, Toxics Release Inventory, Chemical Report—2007. Available at www.epa.gov/triexplorer/.

281 **cancer as homicide:** The environmental analysts Paul Merrell and Carol Van Strum have argued that the concept of acceptable risk is tolerated only because of the anonymity of its intended victims. See P. Merrell and C. Van Strum, "Negligible Risk: Premeditated Murder?" *Journal of Pesticide Reform* 10 (1990): 20–22. Likewise, the molecular biologist and physician John Gofman has argued, "If you pollute when you DO NOT KNOW if there is any safe dose (threshold), you are performing improper experimentation on people without their informed consent. . . . If you pollute when you DO KNOW that there is no safe dose with respect to causing extra cases of deadly cancers, then you are committing premeditated random murder" (J. W. Gofman, memorandum to the U.S. Nuclear Regulatory Commission, 21 May 1994).

281 Rachel Carson's observation: *Silent Spring*, 248. See also M. J. Kane, "Promoting Political Rights to Protect the Environment," *Yale Journal of International Law* 18 (1993): 389–411.

281 **precautionary principle:** European Environment Agency, *Late Lessons from Early Warnings: The Precautionary Principles 1886–2000* (Luxembourg: Office for Official Publications of the European Communities, 2001); N. J. Myers and C. R. Raffensperger, *Precautionary Tools for Reshaping Environmental Policy* (Cambridge, MA: MIT Press, 2005).

283 **precautionary approach:** [illegible], "Cancer Prevention Through a Precautionary Approach to Environmental Chemicals," presentation before the President's Cancer Panel, East Brunswick, NJ, 16 Sept. 2008.

282 **pollution as the result of outmoded technology:** [illegible], "Eight Bits of Green Wisdom for World Environment Day," *China Daily*, 30 May 2008.

284 **mammary gland carcinogens:** R. Rudel, "Chemicals Causing Mammary Gland Tumors and Animals Signal New Directions for Epidemiology, Chemicals Testing, and Risk Assessment for Breast Cancer Prevention," *Cancer* 109 (2007, S-12): 2635–66.

284 **need for vision and courage:** P. Grandjean, "Seven Deadly Sins of Environmental Epidemiology and the Virtues of Precaution," *Epidemiology* 19 (2008): 158–62.

284 **quote by Bradford Hill:** A. B. Hill, "The Environment and Disease: Association or Causation?" *Proceedings of the Royal Society of Medicine* 58 (1965): 295–300.

284 **abnormal changes in juvenile rats:** M-H. Li and L. G. Hansen, "Enzyme Induction and Acute Endocrine Effects in Prepubertal Female Rats Receiving Environmental PCB/PCDF/PCDD Mixtures," *EHP* 104 (1996): 712–22.

acknowledgments for the second edition

Over the years, many people have commented to me on the length of this book's acknowledgment section. My response is that any book that ranges as widely across biological disciplines as this one requires many different eyes to vet it. The original acknowledgments section is reprinted below, along with the original affiliations of my critical readers—some of whom have since moved on to other posts. Of course, responsibility for accuracy continues to belong only to me.

During the process of revising and updating the book for this second edition, as well as during the process of creating the film version of *Living Downstream*, I have accumulated other significant debts. The biggest of these are to my husband, Jeff, and our two children, Faith and Elijah. A book project is like a houseguest who refuses to leave and talks too much. A book project that involves a film crew in the living room at 6 A.M. is like a houseguest who invites all his friends over. As my children freely point out to me, books are rude.

Second only to those who share my street address is Merloyd Lawrence, who has patiently conducted me through innumerable drafts over many years. Having Merloyd as my editor for both editions of *Living Downstream* (as well as for *Having Faith* and my forthcoming book on children's environmental health) makes me—quoting my children again—a lucky duck.

During the months of writing, many farmers and librarians plied me with organic food and research materials. In at least one case, the farmer and the librarian are the same [illegible] librarian at Ithaca College, also sells me salad greens and heirloom garlic at my village farmers' market—sometimes while discussing my latest interlibrary loan request. Thanks to all the farmers at the Trumansburg Farmers Market, as well as to Paul Martin and Evangeline Sarat, who operate the Sweetland Farm CSA—only a half-mile from my home—providing me year-round food security in the form of eggs, beets, and a chest freezer full of green beans and strawberries . . . along with the chance to revise chapters in my head while picking sugar peas. Paul and Evangeline also shared with me their expert knowledge of organic pest control methods.

I am especially grateful to documentary film producer and director Chanda Chevannes of The People's Picture Company, Inc., who found a way to capture on camera the lyric beauty of my home state, my private life as a medical patient, and the scientific argument contained within these pages—and then shaped it all into a coherent cinematic story. From the start, Chanda possessed an uncanny fluency with the language of this book as well as a mastery of the science. As our collaboration continued, I began to see Chanda as a coauthor. Her insights showed me where and how the book could be revised. I thank also her film crew for their boundless patience with me as I struggled to overcome my self-consciousness with cameras and boom mikes: Nathan Shields, Benjamin Gervais, Trent Richmond, P. Marco Veltri, Larissa Shames, Rebecca Rosenberg, Joshua Kraemer, Zachary Pedersen, Bill Pope, Bryant Cardona, Michelle MacLachlan, Garrett Shields, Jill Chevannes, and Liz Armstrong. The film's funders and supporters include The Ceres Trust, the Tides Foundation, the Canadian Independent Film and Video Fund, Canada Council for the Arts, Park Foundation, Canadian Auto Workers–Social Justice Fund, Doris Cadoux and Hal Schwartz, Ya Ya Sistahs & Bruddahs Too! Cancer Prevention Challenge Team, and organizational partners Women's Healthy Environments Network and Insight Productions.

During the making of the film, Marge Melchers, Terra Brockman, Joel Smith, and Henry Brockman provided locally grown, organic food and reintroduced me to the riverbanks, floodplains, and bottomlands of central Illinois. My mother Kathryn Steingraber opened her home and her heart to me and the film crew, as did my cousin John and his wife Emily, who gave up a day of work on a rainless Monday in July to recreate on camera the farm scenes in Chapter Seven. This is no small sacrifice for Illinois farmers. My conversations with John on camera gave me new insights about farm chemicals, which helped guide my revisions of the book. My Uncle Roy and Aunt Ann, whose farm appears in Chapter Ten, likewise offered their enthusiasm and intimate knowledge of agriculture. The Land Connection, which found new owners for the farm after the death of my grandmother in 2006, deserves credit not only for saving our family's farm but for helping rebuild regional food security throughout Illinois by finding land for young organic farmers. My family's farm in Pleasant Ridge Township, which is now a completely organic operation, is only one beneficiary of The Land Connection's good work. By fostering sustainable, community-based food systems, The Land Connection is providing solutions to the problems I identify in Chapters One, Seven, and Ten.

For fact checking and manuscript commentary on this second edition, I thank Joyce Blumenshine, Terra Brockman, Charlotte Brody, Julia Brody, Dick Clapp, Terrence Collins, Kamyar Enshayan, Jeanne Handy, Tim LaSalle, Dave Miller, Carmi Orenstein, Susan Richardson, Ruthann Rudell, Jennifer Sass, Ted Schettler, Joel Smith, John Spinelli, and Gina Solomon.

My knowledge of cancer biology has been deepened by my advisory work with the California Breast Cancer Research Program. As a steering committee member on CBCRP's Special Research Initiative, I had the opportunity to review the literature on the environmental links to breast cancer and, together with an expert strategy team, identify the gaps in this body of research in need of targeted funding. Our 2007 draft report, *Identifying Gaps in Breast Cancer Research: Addressing Disparities and the Roles*

versity of Chicago Medical Center; Susan Matsuko Shinagawa, Asian and Pacific Islander National Cancer Survivors Network; and David R. Williams, PhD, Harvard School of Public Health. Likewise, my commissioned work in 2007 for the Breast Cancer Fund on the causes of early puberty in girls deepened my understanding of the biology of breast development, and insights gleaned from that research project are now woven into these chapters. Thanks especially to Jeanne Rizzo, Tamara Adkins, Brynn Taylor, Janet Nudelman, and Nancy Evans.

I am also grateful to members of Cornell University's Program on Breast Cancer and Environmental Risk Factors, where I served as a visiting scholar from 1999 to 2003. My colleagues there, Suzanne Snedeker, PhD, and Carmi Orenstein, MPH, continue to answer my queries, forward data, critique my writing, and point me in new directions. Their influence is particularly evident in Chapters Six and Eleven.

I have also learned much from my colleagues at the Science and Environmental Health Network, on whose board I sit. SEHN's ongoing work on the Precautionary Principle, ecological medicine, and chemicals policy reform is unsurpassed. For their generous tutelage in all these areas, I thank this organization's staff and board.

At Ithaca College, where I now serve as scholar in residence, a cadre of able librarians has assisted me in myriad ways. Tanya Saunders, PhD, dean of the Division of Interdisciplinary and International Studies, has provided me a wonderful academic base of operations. I am grateful, too, to Marian Brown, PhD, and Susan Allen-Gil, PhD, director of the Environmental Studies Program.

For enabling my work in direct and indirect ways, I thank Adelaide Park Gomer of the Park Foundation and Marni Rosen of the Jenifer Altman Foundation. I also thank Janet Wallace of the Wallace Global Fund;

Pete Myers, PhD, of Environmental Health Sciences; Michael Lerner, PhD, of Commonweal; Liz Armstrong of Prevent Cancer Now; and Janet Collins. And a special thanks to Judy Kern.

For helping me carry the message of *Living Downstream* into universities, town hall meetings, medical schools, public libraries, county fairs, church basements, and international conferences, I thank Jodi Solomon, Bill Fargo, and Stacy Borden at the Jodi Solomon Speakers Bureau in Boston. For bringing its message to these pages, I thank my literary agent, Charlotte Sheedy.

I thank my urologist Sanjeev Vohra, MD, for letting us film my annual cystoscopy, thereby pulling back the curtain of privacy on a cancer-screening procedure that saves lives and yet is unfamiliar to (or feared by) many people. I thank Dr. Vohra again for his compassion and due diligence as my physician.

Finally, I thank members of my extended family who have generously shared with me their knowledge of my biological roots. I am particularly grateful to Doris and Arne Petersen of Palo Alto, California, and to Steve and Alice Kauble of Tulsa, Oklahoma, and their daughter Sarah.

acknowledgments for the first edition

During the years of researching and writing this book, I have enjoyed the encouragement, support, and direct assistance of considerable numbers of individuals and institutions. Foremost among these are the Bunting Institute of Radcliffe College, the Center for Research on Women and Gender at the University of Illinois, and the Women's Studies Program at Northeastern University in Boston. All variously provided me residencies, fellowship money, and communities of inspiring colleagues. For these necessities, I thank their respective directors, Drs. Florence Ladd, Alice Dan, and Christine Gailey. I am grateful also to the Radcliffe Research Partnership Program for sponsoring a coterie of Harvard University students to help me with the formidable task of library research: Rebecca Braun, Christine Chung, Theresa Esquerra, Palmira Gómez, Julie Nelson, Kathryn Patton, and Amy Stevens. I thank each of them for their enthusiastic and dogged assistance.

I owe a considerable debt to many librarians. I am especially grateful to those at the Harvard University Medical School and Widener Libraries, the National Library of Medicine in Bethesda, the Beinecke Library at Yale University, the Snell Library at Northeastern University, and the public libraries of Boston, Somerville, Peoria, and Pekin.

Several right-to-know experts aided in the search for my ecological roots. They are John Chelen at the Unison Institute, Lisa Damon at the

Illinois Hazardous Waste Research and Information Center, Joe Goodner at the Illinois Environmental Protection Agency, [illegible] Kallis, Ed Hopkins of Citizen Action, and Paul Orum at the Working Group on Community Right-to-Know. I am also, once again, indebted to Brian Burt for his redoubtable computer savvy and wholehearted resourcefulness.

Many colleagues in science, medicine, and policymaking contributed their expert knowledge to this project by reading parts of the manuscript. For their invaluable commentary and criticisms, I thank Ruth Allen, PhD, MPH, of the National Cancer Institute and U.S. Environmental Protection Agency; Dorothy Anderson, MD, of Mason City, Illinois; Ann Aschengrau, ScD, of Boston University; Pierre Béland, PhD, of the St. Lawrence National Institute of Ecotoxicology in Québec; Judith Brady of San Francisco; Julia Brody, PhD, of Silent Spring Institute; Leslie Byster of the Silicon Valley Toxics Coalition; Kenneth Cantor, PhD, of the National Cancer Institute; Jackie Christensen of the Institute for Agriculture and Trade Policy; Richard Clapp, ScD, MPH, of Boston University; Brian Cohen of the Environmental Working Group; Penelope Fenner-Crisp, PHD, of the U.S. Environmental Protection Agency; Joan D'Argo of the National Coalition for Health and Environmental Justice; Devra Lee Davis, PhD, MPRH, of the World Resources Institute; Samuel Epstein, MD, of the University of Illinois, Chicago; James Davis, PhD, of St. Louis University; Thomas Downham, MD, of the Henry Ford Medical Center; Jay Feldman of the National Coalition Against the Misuse of Pesticides; Vincent Garry, MD, of the University of Minnesota; John W. Gephart, PhD, of Cornell University; Benjamin Goldman, PhD, of Boston; Joe Goodner of the Illinois Environmental Protection Agency; Ross Hall, PhD, professor emeritus of McMaster University; John Harshbarger, PhD, at the Registry of Tumors in Lower Animals; Monica Hargraves, PhD, of Ithaca, New York; Robert Hargraves, PhD, professor emeritus of Princeton University; Peter Infante, PhD, of the Occupational Safety and Health Administration; Frieda Knobloch, PhD, of St. Olaf's College; Nancy Krieger, PhD, of Harvard University; Philip Landrigan, MD, of Mt. Sinai School of Medicine; Linda Lear, PhD, of George Washington University and the Smithsonian Institution; Ronnie Levin of the U.S. Environmental Protection Agency; June Fessenden MacDonald, PhD, of Cornell University's Program on Breast Cancer and Environmental Risk Factors in New York State; Donald Malins, PhD, of the Pacific Northwest Research Foundation; Robert Millikan, PhD, of the University of North Carolina; Monica Moore of the Pesticide Action Network North America Regional Center; Mary O'Brien, PhD, of Eugene, Oregon; Maria Pellerano and Peter Montague, PhD, of the Environmental Research Foundation; Frederica Perera, DrPH, of Columbia University; David Pimentel, PhD, of Cornell University; Mike Rahe of the Illinois Department of Agriculture; Edmund Russell III, PhD, of the University of Virginia; Arnold Schecter, MD, MPH, of the State University of New York, Binghamton; Paul Schulte, PhD, of the National Institute of Occupational Safety and Health; Carl Shy, MD, DrPH, of the University of North Carolina; Carlos Sonnenschein, MD, and Ana Soto, MD, of Tufts University; William H. Smith, PhD, of Yale University; A. G. Taylor, CPSS, of the Illinois Environmental Protection Agency; Susan Teitelbaum, MPH, of Columbia University; Paul Tessene and Susan Post of the Illinois Natural History Survey; Rebecca Van Beneden, PhD, of the University of Maine; Louis Verner, PhD, of the Antioch New England Graduate School; Tom Webster, PhD, of Boston University; Gail Williamson, MD, of Brookfield, Illinois; Mary Wolff, PhD, of Mt. Sinai School of Medicine; and Sheila Hoar Zahm, ScD, of the National Cancer Institute. Of course, all responsibility for the accuracy and validity of the text rests with me.

Friends in arts and letters also offered their insights as readers and commentators. They are Karol Bennett, Anthony Brandt, Robert Currie, Joellen Masters, Kim McCarthy, Marnie McInnes, John McDonald, and Ann Patchett.

Many other scholars and researchers, too numerous to name here, cheerfully offered advice, shared data, fielded questions, and alerted me to important areas of study. I owe them all a great deal—as I do numerous public servants in an assortment of government agencies. These include the U.S. and Illinois Environmental Protection Agencies; the National Cancer Institute; the National Institute of Occupational Safety and Health; the National Institute of Environmental Health Sciences; the Agency for Toxic Substances and Disease Registry; the National Program of Cancer Registries; the National Center for Health Statistics; the Illinois Geological, Water, and Natural History Surveys; the Illinois Department of Conservation; the Illinois and Massachusetts Departments of Public Health; and the Mason County Health Department. The conclusions and recommendations expressed in this book may not be shared by those who assisted me in the research process nor by the agencies they represent.

All of us are indebted to those working, on local as well as national levels, for cancer prevention and environmental protection. I owe a special thanks to members of the Women's Community Cancer Project in Cambridge, Massachusetts; Nancy Evans of Breast Cancer Action; Cathie Ragovin, MD, of the Massachusetts Breast Cancer Coalition; Barbara Balaban, Geri Barish, and Joan Swirsky of Long Island; Sandra Marquardt of Mothers and Others for a Liveable Planet; the staffs of Pesticide Action Network and the Northwest Coalition for Alternatives to Pesticides; and the citizens and scientists who serve with me in Washington on the National Action Plan on Breast Cancer.

Closer to home, I received assistance from Kevin Caveny, district supervisor of my hometown water utility; Elaine Hopkins of the *Peoria Journal Star*; state representative Ricca Slone; and Dr. Earl and Marge Melchers of Pekin, Illinois. My parents, Wilbur and Kathryn Steingraber, and sister, Julie Skocaj, read drafts, checked facts, and freely offered the authority of their memories and experience.

No writer could ask for a more steadfast pair of advocates than I have found in my literary agent, Charlotte Sheedy, and my editor, Merloyd Lawrence. Both have been unflagging in their enthusiasm and support—even in the face of tribulations and unforeseen delays on my part. Dropping a finished chapter through Merloyd's Beacon Hill mail slot was the most satisfying of gestures. For her patience, caring, and sharp editorial judgment, I am forever grateful.

Finally, I express my abiding appreciation to Bernice Bammann, to the extraordinary writers Valerie Cornell and Karen Lee Osborne, and to Jeff de Castro, whose intellect informs many of these chapters and continues to bless my life.

Index

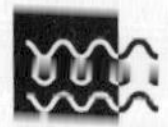

About the Author

Sandra Steingraber, Ph.D. holds a doctorate in biology and master's degree in creative writing. Called "a poet with a knife" (*Sojourner*), Steingraber is the author of *Post-Diagnosis*, a volume of poetry and coauthor of a book on ecology and human rights in Africa, *The Spoils of Famine*. Her memoir, *Having Faith: An Ecologist's Journey to Motherhood*, explored the intimate ecology of pregnancy and was selected by the *Library Journal* as a best book of 2001.

The original edition of Steingraber's *Living Downstream* was the first to bring together data on toxic releases with data from U.S. cancer registries. *Living Downstream* won praise from international media, including *The Washington Post, The Chicago Tribune, Publishers Weekly, The Lancet, The Times Literary Supplement*, and *The London Times*. For this work, Chatham College selected Steingraber to receive its biennial Rachel Carson Leadership Award in 2001. In 2006, Steingraber received a Hero Award from the Breast Cancer Fund and, in 2009, the Environmental Health Champion Award from Physicians for Social Responsibility, Los Angeles.

In 2010, The People's Picture Company of Toronto released the film adaptation of *Living Downstream*. This eloquent and cinematic documentary follows Steingraber during one pivotal year as she travels across North America, working to break the silence about cancer and its environmental links.

Steingraber has served as advisor to the California Breast Cancer Research Program, provided Congressional briefings, and lectured on many college campuses. A columnist for Orion magazine, Steingraber is a scholar in residence at Ithaca College in Ithaca, New York. She is married to artist Jeff de Castro, and they have two children. Please visit her Web site: www.steingraber.com.